GATEWAY TO THE MOON

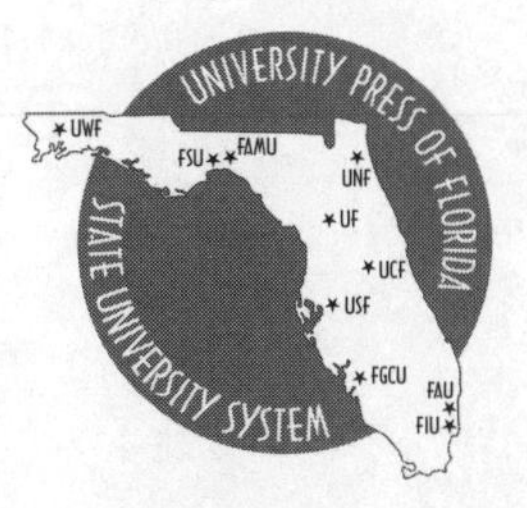

Florida A&M University, Tallahassee
Florida Atlantic University, Boca Raton
Florida Gulf Coast University, Ft. Myers
Florida International University, Miami
Florida State University, Tallahassee
University of Central Florida, Orlando
University of Florida, Gainesville
University of North Florida, Jacksonville
University of South Florida, Tampa
University of West Florida, Pensacola

USA

GATEWAY TO THE MOON

Building the Kennedy Space Center Launch Complex

Charles D. Benson and William B. Faherty

University Press of Florida

Gainesville · Tallahassee · Tampa · Boca Raton

Pensacola · Orlando · Miami · Jacksonville · Ft. Myers

First cloth printing 1978, as first half of *Moonport: A History of Apollo Launch Facilities and Operations,* by National Aeronautics and Space Administration, Scientific and Technical Information Office
NASA SP-4204

First paperback printing 2001 by University Press of Florida
Printed in the United States of America on acid-free paper

06 05 04 03 02 01 6 5 4 3 2

Library of Congress Cataloging-in-Publication data are available.
ISBN 0-8130-2091-3

The University Press of Florida is the scholarly publishing agency for the State University System of Florida, comprising Florida A&M University, Florida Atlantic University, Florida Gulf Coast University, Florida International University, Florida State University, University of Central Florida, University of Florida, University of North Florida, University of South Florida, and University of West Florida.

University Press of Florida
15 Northwest 15th Street
Gainesville, FL 32611-2079
http://www.upf.com

CONTENTS

Gateway to the Moon: Building the Kennedy Space Center Launch Complex was originally published in 1978 by the National Aeronautics and Space Administration as part of the NASA History Series. The original volume was a single book titled *Moonport: A History of Apollo Launch Facilities and Operations*. This volume corresponds to chapters 1–14 and pages 1–351 of that publication.

FOREWORD

By now the grandeur of the achievement of landing men on the moon and returning them to earth has taken its place in our language as a yardstick of human accomplishment—"If we could send men to the moon, why can't we do so-and-so?" The most imposing artifact of that achievement is the Apollo launch facilities at Kennedy Space Center.

When the national objective of landing men on the moon was dramatically announced in May 1961, it quickly became apparent within NASA that the remainder of the decade was little enough time to design, build, and equip the extensive and unprecedented facilities required to launch such missions. Indeed, time was so pressing that for many months the planning, designing, even initial construction of launch facilities had to go forward without answers to some essential questions, such as: How big would the launch vehicle(s) be? How many launches would there be, and how often?

Intense effort by a rapidly growing team of people in government, industry, and the universities gradually filled in the grand design and answered those questions. Land was acquired, ground was broken, pipe was laid, concrete was poured, buildings rose. When the launch vehicles and spacecraft arrived, the facilities were ready and operations could begin. Seldom was the pressure off or the path smooth, but the end of the decade saw the deadline met, the task accomplished.

This history tells the story of the Apollo launch facilities and launch operations from the beginning of design through the final launch. You will meet many of the cast of thousands who took part in the great adventure. You will read of the management techniques used to control so vast an undertaking, of innovation in automation, of elaborate, repetitive, exhaustive testing on the ground to avoid failures in space. You will also learn something of the impact of the Apollo program on the citrus groves and quiet beaches of Florida's east coast.

It is fitting that, as this manuscript was being prepared, these same facilities were being modified to serve as the launch site for Apollo's successor, the Space Shuttle, for at least the remainder of this century.

August 1977

Lee R. Scherer
Director
Kennedy Space Center

Preface

On 28 July 1960, the National Aeronautics and Space Administration (NASA) announced a new manned spaceflight program. Called Apollo, its aim was to put three astronauts into sustained earth orbit, or into a flight around the moon. The timing of the announcement was not auspicious. The next day, NASA's first Mercury-Atlas (MA-1) disintegrated and fell into the ocean 58 seconds after takeoff from Cape Canaveral. This disaster ushered in a bleak four months during which the test rocket Little Joe 5 joined the MA-1 in the ocean, and the first Mercury-Redstone lifted a fraction of an inch and settled back on its launch pad. The last failure, on 21 November, marked the absolute nadir of morale for the engineers working on Mercury. The people at the new NASA headquarters in Washington, coping with financial and administrative problems and facing a change of administration after the national election, were only a little less dispirited than the workers in the field. But the fledgling space agency had an asset that made its announcement of an ambitious Apollo program more than an exercise in wishful thinking—it had the support of the American people.

If there is an American psyche, it had been shaken 4 October 1957 by the news that Russia had launched the first man-made earth satellite—Sputnik 1. To those apprehensive of anything Soviet, the news was a red flag. The military and the President played down Sputnik's significance, but a layman could not but wonder if Sputnik was one of those scientific breakthroughs that could alter the balance of power. The average American was perhaps most concerned because someone else was excelling in technology—an area in which the U.S. was accustomed to leading.

There was an almost unanimous determination to get into the space race and win it. Three Presidents, with firm support from Congress, channeled the public will into an answer to the Russian challenge. Lyndon B. Johnson, the Senate majority leader, pushed the Aeronautics and Space Act through Congress in 1958. Under its authority, President Eisenhower set up NASA and transferred the armed services' non-military space activities to the new civilian agency. The following year NASA received a vital asset—the Army team of former German V-2 experts who were working up plans for Saturn, a large rocket. Assigned the task of manned spaceflight, NASA's immediate goal was the successful orbiting of a man aboard a Mercury spacecraft. NASA's Ten Year Plan of Space Exploration, revealed to Congress in early 1960, called for nearly 260 varied launches during the next decade, with a manned flight to the moon after 1970. The House Committee on Science and Astronautics considered it a good program except that it did not move ahead fast enough.

Meanwhile, the Russians were not idle. On 12 April 1961, they put Major Yuri A. Gagarin into orbit around earth. The Soviet Union and the United States were locked in a confrontation of prestige in Cuba, in Berlin—and in space. Convinced it was necessary to show the world what America could do, President Kennedy told Congress on 25 May 1961:

> Now it is time to take longer strides—time for a great new American enterprise—time for this nation to take a clearly leading role in space achievement which in many ways may hold the key to our future on earth . . . . I believe that this nation should commit itself to achieving the goal, before this decade is out, of landing a man on the moon and returning him safely to earth. No single space project in this period will be more exciting or more impressive to mankind or more important for the long-range exploration of space and none will be so difficult or expensive to accomplish . . . . In a very real sense, it will not be one man going to the moon—it will be an entire nation. For all of us must work together to put him there.

If President and people were agreed on the end, what about the means? Kennedy's proposal was not made lightly. Before coming to a decision, he had taken counsel with advisors who believed that the moon project was feasible, largely because it could be accomplished without any new scientific or engineering discoveries. It could be done "within the existing state-of-the-art" by expanding and extending the technology that existed at that time.

What was the "existing state-of-the-art" as of 25 May 1961? Since December 1957, when the first Vanguard orbital launch attempt had collapsed in flame before a television audience, the United States had tried to put 25 other scientific satellites into earth orbit; 10 had been successful. Two meteorological satellites had been placed into orbit, and both had operated properly. Two passive communications satellites had been launched, but only one had achieved orbit. Nine probes had been launched toward the moon; none had hit their target, although three achieved a limited success by returning scientific data during flight. After its 1960 failures, NASA had put a Mercury with Alan B. Shepard aboard into suborbital flight on 5 May 1961.

Just 18 months before the Kennedy recommendation, the Atlas military missile, at that time America's most powerful space booster, had made its first flight of intercontinental range—some 10000 kilometers. Not three years had gone by since the smaller intermediate range ballistic missiles, Jupiter and Thor, had made their first full-range flights. Yet by May of 1961 none of these military rockets had reached a high degree of reliability as space carriers.

When the President laid his proposed goal before the Congress, the spacecraft that would carry man to the moon existed only as a theoretical concept tentatively named Apollo.

The powerful rocket that would be necessary to launch the spacecraft with sufficient velocity to escape earth's gravity was only a few lines on an engineer's scratch pad. Conceivably, it would be one of a family named Saturn: specially designed space carrier vehicles, each generation larger and of greater power than the preceding one. The first Saturn would not make its maiden flight for another six months.

The vast support, checkout, and launch facilities of the earthbound base whence men would launch other men on their journey did not exist. The moonport had yet to be located, designed, built, and activated—and this book tells that story.

Other books now being prepared for NASA deal with the other aspects of the program—the Saturn launch vehicles, the Apollo spacecraft, astronaut training and the missions. Another volume, a history of NASA administration, 1963–69, will include the headquarters story of Apollo.

The central feature of this book is launch complex 39 (LC-39), where American astronauts were launched toward the moon. Its story begins in early 1961 with the earliest plans for a mobile launch complex and proceeds through design and construction to the launching of Apollo 11 and subsequent lunar missions. The construction story is a big one—the building of the Apollo launch facilities was the largest project of its time. In many ways, however, the operations at LC-39 were an even greater challenge. As an Apollo program manager has noted, the Kennedy Space Center was at the "tail end of the whip." There all the parts of the Apollo program came together for the first time. The launch team ensured that the space vehicle would work.

While LC-39 is the principal focal point, it is not the only one. Two other Apollo-Saturn complexes on Cape Canaveral, LC-34 and LC-37, launched the program's early flights; at LC-34 the program's great tragedy occurred. The Apollo spacecraft were tested in the operations and checkout building in the Merritt Island industrial area. Vital telemetry equipment was located nearby in the central instrumentation facility. Moreover, the size and shape of the launch facilities were largely determined by the Saturn family of launch vehicles, which were produced under the direction of Marshall Space Flight Center at Huntsville, and by the Apollo spacecraft, under the Manned Spacecraft Center at Houston. An understanding of launch facilities and operations requires, to some degree, an appreciation of program-wide activities.

The history is complicated because planning, construction, and launch operations were conducted concurrently during much of the program. Three topics take up most of the first ten chapters: the construction of launch complexes 34 and 37 and the subsequent Saturn I tests; the planning of a moonport on Merritt Island and the purchase of that area; and the buildup of the launch team. Chapters 11–15 relate the design, construction, and activation of launch complex 39. Chapters 16–23 describe the Apollo launch operations from early 1966 through the launch of Apollo 17 in December 1972. Chapter 24 is a tentative summing-up.

The work comprehends three kinds of history: official, contemporary, and technological. The technology of the moonport crossed many scientific and engineering disciplines from microelectronics to civil engineering; expertise was needed in telemetry, fluid mechanics, cryogenics, computers—even lightning strikes. Although NASA engineers gave us a great deal of help, it was our task to make the technical terms comprehensible. Another problem stems from NASA's requirement that its authors use the new international system of units. One obvious way to comply, without losing most of our readers, would have been to give all measurements in both international and old-fashioned units. Unfortunately, with that solution the prose immediately bogs down. We have therefore proceeded as follows. First, where physical units were not essential, we have eliminated them. Second, the more familiar of the international units, such as meters and kilograms, we have used alone. Third, only the more esoteric terms, such as newtons, have we translated in the text.

The contemporary historian's task is to walk into a virgin forest of unsifted materials, with no clearings made by destruction of the unimportant and no trails blazed by prior researchers. Yet the journey can be propitious: we were able to interview hundreds of eyewitnesses who told it as they saw it. They recalled personality conflicts that sometimes affected major decisions. They narrated events never put down in writing and reached into personal files for documents not available in the archives. The use of eyewitnesses naturally required the resolution of some conflicting evidence, and their additional material increased the problems of selection. The insights gained, however, more than compensated for the trouble.

The great weakness of contemporary history, a want of perspective, is irremediable. Until the Russian story is on the record, our view of the space race is limited. Future judgments of the Apollo program will reflect further developments in space exploration. Thus, with respect to the launch facilities, the wisdom of building the moonport in the way it was done depends in part on the programs to be launched henceforth. The moonport was funded, designed, and built on the assumption that the lunar landing was only a beginning. With these considerations in mind, we defer to 21st-century historians a definitive evaluation of the effort.

Under the contract with the University of Florida, NASA enjoyed the rights to final review and publication of this book. We worked largely from NASA documents and with NASA officials. This may have tempered some of our conclusions, consciously or not, but we are satisfied that this is not a court history. Criticisms directed at the Kennedy Space Center (KSC) team and mistakes in the launch operations are treated in detail. Contrary to the wishes of some participants, conflicts within the program are aired. A greater fault may lie in our dependence on NASA documents. Although we tried to balance the account with corporation documents and interviews, the history inevitably focuses on NASA's direction of Apollo launch operations. The Apollo contractors and other support agencies, such as the Air Force, may receive less than their due.

Understandably, our treatment of certain events will not satisfy everyone. For example, too much controversy still surrounds the Apollo-Saturn 204 fire. We have largely avoided two other controversial questions. Was the KSC operation more or less efficient than other governmental projects of the 1960s? There was undoubtedly waste in the construction of the Apollo launch facilities and in the launch operations, but we are not in a position to judge the cost efficiency of the KSC team against similar projects, such as a large defense contract. The second question—the worth of the Apollo program—will be, as previously stated, left to future historians. In our personal view it was a noble goal, nobly achieved.

A word is in order with regard to Kennedy Space Center speech usages, especially acronyms. The scientists and engineers at KSC do not use a peculiar tongue to mystify the layman—but as a matter of fact, that is one result. When an LCC man says "the crawler is bringing the bird back from pad 39 to the VAB," he is understood by anyone at the space port. Every discipline has its technical language, which sometimes goes too far. We believe we reached the nadir in space jargon when we uncovered the record of a "Saturn V Human Engineering Interstage Interaction Splinter Meeting of the Vehicle Mechanical Design Integration Working Group."

Apollo scientists and engineers were establishing a terminology for new things; no one had defined them in the past because such things did not exist. *Module* is an example. As late as 1967, the *Random House Dictionary of the English Language* gave as the fifth definition of *module* under *computer technology:* "A readily interchangeable unit containing electronic components, especially one that may be readily plugged in or detached from a computer system." The space world was well ahead of the dictionary because, as every American television viewer knew, a module—command, service, or lunar—was a unit of the spacecraft that went to the moon. *Interface* is another word that was recast at the space center. Defined in the dictionary as "a surface that lies between two parts of matter or space and forms their common boundary," it grew to encompass any kind of interaction at KSC. Perhaps this was subliminal recognition that Kennedy Space Center was the Great Interface where the many parts and plans that went into the moon launch had to be fitted together.

Like all government agencies since 1950, NASA made extensive use of acronyms. In February 1971, the Documents Department of the Kennedy Space Center Library compiled a selective list of acronyms and abbreviations. It contained more than 9500 entries. We have tried to avoid acronyms as much as possible; when used, the acronym is coupled with its full and formal terminology on its first use.

The astronauts were quick to acknowledge that Apollo was a team effort. Appropriately enough, the same can be said for this history of the Apollo launch operations. We drew extensively upon the work of previous researchers. Dr. James Covington and Mr. James J. Frangie prepared material on the design and construction of the launch facilities. Dr. George Bittle and Mr. John Marshall performed helpful research on launch operations. Mr. William A. Lockyer and Mr. Frank E. Jarrett of the KSC historical office provided much reliable criticism. Dr. David Bushnell, the University of Florida's project director for the history, rendered administrative and editorial assistance. Finally, thanks are due to scores of KSC personnel who provided recollections, documents, and patient explanations on the workings of Apollo.

1

The First Steps

Genesis of the Saturn Program

America took its first step toward the moon in the spring of 1957, four years before President Kennedy declared the lunar expedition a national mission. While still preparing for the launch of its first Jupiter (31 May 1957), the Army rocket team at Huntsville, Alabama, began studies of a booster ten times more powerful than the 667 200-newton (150 000-pound-thrust) Jupiter. The tenfold increase in thrust could put a weather and communications satellite into orbit around the earth, or propel a space probe out of earth's orbit.

The change of emphasis from intermediate range and intercontinental ballistic missiles (Jupiter, Thor, Atlas) to a super-rocket capable of space exploration signified a change of attitudes at the Department of Defense. The change was also grounded in interservice politics: the previous November, Secretary of Defense Charles Wilson had assigned responsibility for all intermediate and long-range missiles to the Air Force. If the Army was to stay in the big-rocket business, it would have to find new tasks for its Wernher von Braun team of rocket experts at the Redstone Arsenal in Huntsville.* Maj. Gen. John B. Medaris, commander of the Army Ballistic Missile Agency (ABMA), set his sights on the new super-rocket, subsequently to be named Saturn.[†1]

Medaris's effort to gain Defense Department support for the big rocket was bolstered by the Soviet Union's accomplishments in the fall of 1957. The contrast between the 500-kilogram Sputnik 2 and America's

---

*In the collapse of the Third Reich in 1945, United States Army Ordnance seized 300 carloads of V-2 components—the operational rocket used by Germany in the last winter of the war. In addition, 115 German rocket specialists, led by Wernher von Braun, senior civilian scientist at the V-2 rocket station at Peenemünde, signed contracts to work in the United States. First located in Fort Bliss, Texas, and White Sands, New Mexico, the group was moved in 1950 to Redstone Arsenal, Huntsville, Alabama, headquarters for the Army Ballistic Missile Agency.

†Originally termed the Juno V, the super-rocket was renamed Saturn in Huntsville work papers of mid-1958, and the new name received official status in early 1959. From the beginning it had a dual connotation: (1) a clustered booster, and (2) a multistage rocket in which the clustered booster would serve as the first stage.

8-kilogram Explorer 1 was persuasive. In December von Braun's group (officially known as the Development Operations Division of the ABMA) set out arguments for the new booster program. The super-rocket would develop 6 672 000 newtons (1 500 000 pounds of thrust) and serve as a steppingstone to an even larger rocket capable of manned lunar missions. Its early development and adaptation in a multistage vehicle could accomplish a number of space objectives pointing toward a landing on the moon in 1967.[2]

Although the ABMA proposal was reinforced by the public's embarrassment over Sputnik, approval for the Huntsville project was delayed for several months. Medaris's program faced two obstacles: the Eisenhower administration's fiscal conservatism and the priority given to intercontinental missiles. While Medaris pressed his campaign, the von Braun team was far from idle. Between April 1957 and August 1958, ABMA logged 50 000 man-hours on the project. Finally, in July 1958, the Advanced Research Projects Agency, established earlier that year to coordinate Defense Department space activities, announced its intention to develop a super-rocket. The following month ABMA was directed to start on the Saturn.[3]

In September 1958, General Medaris and Roy Johnson, the Director of the Advanced Research Projects Agency, established a flight-test schedule of four Saturn launches. The first was set for September 1960. The third, eight months later, would employ an upper stage to place limited payloads in orbit. The written agreement between the two men was still shadowed by the Eisenhower administration's reluctance to spend money on non-military space ventures. Johnson promised to provide $72.3 million over a three-year period. (The Saturn I program would eventually cost more than a billion dollars.) The size of the commitment meant that, at least in the beginning, Saturn would operate on a shoestring.[4]

The original Saturn design reflected a concern to save time and money, and to employ components that could be moved by air transport. The booster made extensive use of available Army hardware. It used eight engines and a cylindrical center tank copied after the Jupiter, a single-stage rocket with a range of 2700 kilometers. For its eight clustered tanks, the von Braun team went back to their favorite Redstone rocket. The propellants would be RP-1 (kerosene) and liquid oxygen.

Early plans included a stipulation that no component could exceed 11 340 kilograms or a cross-sectional dimension of 3 meters, the maximum limits of aircraft transport at the time. To meet these limitations, the booster was initially designed with the center and eight outer tanks separate from the frame and engine assembly. The fuel tanks were to be mated with the frame on the launch pad. The idea was discarded in early 1959 for two reasons. Huntsville engineers agreed that flying out a disassembled thrust unit and rebuilding it on the pad would reduce reliability; and transportation studies indicated that air freight by 11 C-124s would cost more than construction of a cradle to carry the Saturn down the Tennessee River by barge.[5]

A Saturn Launch Site

With better than 20 years' experience, the von Braun team preached and practiced that rocket and launch pad must be mated on the drawing board, if they were to be compatible at the launching. The new rocket went hand in hand with its launching facility. The short-lived plan to transport the Saturn by air was prompted by ABMA's interest in launching a rocket into equatorial orbit from a site near the Equator; Christmas Island in the Central Pacific was a likely choice. Equatorial launch sites offered certain advantages over facilities within the continental United States. A launching due east from a site on the Equator could take advantage of the earth's maximum rotational velocity (460 meters per second) to achieve orbital speed. The more frequent overhead passage of the orbiting vehicle above an equatorial base would facilitate tracking and communications. Most important, an equatorial launch site would avoid the costly dogleg technique, a prerequisite for placing rockets into equatorial orbit from sites such as Cape Canaveral, Florida (28° north latitude). The necessary correction in the space vehicle's trajectory could be very expensive—engineers estimated that doglegging a Saturn vehicle into a low-altitude equatorial orbit from Cape Canaveral used enough extra propellant to reduce the payload by as much as 80%. In higher orbits, the penalty was less severe but still involved at least a 20% loss of payload. There were also significant disadvantages to an equatorial launch base: higher construction costs (about 100% greater), logistics problems, and the hazards of setting up an American base on foreign soil. Moreover in 1959 there was a question as to how many U.S. space missions would require equatorial orbits. The only definite plans for equatorial orbits were in connection with communications and meteorological satellites operating at 35000 kilometers.[6]

While there was disagreement over the merits of an equatorial base for future Saturn operations, the Atlantic Missile Range was the clear choice for the developmental launchings. At the range's launch site, Cape Canaveral, the Air Force Missile Test Center provided administrative and logistical support. The range's ten tracking stations, stretching into the South Atlantic, gave good coverage of test flights. Moreover, ABMA's launch team, the Missile Firing Laboratory (MFL), had launched missiles from Cape Canaveral since 1953. Cost and time considerations agreed. As an MFL study noted, the Atlantic Missile Range met "the established [launch] criteria in the most efficient, timely manner at a minimum cost."[7]

The Making of "the Cape"

Cape Canaveral, better known as "the Cape," had been earmarked as a missile testing range in 1947.* An elbow of land jutting out into the Atlantic midway between Jacksonville and Miami, the Cape covers about 60 square kilometers. Early Spanish sailors, marking it down as the only major feature of the long Florida coast line, named it for its abundance of cane reeds. Its choice as a missile range was dictated by several factors: the planners could set up a line of tracking stations stretching southeasterly over the Atlantic to provide the longest range necessary for missile testing; the Banana River Naval Air Station could serve as a support base; and the launch area was accessible to water transportation. The Air Force took over the Banana River Naval Air Station on 1 September 1948, contemplating its use as a headquarters for a Joint Long Range Proving Ground. The Coast Guard opened its 2.5 square kilometers on Cape Canaveral to missile use in February 1950. The government obtained the remainder from private owners by negotiation or condemnation.

Cape Canaveral was a scenic but comparatively unsettled place—beautiful beaches, excellent fishing areas, a lighthouse, scattered private residences, an inn that became the Cape Canaveral Auxiliary Air Force Base Headquarters, a few unpaved roads or trails, a dock used by shrimpers, and welcome and unwelcome wildlife including deer, alligators, rattlesnakes, and many millions of the pests that gave their name to Mosquito Lagoon to the north. In a clearing, made by burning the underbrush and uprooting the palmettos with bulldozers, construction workers completed a concrete pad on 20 June 1950. They also cleared all land within 1.6 kilometers of the pad.

Few pictures reflect the state of American rocketry in 1950 so accurately as the first launch pad at Cape Canaveral. It was a 30-meter-wide layer of concrete, poured on top of sandy soil a little more than a kilometer north of the lighthouse. When a dozen jeeps and delivery trucks sank to their axles on the sandy paths that passed for roads, a layer of gravel was laid over the

---

*The selection was made by a Joint Chiefs of Staff committee. When the armed services went into rocketry in 1945, the Army stationed its launch team of German V-2 experts at White Sands, New Mexico—near the scene of Robert Hutchings Goddard's pioneering work in the 1930s. The southwestern desert proved too small for rockets. On 29 May 1947, a modified V-2 went the wrong way and landed in a cemetery south of Juarez, Mexico—one of the factors that decided the Joint Chiefs to move rocket experiments to the east coast of Florida.

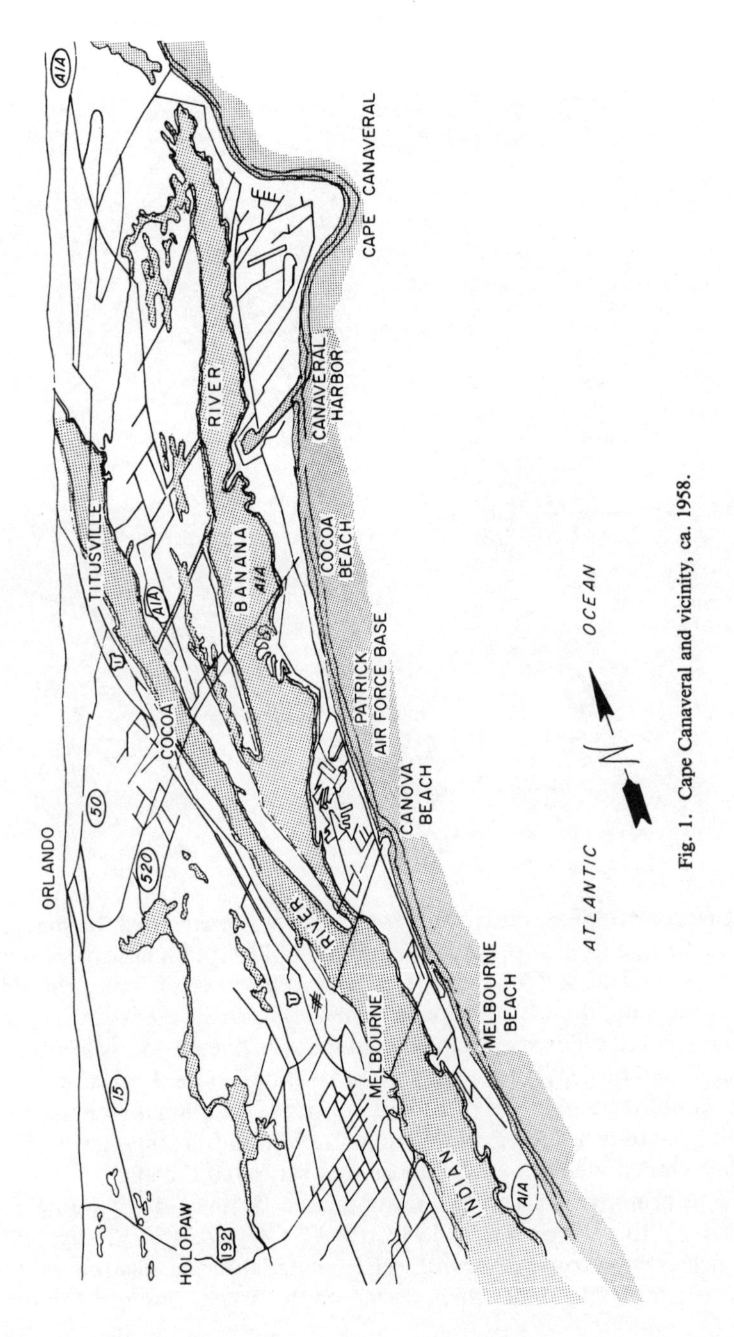

Fig. 1. Cape Canaveral and vicinity, ca. 1958.

Fig. 2. Cape Canaveral. View south from the lighthouse, ca. 1950.

sand. Steel scaffolding, purchased from painters, surrounded the missile to form the first gantry, or service support tower. Plywood platforms stood at various levels of the scaffolding. If more than ten workers climbed the piping at the same time, the whole rickety framework seemed ready to fall down. The crew stacked sandbags around an old shack, a onetime dressing room for swimmers, and turned it into a launch control blockhouse. It stood a scant 91 meters from the pad. A row of trailers contained additional facilities to coordinate countdown, information, and reports from tracking sites. Heat and humidity sapped men's energy. Mosquitoes saturated the air.

The primitive spaceport was inaugurated 19 July 1950 by Bumper 7, a modified V-2 first stage combined with a WAC Corporal second stage. While the launch crew—Army, General Electric, and California Institute of Technology people—and 100 newsmen waited on the beach, Bumper 7 sputtered and fizzled at countdown. An autopsy revealed that salt air had corroded some of its elements. Five days later, the launch crew tried again with Bumper 8, a sister missile. The missile rose steadily into the air while a thundering roar rolled across the Cape. At 15 500 meters, the WAC Corporal second stage ignited and accelerated to 4350 kilometers per hour before dropping into the sea. Thereafter, the Cape was in almost continuous use as the armed services brought missiles to Florida for testing—the Lark, Matador, Snark, Bomarc.[8]

The Cape had its share of growing pains. The Korean War diverted funds. The multi-service operation posed problems. On 30 June 1951, the Defense Department changed the official title of the Air Force unit managing the Cape from Headquarters, Joint Long Range Proving Ground Division, to Headquarters, Air Force Missile Test Center, with the Air Force in sole charge. The Cape was designated the Cape Canaveral Missile Test Annex. The Navy had Point Mugu, California, and the Army had White Sands, New Mexico. But soon after the Army's rocket team moved to Huntsville, a representative was knocking on the door at the Cape, asking for launch facilities.

In the meantime, negotiations with Great Britain resulted in the Bahamas Long Range Proving Ground Agreement on 21 July 1951. This pact and subsequent agreements gave the United States the use of a 1600-kilometer range through the Bahamas with tracking stations at Point Jupiter, Florida; Grand Bahamas Bank; and Grand Turk Island. Subsequent negotiations extended the range to Ascension Island, more than 8000 kilometers southeast of Cape Canaveral.[9]

While working out the downrange bases, the Air Force had to cope with a communications problem at home. The division of operations between the administrative headquarters at Patrick Air Force Base and the launch site at Cape Canaveral, 29 kilometers to the north, resulted in a costly duplication of effort. In the summer of 1953 Pan American World Airways, an old hand at operating bases around the world, convinced the Air Force that it could reduce the costs of running the range. Pan American was awarded a contract for day-to-day operations and was soon engaged in many activities from setting up cafeterias to providing security on the pads. The Radio Corporation of America received a subcontract for the technical aspects of range operations.

With the launch of Redstone #1 in August 1953, the Missile Firing Laboratory inaugurated the testing of ballistic missiles. In those days, launch procedures were unsophisticated. Albert Zeiler, one of the Peenemünde veterans, had to decide within a split second whether to shut off the engine immediately after ignition, basing his decision upon the color of the flames. An off-color indicated an improper mix of the propellants. A couple of minor delays had occurred earlier, but on the morning of 20 August 1953 the flame color met Zeiler's approval, and the Redstone rose. The powered flight lasted only 76 seconds and fell far short of the anticipated 257-kilometer range. Still the missile met most of the test objectives, its structure proved sound, and the propulsion system worked well.[10]

Building a Launch Complex

By the late 1950s, the Cape Canaveral skyline already had distinctive features. Towering gantries rose along "ICBM Row." The various missiles had certain similarities in ground environmental needs and operational requirements. In the test phase, each required an assembly and checkout building, transport from assembly area to launch complex, a launch pad, a gantry service tower, a blockhouse for on-site command and control of the launch, and a network of power, fuel, and communication links that would bring it to life. For a long while, the complexes resembled each other. Igloo-shaped blockhouses stood 230 meters from the pads and looked like the pillboxes of World War II. They provided protection for the launch crew and the control consoles and instrumentation. In the case of complexes 11, 12, 13, and 14, designed for the Atlas ICBM, the inside walls of the 12-sided domed structures were 3.2 meters thick at the base with 12 meters of sand around them.

Besides the blockhouse or launch control center, the essential features of a fixed-pad complex included a concrete or steel pedestal on which to erect and launch the vehicle, a steel umbilical tower to provide fluid and electrical connections to the vehicle, a flame deflector, and a mobile service structure that moved around the vehicle so ground crews on platforms could service and test various components. Other features of the complex included an operations support building, storage facilities for kerosene and liquid oxygen, a tunnel for instrumentation and control cables, roads, camera sites, utilities services, and security.

Three factors largely determined the choice of sites for the launch complexes: explosive hazards, the dangers of overflight, and lines of sight. In 1959 the launch planners assumed that the first five or ten missiles in a new program would have a high rate of failure on the pad or shortly after launch. Approximately 5% of the Cape's previous developmental launches had exploded a few seconds after takeoff, most of them in an area 10° to either side of the intended azimuth (direction) of launch. Experience thus showed the wisdom of locating a pad in an area where there were no permanent facilities immediately downrange. Likewise, the frequency of accidents during test

Fig. 3. ICBM row, December 1967.

programs made backup pads desirable. The explosive hazard further influenced the placement of facilities within the launch site to minimize damage to "long-lead-time" equipment. Planners also had to maintain a clear line of sight from the launch vehicle to the launch control center, and to electronic and optical instrumentation sites.[11]

To meet the constantly expanding needs of the many missile groups, the Corps of Engineers eventually built 21 missile assembly buildings patterned after Marine Corps hangars at El Toro, California. Shop, office, and assembly area met the requirements of the early missiles; inside, a maze of power and instrumentation circuits ran through covered trenches. Criteria prepared by the Facilities Division of the Joint Long Range Proving Ground standardized the basic framework of the last 18 of these assembly buildings and developed overhead cranes that were interchangeable in all structures.[12] As missiles grew more complicated over the years, the assembly buildings began to reflect the characteristics of the individual vehicles they would service.[13]

Missions for Saturn

In the fall of 1958, the Army Ballistic Missile Agency's Missile Firing Laboratory (MFL), after five years at Cape Canaveral, was concluding its Redstone research and development program; the launch on 5 November was the last in a series of 38. A parallel program, training field artillery units to launch Redstone, was also nearing completion. With Redstone attaining operational status, MFL's Cape activities would center around Jupiter launches and the preparation of Pershing facilities. Big on the horizon was its greatest challenge—Saturn. Although Defense Department officials had approved the Saturn rocket and its Cape Canaveral launch site, wheels at Washington would grind another 18 months before the program was (to indulge in government jargon) finalized. The rocket teams at Huntsville and Cape Canaveral had to work, if not in the dark, at least in a twilight zone where there were few certainties. What was the United States going to do in space? What part would the Saturn have in the space program? What governmental agency would handle its development? How much money would be available? It was the beginning of the if-and-when planning that would bedevil the program for five years.

Even as initially set up by General Medaris and Roy Johnson, the project was dotted with question marks. Some were in the technological area, involving the working out of the overly simplified reference in the Medaris-Johnson pact to "booster flights which, without sophisticated upper stages, would be capable of placing limited payloads in orbit" (page 2). More questions developed from the involved process of transferring the Saturn project from the Army to NASA. In 1958, the Defense Department's Advanced Research Projects Agency (ARPA) was dealing with the Army Ballistic Missile Agency (ABMA) concerning the Development Operations Division's Saturn, and its Missile Firing Laboratory's Saturn launch facilities. By 1960 NASA's Office of Launch Vehicle Programs was handling the same subject matter with the Marshall Space Flight Center (MSFC) and its Launch Operations Directorate (LOD). All of this called for much clearing of the lines of authority.

Meanwhile, the space experts debated the use of the new booster in multistage vehicles. In December 1958, with Saturn still an Army project, ARPA ordered ABMA to study future Saturn configurations with second and third stages. Herman Koelle, chief of the Future Projects Office, directed a task group in an examination of 1375 configurations during the next three months. The study concluded that a modified version of the Atlas, the 3-meter-diameter Titan, or the 4-meter Titan could be used as a second stage on top of the Saturn booster already on the drawing boards at Huntsville. The Centaur was recommended as the logical choice for the third stage.* An ARPA evaluation committee, composed of NASA and Defense Department members, accepted the study findings and selected the 3-meter Titan for the second stage. In May 1959, ABMA was directed to develop the three-stage Saturn.[14]

Within days after completing the Saturn systems study, the Koelle group was attempting to devise an appropriate mission for the super-rocket. A 24-hour communications satellite, the only firm requirement for Saturn, did not justify ABMA's large expenditures. Koelle's answer was Project Horizon, a plan to place a military colony on the moon. The summary of the five-volume Horizon study appeared in June 1959. The report proposed a manned lunar landing in 1965, with establishment of a 12-man lunar outpost the following year. As logistical support for a lunar base would require the launching of 64 Saturns annually, approval of the Horizon project would secure ABMA's position for at least a decade.[15]

While ABMA and the Army examined ways to employ the Saturn, NASA was drawing up its own plans for programs beyond Mercury.† Suggestions included an earth-orbiting manned space station, manned circumlunar flights, manned lunar landings, and ultimately interplanetary flights.

---

*The Air Force began work on the Titan I missile in May 1955 as a backup to the Atlas. The missile was 30 meters long, burned LOX and RP-1, and relied on radio guidance. It first flew at AMR on 5 Feb. 1959. The Centaur, the earliest hydrogen-fueled stage, was built by Convair and achieved 133 440 newtons (30 000 pounds of thrust).

†Mercury was the first U.S. manned spaceflight program. Its objectives—orbital flight and successful recovery of a manned satellite, and a study of man's capabilities in a space environment—were achieved in a series of flights, 1961-63. See Loyd S. Swenson, Jr., James M. Grimwood, and Charles C. Alexander, *This New Ocean: A History of Project Mercury*, NASA SP-4201 (Washington, 1966).

NASA appointed the Research Steering Committee on Manned Space Flight, chaired by Harry J. Goett of Ames Research Center, to study those suggestions. On 25 May 1959, the committee recommended manned interplanetary travel as NASA's ultimate goal. As a more immediate objective, some members wanted manned flights around the moon; others wanted to land on the moon. George Low of Space Flight Development strongly urged the latter objective. He believed that, among other advantages, Congress would more readily fund this package. He further urged using existing vehicles, such as the Army's Saturn booster, rather than developing a completely new and larger launch vehicle.[16]

Meanwhile, NASA's Office of Program Planning and Evaluation, under the direction of Dr. Homer Joe Stewart, whose specific task was to formulate an overall program, set up a Long Range Objectives and Program Planning Committee. With the assistance of the Goett Committee, the Planning Committee submitted a working draft on 1 June 1959, spelling out the problems, costs, and equipment required for landing one or two men on the moon and returning them safely to earth after a period of exploration.[17]

A Marriage of Convenience

At this point the Army had a Saturn vehicle for which it was seeking a mission, and NASA had a mission for which it was seeking a vehicle. A marriage of convenience was indicated. Dr. T. Keith Glennan, first NASA administrator, had attempted to bring half of the von Braun team into his new organization on 15 October 1958. Secretary of the Army Wilbur Brucker and General Medaris successfully rebuffed that effort; the Army still had military projects to supervise (Jupiter and Pershing) and did not want to break up the von Braun team. Brucker suggested, as a compromise, that NASA place a liaison group at Huntsville and plan to use the Redstone Arsenal facilities for certain programs. Coveting the Saturn program, NASA accepted Brucker's proposal as the best of a bad bargain. In January 1959, ARPA and NASA representatives established a National Space Program. NASA would concentrate on smaller vehicles while the Defense Department developed larger ones including the Saturn. Although this understanding appeared to secure a role for Saturn, it actually spelled trouble for ABMA. The Huntsville organization had hoped that NASA would provide financial assistance for Saturn since the new space agency would likely use the big booster. NASA, however, unable to direct the Saturn program, refused to underwrite any of its costs.[18]

Saturn's prospects worsened after a key Defense Department official opposed the Army program. In the spring of 1959, Dr. Herbert F. York, newly appointed Deputy Secretary for Research and Engineering, assigned responsibility for future military space activities to the Air Force. Having previously disclaimed any Defense interest in moon exploration, York in April indicated a desire to cancel the Saturn.* He could see no military justification for the big rocket. ARPA, perhaps influenced by York, suspended studies of the second stage on 31 July, directing ABMA to conduct a new series of cost and time estimates based on a 4-meter Titan. The larger Titan offered several advantages, including compatibility with the Air Force Dyna Soar, a manned space-glider program.[19]

Two decisions in September reaffirmed the Saturn program. An ARPA-NASA Large Booster Review Committee, after examining Army, Air Force, and industry programs, recommended the clustered Saturn booster as "the quickest and surest way to attain a large space booster capability in the million-pound thrust [4 448 000-newton] class."[20] York and Dr. Hugh Dryden, NASA's Deputy Administrator, reached a similar conclusion in their comparison of the Saturn and the Air Force's Titan C proposal. (The latter would have employed a cluster of upgraded Titan I engines to provide a thrust comparable to the Saturn.)[21] The York-Dryden committee also recommended that ABMA conduct a new study of second and third stages.

ABMA presented a second Saturn systems study to a Defense Department conference in Washington 29–30 October 1959. The report offered four alternative configurations, ranging from a Titan second stage and Centaur third stage to an optimum vehicle with a new 5.6-meter-diameter conventional second stage (burning RP-1), a new hydrogen-fueled third stage, and a Centaur fourth stage. Knowledge that President Eisenhower had decided to transfer Saturn and the Development Operations Division to NASA lessened the study's impact. After assuming technical direction of the Saturn in November, NASA initiated still another study of upper stages. Dr. Abe Silverstein, NASA's Director of Space Flight Development, headed a committee representing the Air Force, NASA, ARPA, and ABMA.[22]

Upper Stages

The Silverstein Committee established two criteria for a successful Saturn program: development of a rocket with an early launch capability as well as growth potential. The group listed three missions for the initial Saturn

*In a letter to the authors, York elaborated on his motivation. In early 1959 York viewed the U.S. space program as a "mess" and thought the transfer to NASA of the von Braun team and its big booster would improve matters. Neither the Army nor the Navy needed large rockets, and the Air Force was developing the Titan. NASA, on the other hand, required large boosters in future space programs. York wrote, "While ARPA did have other legitimate roles in Defense R&D, I concluded it was really just one more unnecessary layer in the management of large rocket and space programs, and so I recommended its role in Space be cancelled."

vehicle: unmanned lunar and deep space missions with an escape payload of about 4500 kilograms; 2250-kilogram payloads for a 24-hour equatorial orbit; and manned spacecraft missions in low orbits, such as Dyna Soar. The committee matched a number of configurations against these missions. Current ICBMs such as the Titan were adjudged unsatisfactory; they would not generate sufficient thrust for the lunar mission. A larger, conventionally fueled second stage—5.59-meter diameter—met mission requirements, but time and cost seemed excessive for a rocket stage with little growth potential. The solution lay with the early development of high-energy (liquid hydrogen) propellants for all stages above the first. In defense of this rather bold position the committee noted: "If these propellants are to be accepted for the difficult top-stage applications, there seems to be no valid engineering reasons for not accepting the use of high-energy propellants for the less difficult application to intermediate stages." The committee also recommended a building block concept stating that "vehicle reliability will be emphasized . . . through a continued use of each development stage in later vehicle configurations." The Saturn C-1* would consist of the clustered booster, a new Douglas Corporation second stage with four hydrogen-burning Centaur engines of 66 720–88 960 newtons (15 000–20 000 pounds of thrust) per engine, and a modified Centaur as a third stage. The C-1 would become the C-2 upon insertion of a new oxygen-hydrogen second stage with two 667 200–889 600-newton (150 000–200 000 pounds of thrust) engines. The top two stages of the Saturn C-1 would then become stages three and four on the C-2 version. The committee proposed to launch ten C-1s starting in the fall of 1961.[23]

On the last day of 1959, Glennan approved the Silverstein recommendations, and Saturn got its upper stages. Chances of meeting the new schedule improved with two Eisenhower administration decisions in January 1960. The Saturn project received a DX rating, which designated a program of highest national priority. Besides reflecting the administration's support, the rating gave program managers a privileged status in securing scarce materials. More important, the administration agreed to NASA's request for additional funds. The Saturn FY 1961 budget was increased from $140 million to $230 million.[24] On 15 March 1960 President Eisenhower officially announced the transfer of the Army's Development Operations Division to

*Until 1963 Saturns were classified by a *C* and an arabic numeral. People generally assume that *C* stood for configuration; but according to Kennedy Space Center's *Spaceport News* (17 Jan. 1963), MSFC engineers used it to designate vehicular "concepts." Saturn C-1 denoted the concept of the S-I booster topped with upper stages using liquid hydrogen as a propellant. C-2, C-3, and C-4 were drawing-board concepts that preceded the C-5 (Saturn V) moon rocket. For additional information on the origins of Saturn, see John L. Sloop, *Liquid Hydrogen as a Propulsion Fuel, 1945–1959,* NASA SP-4404, in press, chap. 12.

NASA. He took the occasion to name the Huntsville installation the Marshall Space Flight Center, for his wartime commander, General George C. Marshall. The DoD's Missile Firing Laboratory at Cape Canaveral became the Launch Operations Directorate of the new organization.

2

Launch Complex 34

The Director

The Missile Firing Laboratory's director, Dr. Kurt Debus, had wasted no time in getting a launch pad ready for the new rocket. Early on the morning of 26 September 1958, four days after the Medaris–Johnson Agreement to launch four Saturn boosters, a small group of MFL members left Huntsville Airport for Patrick Air Force Base. They joined the Cape Canaveral members of the MFL team in Debus's office for a discussion of ways and means of putting the proposed super-rocket into space.

After the Army had relocated its missile team from Fort Bliss to Huntsville in 1950, Wernher von Braun became Technical Director of the Ordnance Guided Missile Center. Debus, who had worked with von Braun at Peenemünde, was Assistant Technical Director. When von Braun established an Experimental Missile Firing Branch, Debus was placed in charge. The name was changed to Missile Firing Laboratory in January 1953, with Debus remaining at the helm. MFL maintained offices at Huntsville, although Debus spent much of his time at Cape Canaveral. During his early years at the Cape, Debus wrestled with a gamut of problems. One was a shortage of experienced people; a year after its formation, his team had only 19 members. The launch team for Redstone 1 in the summer of 1953 numbered 82, but only 37 were permanently assigned to the Missile Firing Laboratory. As the Redstone and Jupiter programs burgeoned, MFL grew also and by 1960, on the eve of its transfer from the Army to NASA, numbered 535 people. Thus, while the well-known "von Braun team" operated in Alabama, a less known and initially subsidiary "Debus team" was growing up at Cape Canaveral.[1]

Slowly the qualities of Dr. Debus became evident as he moved out of the shadow of the more charismatic von Braun. A doctor of philosophy in engineering from Darmstadt University, Debus had been headed for a professor's chair when he was recruited into the Peenemünde group. Debus was a systematic man; he kept a daily journal and believed a well-ordered desk was a sign of an orderly mind. On his monthly inspections, he might help a subordinate clear his desk of nonessentials; or he would do it himself if the man was away at the time. He purged his own files regularly.

Totally committed to his work, Debus expected total commitment from those with him. Thus he would have less respect for a happy-go-lucky individual, no matter how well that man might do his job, than for one who shared his own seriousness of deportment. He set his goals and brooked no opposition to them. But he allowed his subordinates a choice of methods in reaching those goals. He relied more on his personal experience of a man's capabilities than on records or written recommendations—a penchant he could not indulge in later years as the operation expanded. While not outgoing in manner, he had a deep concern for others. He showed the same reserved courtesy to the electrician who interrupted his busy day to replace a burned-out fluorescent tube as to the congressional leader who came to his office to discuss launch operations. While his team was small, he remembered birthdays with letters and cards. Straightforward in approach, he let his achievements speak for him—not always the most effective means of getting ahead. He was a man to get the job done. Now his job was to put a Saturn into space.

The proposed super-rocket dwarfed anything heretofore handled by the Army Ballistic Missile Agency (see table 1). The problems caused by the clustered engines were particularly significant. To guarantee proper ignition of all engines, the booster would have to be held on the launch pad for a few seconds. A complex mechanism to do this had to be developed. There was also a psychological factor, related to the Saturn's great expense. With

TABLE 1. COMPARISON OF ROCKETS LAUNCHED BY MFL/LOD/LOC, 1953–1965

(Vehicle characteristics varied during rocket development; figures represent an approximate average.)

	Redstone	Jupiter	Saturn I (Block I) SA-1–SA-4	Saturn I (Block II) SA-5–SA-10
Height (meters)	21	18	50	58
Diameter (meters)	1.75	2.63	6.40	6.40
Propellant weight (kilograms)	18 000	38 500	290 000	450 000
Total weight at liftoff (kilograms)	28 000	50 000	430 000	515 750
Total thrust (newtons)	333 600	667 200	5 871 400	1st stage, 6 690 000 2d stage, 400 300
RF links	2	4	8	13
Telemetered measurements	116	215	560	1180
Pad time (days)	15	25	61	103

previous military missiles, launch equipment failures had been relatively inconsequential. Each program called for a number of tests; the MFL staff learned from mistakes. The millions of dollars tied up in each Saturn, however, meant that launch facility failures could not be tolerated. Finally there was the problem of time. With the first launch only two years away, there could be no serious delay in determination of criteria, in design, or in construction.[2]

Conversations with the Air Force

The purpose of the MFL staff meeting on 26 September 1958 was to determine the support requirements needed from the Air Force. A number of topics were discussed including safety zones, construction costs, fuel requirements, instrumentation, a service structure, and a launch site. The matter of a site, for what would eventually be launch complex 34, received further attention that Friday afternoon when Debus introduced the Saturn project to Maj. Gen. Donald N. Yates, Air Force Missile Test Center Commander. Debus suggested placing the new complex in the central part of the Cape near pad 26. That pad was presently in use for the Jupiter program, but would be phased out in 1960. Yates believed that construction near LC-26 would interfere with other contractors and pose safety problems. He suggested the use of areas near complex 20 (Titan), complex 11 (Atlas), or at the north end of the launch area, which had been tentatively reserved for large boosters.[3]

During the next two weeks an MFL facilities team made a preliminary survey of five possible sites. James Deese drew upon eight years of Cape experience in directing the survey. The team focused much of its attention on ground safety. The potential blast effect of an explosion on the pad established a ground safety zone and a minimum intraline distance. The safety zone, marking the danger area for exposed personnel, would be cleared of all persons 30 minutes prior to launch. The minimum intraline distance delimited the area within which a pad explosion would cause damage to adjacent pad structures or vehicles. Deese estimated that the fuel would have half the explosive force of TNT. With an estimated fuel load of 476 tons (equivalent to 238 tons of TNT), the three-stage Saturn would require a ground safety radius of 1650 meters and intraline distance of 400 meters. The proposed firing azimuths (44° to 110°) excluded sites that would result in overflying permanent launch facilities already constructed to the east.[4]

The Deese team recommended only one site, an area approximately 300 meters north of complex 20. By using the existing Titan I blockhouse (launch control center) at LC-20, costs and construction time would be minimized. The Air Force Missile Test Center objected to this location, contending that the Saturn pad should be at least 610 meters from other structures. This precluded joint use of the Titan blockhouse, because the data transmission equipment used in checkout of the Saturn would be adversely affected by voltage drops over a 610-meter circuit.* MFL arguments that the Air Force recommendation would increase facility costs by 30% and construction time by four months proved to no avail. In mid-January, after a six-week delay, the Advanced Research Projects Agency sited the Saturn complex 710 meters north of pad 20.[5]

Writing the Criteria Book

Criteria development for the Saturn complex proceeded more cordially. Close coordination was required between four groups: MFL, the Systems Support Equipment Laboratory of the Development Operations Division at Huntsville, the Jacksonville District Office of the Army Corps of Engineers, and an architect-engineering firm. Their goal was to collect and organize all the data necessary for satisfactory design and construction. The procedures used in developing Saturn launch criteria followed a pattern set in earlier programs. MFL and the Systems Support Equipment Laboratory prepared basic data on all launch facilities and equipment. The architect-engineer then formalized the data in a criteria book. The Army Corps of Engineers reviewed this document for cost, utility, and compliance with federal and Atlantic Missile Range codes. The launch criteria book provided a general description of facilities, proposed methods of construction, the placement of utilities and equipment, facility dimensions, distances between facilities, cost estimates, and preliminary drawings.[6]

The blockhouse for LC-34 was patterned after the control center at complex 20. The reinforced concrete design permitted the planners to locate the structure 320 meters from the launch pedestal. A domed roof would be built up in three layers: an inner layer of reinforced concrete 1.5 meters thick; a middle layer of earth fill 2.1 to 4.2 meters in depth; and a 10-centimeter cover of shotcrete. The last, a concrete with a high cement content, was pressure-driven through a 15-centimeter tube onto a reinforced

*Some MFL officials believed the Air Force simply did not want to share blockhouse 20. The Air Force, however, consistently gave range safety a high priority. As General Yates recalled, the Air Force received numerous complaints from contractors because of concessions the Missile Test Center made to MFL.

mesh screen. The 930 square meters of floor space provided room for 130 persons, with test and launch consoles, instrumentation racks, remote control fueling devices, and television and periscope equipment for the observation of activities on the launch pad. Blockhouse operations required substantial air conditioning for such equipment as computers, as well as for the people. Should a delay in firing occur after the rocket was fueled, the blockhouse could be buttoned up for 20 hours. Two tunnels provided escape routes in case an explosion sealed the door.[7]

Two Cape veterans, R. P. Dodd and Deese, drew up preliminary criteria for the launch complex. Their plans called for a two-pad complex with only the northern pad (pad A) constructed initially. A raised concrete circle 130 meters in diameter would form the base of the pad. The central area's slight depression facilitated replacement of refractory brick after a launch. Dodd included a water deluge system to reduce the intense heat and wash away spilled fuel, which would be channeled toward a perimeter trench. A skimming basin would prevent kerosene from entering the area's drainage ditches. Beneath the pad, a series of rooms provided space for mechanical and electrical checkout and firing equipment such as terminal boards, instrumentation racks, electrical cables, and generators.

Three facilities along the south edge of the complex would service the Saturn's propellant needs. In the southeast corner near the ocean stood tanks for RP-1, a grade of kerosene, to fuel the Saturn I booster (first stage). The liquid oxygen (LOX) tank in the middle of the southern boundary stored the oxidizer for all Saturn stages. This tank was insulated; in its liquid state, oxygen is cryogenic—super cold—with a boiling temperature of 90 kelvins (−183°C). Dodd and Deese placed a high-pressure-gas facility in the southwest corner of the complex, near the blockhouse. The tanks in this storage area held two gases, nitrogen and helium, used in launch operations. Large amounts of nitrogen were used to purge and dehumidify the cryogenic lines that ran from the LOX tanks to the Saturn vehicle. The nitrogen also actuated LC-34's pneumatic ground support equipment. On later launches, gaseous helium would be used to purge the hydrogen fuel lines to the Saturn upper stages. With an even lower temperature than liquid oxygen, liquid hydrogen boils at 20 kelvins (−253°C). Since nitrogen would solidify in the presence of liquid hydrogen, helium was substituted. A few bottles of nitrogen and helium went aboard the launch vehicle to pressurize some of the subsystems.

In the final plans, the flame deflector and its spare were parked north of the pedestal. The service structure pulled away on rails running from the pad to a parking area 185 meters west. The designers placed the umbilical tower on the northeast side of the launch pedestal. Eventually 70 meters

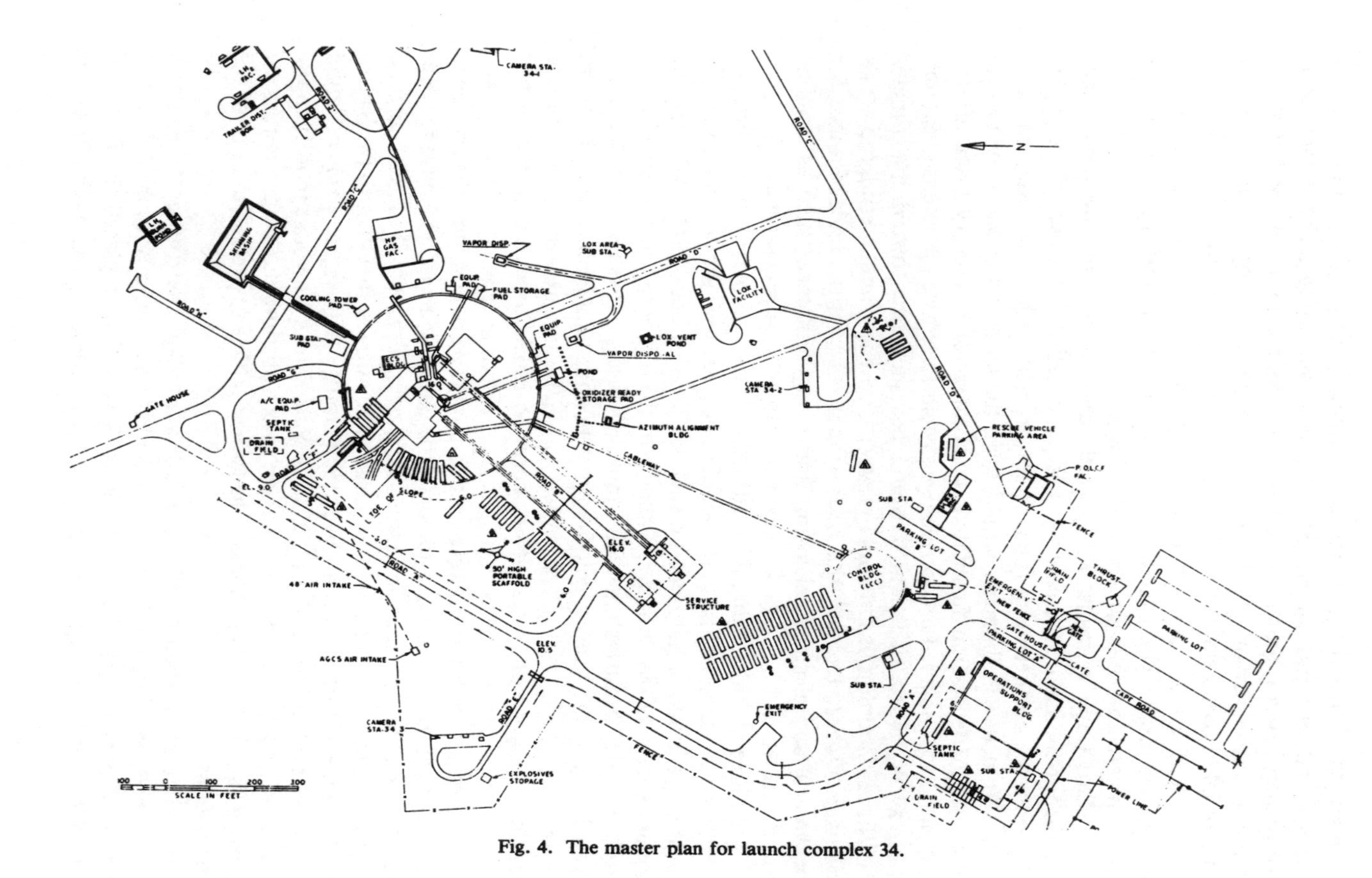

Fig. 4. The master plan for launch complex 34.

high, it would carry fuel lines and other connections to the Saturn before liftoff. Two requirements governed the location of the umbilical tower and the service structure: the need for clear lines of sight from the erected launch vehicle to radar and telemetry stations in the industrial area 3 kilometers to the southwest, and an anticipated launch azimuth of 75° to 90°.[8]

Problems in Design

At Huntsville the Systems Support Equipment Laboratory designed the ground support equipment, a term applied to components used in the preparation, testing, monitoring, and launching of a rocket. The interface, or fit, of the launch vehicle and the support equipment largely determined the design of the latter. Accordingly, work in the Systems Support Equipment Laboratory paralleled Saturn development and was very much a research and design effort. Five design problems, in particular, challenged the laboratory: the launch pedestal, the hold-down and support mechanisms, the deflector, the cryogenic transfer equipment, and the umbilical tower.

Initial launch pedestal plans called for a hexagonal structure of tubular steel. George Walter, the laboratory's expert on structures, suggested a reinforced concrete design, which was eventually adopted. Walter's pedestal, 13 meters square and 8 meters high, was supported by corner columns and opened on all four sides to allow use of a two- or four-way flame deflector. A torus ring of large water nozzles, designed by Edwin Davis, encircled the 8-meter-wide exhaust opening. During launch and for some seconds thereafter, the nozzles would spray water on the pedestal, across the exhaust opening, and down the opening's walls, cooling the deflector and pedestal.[9]

Designing the eight vehicle support arms to be located on top of the pedestal proved a long and difficult task. Four of the arms, cantilevered at the Saturn's outboard engines, would retract horizontally after ignition, providing clearance for the engine shrouds at liftoff. Should one of the engines fail during the first three seconds following ignition, these four arms could return to the support position. The possibility of damaging the rocket as it settled back on its supports complicated the design of the arms. The Systems Support team developed a nitrogen-fed pneumatic device that brought the support arms safely back under the launch vehicle within 0.16 second. The remaining four support arms were designed to hold the vehicle on the pad for three seconds after ignition so that blockhouse instruments could test engine thrust. Donald Buchanan's design section considered more than 20 different proposals before selecting one suggested by Georg von Tiesenhausen, Dep-

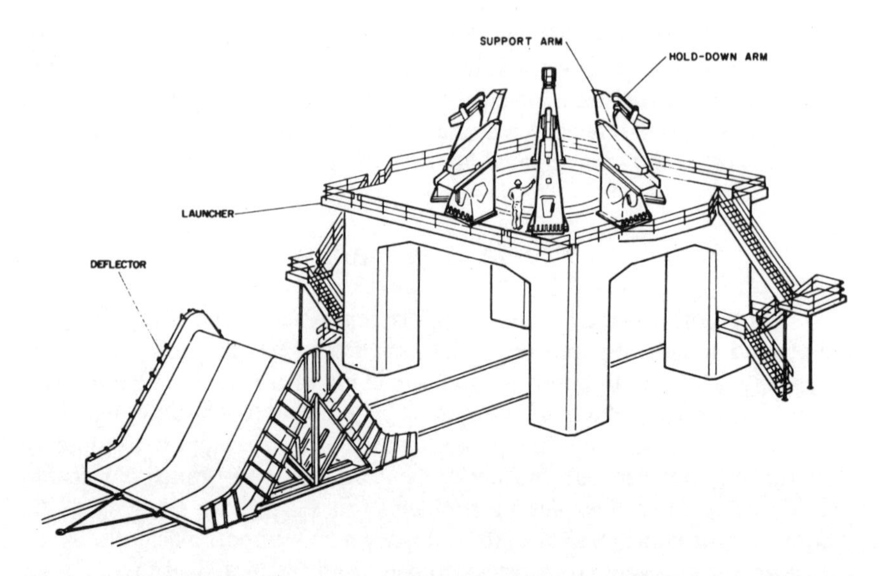

Fig. 5. A 1962 drawing showing the pad at LC-34, including the flame deflector, support arms, and hold-down arms.

uty Chief of the Mechanical Branch. Von Tiesenhausen's concept, modeled after an old German bottle top, had been planned for use in securing a Jupiter seaborne model.* The hold-down arms employed an over-center toggle device to achieve the necessary leverage and rapid release capability.[10]

The flame deflector design stirred debate within the laboratory: Should it have two or four sides? Should it be dry or wet (with cold water circulating through pipes beneath the metal shield)? The Huntsville engineers ruled out the four-sided deflector, previously used for Redstone and Jupiter missiles. The flame, spewing in all directions, would obstruct vision from the blockhouse and endanger equipment at the base of the umbilical tower. Both the size and cost of a wet deflector were unacceptable; one similar to those used on the test stands at Redstone Arsenal would cost ten times more than an uncooled deflector. Its size would increase the height of the launcher platform above ground, a dimension MFL wished to minimize. Despite doubts that a dry deflector could survive a single launch, a two-way uncooled deflector was selected.[11]

Fueling the Saturn promised to be another problem. The booster required 182 200 liters of liquid oxygen (LOX), six times the amount expended

*In November 1955, Secretary of Defense Charles E. Wilson directed the Navy to adopt the Jupiter as a shipborne IRBM. Navy leaders, unenthusiastic about seagoing liquid-fueled rockets, subsequently were able to replace the Jupiter with the solid-propellant Polaris missile.

Fig. 6. The pad under construction, 1960.

by the Jupiter missile. The LOX would evaporate at a rate of 163 liters every minute during fueling and up until launch; some provision for replenishing this loss was required. Explosive hazards dictated placement of the LOX facility a minimum of 200 meters from the launch vehicle. Orvil Sparkman, a Huntsville native who had been working on propellants since 1953, was responsible for designing the cryogenics equipment.

The main storage tank would be an insulated sphere with a diameter of 12.5 meters; it could hold 473 125 liters of liquid oxygen. A centrifugal pump would deliver the LOX through an uninsulated aluminum pipe to the filling mast on the launcher. This was the "fast fill" and operated at 9460 liters per minute. With some of the LOX boiling off as its temperature rose during the filling process, a smaller (49 205-liter) tank would send additional LOX through a vacuum-jacketed line to replace the boil-off, thus keeping the vehicle tanks full. Since the launch team wanted to automate LC-34 fueling, remote controls were designed for the launch control center. Early plans called for a differential pressure sensing system in the rocket's LOX and RP-1 tanks to control propellant flow (much as a washing machine controls flow by measuring the difference in pressure between the top and bottom of the tank). At Debus's request, the system was later replaced by an electrical capacitance gauge. The LOX tank's fuel level sensor also actuated a pneumatic valve on the replenishing line.[12]

A Service Structure for Saturn

MFL's Mechanical Branch, meanwhile, considered the assembly, transport, and service of the launch vehicle. The March 1959 criteria book called for checkout of the Saturn stages in hangar D in the Cape's industrial area, transfer to the pad, and erection and mating of the stages on the pad. The plan required extensive modifications to hangar D, as the booster's size necessitated an increase in hook height from 8 to 13 meters. This additional space could be provided by cutting the roof structure from its columns, jacking the entire roof up as one assembly, and building up the columns.[13]

Some of the Development Operations Division's plans for a Saturn service facility, drawn in terms of a 25-meter booster and limited funds, seem primitive in contrast with the eventual structure. One early study proposed to eliminate service platforms by designing the upper stages with sufficient work space inside the rocket. Another short-lived scheme lowered platforms down over the launch vehicle, attaching them to the rocket's outer surface at the required working levels. Workers would ride elevator stands up to the work platforms. In a November 1958 memorandum, Albert Zeiler scoffed at the notion of men servicing a rocket from a little platform, high above the ground. He said it would be "practically impossible" to perform assembly and checkout tasks, especially in bad weather. In addition, the rocket would be exposed to rain and high winds; in the event of the latter, it would have to be secured by guy wires.[14] He recommended instead a large stand with lifting equipment to assemble and erect the booster and upper stages. Service platforms to support personnel and equipment, elevators, and weather protection would be incorporated in the stand.

MFL awarded the criteria studies for the launch complex and service structure to the Miami firm of Maurice H. Connell and Associates. The Miami architects, veterans of the Redstone program, completed both studies by mid-March 1959. At a conference later that month, Saturn engineers agreed to complete design work by 1 August 1959. The conference set a 1 July 1960 target date for construction of the complex, excepting the blockhouse and service structure.[15]

Connell and Associates had completed the criteria studies and moved into the design phase when MFL decided on major revisions in the assembly and service concepts. Prior to the criteria review, IDECO, a Dressler Industries division, had approached MFL with a proposal for a tubular steel service structure. It was designed in the shape of an inverted U, open at both ends. IDECO's design offered several advantages: greater accessibility to the booster, a minimum hook height of 13 meters, and more flexible service platforms (the platforms telescoped in and out of the main frame design, were vertically adjustable, and could match up with various booster or upper stage diameters).[16]

An additional attraction of the IDECO proposal stemmed from a new MFL proposal to assemble and check out stages of the launch vehicle in a building 180 meters from the launch pedestal. The facilities people at the Cape had never really liked the idea of modifying hangar D; many thought the raised roof would collapse in a hurricane. In the new plan the staging building shared a bridge crane with the IDECO service structure. Rails at the ceiling level of the staging building matched up with the bridge crane rails at the 13-meter level in the service structure. A cutout portion in the center front of the staging building roof provided space for the bridge crane roof. The dual use of the bridge crane allowed the transfer of the stages from the staging building to the launch pedestal in one operation.[17]

General Medaris was impressed with the IDECO proposal and ordered an extension of the service structure study. At a 13 April meeting, MFL directed Connell representatives to prepare a new design incorporating 14 function capabilities of the IDECO proposal. The Miami firm satisfied this requirement in ten days. The design called for a structure of girders and platforms shaped like an inverted U 95 meters high. The service structure was 40 meters wide, including the 17-meter open space where the structure extended over the launch pedestal. A bridge crane supported 40- and 60-ton hoists at a 75-meter hook height. Each hook had a forward reach of 9 meters and a lateral reach of 6 meters. Seven fixed platforms were housed within the tower legs, each providing 73 square meters of working area. Each half of the six enclosed retractable platforms had a capacity of 12 persons and 272 kilograms of equipment. The platforms were vertically adjustable from the 25-meter to the 68-meter level. Three elevators provided a 227-kilogram lift.[18]

Construction bids followed in June. Since assembly and service methods were still not firm, the contract called for additional design work. Kaiser Steel Corporation's $3.9 million bid, $400 000 less than an IDECO proposal, won the contract. Kaiser formally began work on 14 August 1959, but construction did not start until the following summer.[19]

Brick and mortar work on the new complex proceeded satisfactorily, slowed little if at all by the still meager Army appropriations and the prospects of major administrative changes taking form in Washington. In early June 1959 the Western Contracting Company began hydraulic-fill operations

Fig. 7. The service structure, LC-34.

at the pad. A proprietary process, vibroflotation, was used to compact the fill. The Vibroflot machine consolidated the marshy soil by simultaneous vibration and saturation; the machine vibrated the sand with ten tons of centrifugal force as it pumped in more water than the surrounding soil could absorb. The sand formed a dense mass, the excess water floating fine particles to the surface. Workmen shoveled in backfill (roughly 10% of the total volume compacted) to increase the density. Vibroflotation on LC-34 required 5350 cubic meters of fresh sand to provide compact soil 8.5 meters deep. In late November, the Henry C. Beck Construction firm started work on the pad facilities. Three hundred and twenty meters to the southeast, the blockhouse was taking shape.[20]

A Money Transfusion

In the meantime, Debus received good news. With the Saturn's metamorphosis from Army orphan into NASA prima donna, cost estimates at the launch facility could be revised upward to what were considered more realistic levels (see table 2). Limited Army financing had constrained the Development Operations Division to view the Saturn program as a minimum operation to demonstrate the feasibility of the clustered booster, and funding for its Cape Canaveral launch facility during the first nine months was piecemeal and unrealistic. As a later MFL study noted: "Prior to this date [31 July 1959] no budget submissions could be considered an estimate of re-

TABLE 2. LAUNCH COMPLEX 34 COST ESTIMATES

(in millions of dollars)

	9 March 1959	31 July 1959	August 1960
Blockhouse	1.3	1.1	1.1
Service structure	3.0	4.6	5.1
Launch pad and area development	3.6	5.4	5.4
Capital equipment (high-pressure-gas systems, instrumentation)	.3	2.0	2.5
Ground support equipment		8.0	23.1
Operations support building			.9
Industrial facilities	.5		
Totals	8.7	21.1	38.1

Source: J. P. Claybourne, Saturn Project Office, LOD, memo for record, "Cost of Saturn Launch Facilities and Ground Support Equipment," 13 Sept. 1960.

quirements, merely a series of proposals on how to apply initial inadequate funding with the promise of additional operating funds to come."[21] Rough estimates in September 1958 placed the cost of the launch complex at $4.5 million. The original project request, made on 9 March 1959, called for a total expenditure of $8.7 million. The price of the service structure alone had increased from $400000 to $3000000. By 31 July 1959, revised estimates had increased the figure for MFL expenditures to $13.1 million, with an additional $8 million requested for ground support equipment. One year later MFL officials would be justifying a $38 million price tag for LC-34. Their explanation would offer a number of reasons: underestimates, inflation, organizational changes, vehicle design alterations, and the Saturn program's changing guidelines and objectives.[22]

Besides the rising costs of LC-34, MFL faced the need for a backup Saturn launch complex. While the Silverstein Committee report was pending in late 1959, MFL began its own investigation of hydrogen-filled upper stages. A committee, headed by Charles Hall, examined equivalent TNT forces and concluded that an explosion would render LC-34 useless for a year. MFL reassessed its Saturn launch capability in light of that report. The LC-34 staging building, tentatively located near the pad, was moved back to the industrial area and the service structure was fitted with blow-out panels around the base. In January 1960, Debus notified Eberhard Rees, Deputy Director at Huntsville, of the Hall Committee findings and strongly recommended a second Saturn complex. Construction of LC-34's second pad would not do since the 730 meters separating the LC-34B site from LC-20 was too short for safety, with the new Saturn configuration. The Development Operations Division gave its approval and MFL was soon planning for what would become launch complex 37.[23]

With the transfusion of new money, construction of LC-34 proceeded apace. Reminiscent of Florida's Seminole Indian Wars of the 1830s, the first structure to take form was the blockhouse (launch control center). The dangers had changed and so, too, the design of the blockhouse. The interior diameter of the igloo-shaped building at LC-34 was 24.4 meters, its maximum height 7.9 meters. Two stories provided space for control instrumentation, measuring racks, and firing consoles. Construction took 13 months; the blockhouse was ready for occupancy in July 1960.

The Ground Support Equipment

At Marshall Space Flight Center the development of ground support equipment proceeded under a new office. With the reorganization of ABMA, on the takeover by NASA in March 1960, the Systems Support Equipment

Fig. 8. The blockhouse at LC-34, with the service structure rising behind it, November 1960.

Laboratory disappeared. Most of the laboratory's personnel joined LOD as members of Theodor Poppel's Launch Facilities and Support Equipment Office (LFSEO). Poppel's experience with ground support equipment dated back to World War II. A native of Westphalia, Poppel had begun work at Peenemünde in 1940, following graduation from an engineering school in Frankenhausen. He had been among the first Germans to enter the United States in 1945. The deputy to Poppel, Lester Owens, like many American members of the von Braun team, was Alabama-born and Auburn-trained. Five of the six suboffices worked in Huntsville; the other line unit, R. P. Dodd's Launch Facilities Design Group, was based at the Cape.[24]

By May 1960, LFSEO leaders felt sufficiently confident about the development of ground support equipment to discuss responsibilities for installing equipment at LC-34. A conference on the 17th set beneficial occupancy dates (the dates on which each facility within the complex would become available to MSFC and its contractors for installation of collateral equipment). MSFC plans called for a rudimentary high-pressure-gas (nitrogen and helium) facility. Two flatbed trailers, each with a 3785-liter storage tank, pump, vaporizer, and associated equipment, would bring liquid nitrogen to the pad. After being routed through a vaporizer and warmed to 294 kelvins (21 °C), the gaseous nitrogen would be stored under high pressure in a cluster of bottles protected by a concrete vault. Two booster compressors located in the storage area would pressurize the helium. Since the introduction of upper stages after the third launch would increase high-pressure-gas requirements, Chester Wasileski's Propellants Service Design Group was planning a central compressor-converter for LC-37. The basic LC-34 arrangement, however, would be ready by 1 August 1960.

Discussion moved next to RP-1 facilities. LFSEO planned to test the integrated fuel storage and transfer system at Huntsville before 10 August. During the fall, the General Steel Tank Company would install the two 113 550-liter storage tanks at the pad. Hayes Aircraft personnel would then clean, install, assemble, and pressure-test the entire transfer system from tank to fuel mast. The two 3785-liter-per-minute pumps were standard equipment and posed no problems. The propellant operations and status panel required additional testing at Huntsville, but would be ready for blockhouse installation in the fall. Wasileski thought the entire system could be operational by 1 February 1961.

Wasileski envisioned a similar schedule for LOX facilities: completion of testing at Huntsville on 10 August; installation of storage and replenishing tanks by Chicago Bridge Company between September 1960 and 15 January 1961; cleaning, assembling, and installing the transfer system (the pipes that carried LOX from the tanks to the Saturn) by Hayes before 15 January 1961; and operational status as of 1 February 1961. During the installation, the Chicago Bridge Company would conduct a flow test between the storage tank and replenishing tank to qualify the vaporizer design.

Work on the cable masts was proceeding satisfactorily. The Saturn booster required a 21-meter aluminum boom to service the instrument compartment with pneumatic pressure, electrical power, and coolants. LFSEO expected to deliver the long mast 1 March 1961. Shorter cable masts, mounted on the launcher support arms, would provide pneumatic and electrical connections to the tail section of the booster. The electrical connections powered and monitored the propulsion system, while the pneumatic lines purged and pressurized the fuel systems. The conference set 1 February 1961 as the installation date for launcher support arms, hold-down arms, and short cable masts.[25]

Development of the flame deflectors took Poppel's office about six months longer than expected. LFSEO had begun testing angles of impingement in early 1960 to establish the flow pattern for the Saturn's exhaust. If

Launch complex 34

Figure 9

Figure 10

Figure 11

Fig. 9. The high-pressure-gas facility. Fig. 10. LOX facility (L), service structure (C), and pad (R), seen from the top of the RP-1 facility. Fig. 11. The LOX facility. The pad is to the left.

the deflector was set at the wrong angle, a detached shock wave would form, choke the exhaust flow, and raise the heat and pressure on the launch vehicle beyond tolerable limits. Donald Buchanan's Launcher System and Umbilical Tower Design Section experimented with several angles before contracting with the Hayes Corporation of Birmingham to construct two 80° deflectors. Delivery was set for November 1960. During subsequent tests in August 1960, Buchanan and Edwin Davis determined that a 60° deflector (30° angle of impingement) would further reduce the backflow. Although the 80° deflector still met launch vehicle requirements, rocket designers prevailed upon von Braun to modify the Hayes contract. The Birmingham firm shipped the first deflector to the Cape in April 1961. Due to its size, 6×8×13 meters, and weight, 99 tons, the deflector was shipped in seven sections.[26]

The change in the deflector design was not a unique event. The office frequently altered its designs to fit requirements of other MSFC groups; few concessions were made to LFSEO. For one example, the support arms could have been simplified by strengthening the booster frame. This LFSEO recommendation was rejected because it meant adding 1100 kilograms to the launch vehicle weight. Eventually, the weight increased several times that amount, but for other reasons. MSFC officials, fearful of a rocket collision, restricted the size of the umbilical tower. An LFSEO engineer, believing the final design of the tower base was unsatisfactory, surreptitiously increased its dimensions. Speculation about German-American friction at Huntsville was largely unfounded, but disagreements between vehicle designers and ground support engineers were common.

By mid-1960, costs of the two Saturn launch facilities were burgeoning. A July Saturn Project Office memorandum noted that "rising costs, the influence of the committee on Saturn blast potential [Charles J. Hall Committee], and the full impact of [the Saturn] C-2 on the VHF-37 complex . . . indicated that approximately 44 millions were required in lieu of the available 31 millions [for FY 1961]."[27] The Launch Operations Directorate established priorities to complete the essential portions of LC-34 and LC-37 while the remaining facilities awaited adequate funding: first, prepare LC-34 for the three Saturn booster shots without hydrogen capability or umbilical tower; second, prepare LC-37 as a backup pad for the second launch; third, complete LC-37 for launch of a two- or three-stage Saturn C-1 or C-2; fourth, complete LC-34 for Saturn C-1 configuration.* Construction of

*In July 1960, the Saturn launch schedule called for the first three booster shots to carry dummy upper stages. This was eventually changed to four booster shots (the block I series) and six two-stage launches (block II series). The block I series ran from 27 Oct. 1961 to 28 Mar. 1963.

Launch complex 34

Figure 12

Figure 13

Fig. 12. The flame deflector in position beneath the launch pad. Fig. 13. The top of the pedestal. The metal grating, a work platform, was removed before launch. The spray nozzles can be seen beneath the torus ring. The rocket rested on the hold-down arms, which are under protective covers. The rectangular ducts (one of which is in front of the workman) removed exhaust gases.

LC-34's umbilical tower began in September 1960 but stopped at the 8.2-meter level; the long cable mast would provide umbilical connections for the booster launches.[28]

Labor Difficulties

Construction on several of LC-34's most prominent features including the service structure, launch pedestal, and umbilical tower was just getting under way when a serious labor dispute broke out at the Cape. On 5 August 1960, members of the electricians' union (International Brotherhood of Electrical Workers) informed a Corps of Engineers representative that, "It's too hot and ABMA is making it hotter . . . . We're going fishing."[29] The Launch Operations Directorate had generated the heat earlier that morning when some of its personnel unloaded a dozen firing consoles at the launch control center. The incident touched off four months of conflict between LOD and labor unions at the Cape, and eventually received the attention of Congress and the Secretary of Labor.

Involved were jurisdictional issues between unions, as well as the role of labor unions in research and development work. During the late 1950s, the building trades unions had achieved jurisdiction over a large share of the construction of ground support equipment for missiles. They feared loss of such jobs to the aircraft industry union. LOD officials believed that the building trades unions had won a number of concessions at the Cape because the Air Force normally yielded to labor demands. While the urgency of military programs made the Air Force position understandable, Debus refused to take the same course.* LOD articulated its philosophy in a 6 September presentation to General Davis, Air Force Missile Test Center Commander:

> All ground equipment including measuring, launch controls, plumbing, instrumentation which are directly connected to the missile are a very integral part of the missile system. In the early phase of any program, the missile constitutes a flying laboratory for the purpose of gathering *data* and testing feasibility on design concepts, operational techniques . . . . Thus the ground equipment is just as important to the success of the mission as is the actual flight of the missile . . . and must come

*Gen. Donald Yates contends that Air Force policy was a better approach to labor relations. Non-union contractors did work at the Cape, but the Air Force never placed a non-union contractor on the same job with a union contractor. Furthermore, Yates felt that LOD leaders tended to challenge union labor with their new rules.

> under the direct control, from installation to final use, of the LOD *missile* people. All our firings will be R&D in nature, not operational prototypes.[30]

General Davis was impressed with the LOD arguments, but not so the unions. When LOD personnel returned to the launch control center on 10 October to install more panels, 47 electricians walked out again. Ten days earlier, 27 ironworkers had left work on the service structure complaining of excessive supervision; on 4 October, 17 carpenters stopped work in a jurisdictional dispute with electricians over the installation of static ground lines.[31]

These walkouts were brief and contractors lost only 800 man-days from August to November. Then on 14 November LOD resumed its activities at LC-34, with civil service personnel installing cables and consoles. When the electricians struck again, LOD initiated injunction proceedings. The other trade unions retaliated with a mass walkout at the Cape. By Thanksgiving 650 union members were on strike. With the problem attracting national attention, Secretary of Labor James P. Mitchell intervened. His appointment of a fact-finding committee placated the unions and work resumed 28 November. The committee's findings, released after the New Year, included recommendations that LOD improve its communications with the unions and that both sides reexamine the controversial interface points (between rocket and ground support equipment). While the basic issue remained and work stoppages continued, relations never again reached the low ebb of November 1960.[32]

Work moved ahead rapidly on LC-34's major structures in early 1961. By February the inverted U shape of the service structure's rigid box truss frame was clearly recognizable. At the pad, four reinforced concrete columns, 7 meters high and more than 2 meters thick, stood at the corners of the 13-meter-square launch pedestal. Nearby rose the steel frame of the abbreviated umbilical tower. The walls of the 7 × 7 × 8-meter base would incorporate blowout panels to reduce structural damage from a pad explosion.

At its formal dedication 5 June 1961, LC-34 represented the largest launch facility in the free world. Although complexes 37 and 39 would soon overshadow it, LC-34 was destined to play an important and tragic role in the Apollo history. Its inaugural would come in four months with the first Saturn I launch as the United States tried to recover lost ground in the space race.

Fig. 14. LC-34 soon after its dedication. Looking north, the pad is in the center; the service structure has been removed along its parallel tracks to the parking position. The control center (blockhouse) is on the near side and left of the service structure. In the background, land is being cleared for LC-37.

Fig. 15. LC-34, looking southwest. LC-20 is in the background. The white rectangle in the foreground is the skimming pond. The RP-1 facility is at the extreme left.

3

Launching the First Saturn I Booster

The Magnitude of the Task

Just as launch complex 34 dwarfed its predecessors, Saturn checkout represented a new magnitude in launch operations. The Saturn C-1 stood three times higher, required six times more fuel, and produced ten times more thrust than the Jupiter. Its size, moreover, was only a part of the challenge to the Launch Operations Directorate (LOD) at Cape Canaveral. The costs and complexity had also increased markedly. Because of the costs (eventually $775 million for the Saturn I program's research and development alone), there would be fewer test flights. This meant the engineers at Marshall Space Flight Center (MSFC) had to have more test data per flight—such measurements as the temperature of the flame shield, the pressure in combustion chambers, the rocket's angular velocity in pitch and roll. Whereas two telemetry links (radio transmitter-receiver systems) sending 116 measurements had been adequate for Redstone testing, the first Saturn booster employed eight telemetry links to report 505 measurements. The rocket's overall complexity necessitated a longer checkout: Saturn C-1 launch preparations averaged 9 weeks, almost three times longer than for a Jupiter missile.[1]

Ultimately the new procedures were to work a major change in the human role on the launch pad. Until the Saturn, the Debus team had been on a first-name basis with the rockets. LOD members who were not crawling around inside the Jupiter worked within a few yards of the pad. The Saturn brought little change initially; checkout for the first Saturn C-1 remained largely a manual operation. In the blockhouse, a console operator with a test manual threw a switch connected to a rocket component and checked the results on a meter or strip chart. Automation on the first Saturn booster was rudimentary, limited to relay logic during the last minutes of countdown. It increased as the Saturn grew more complicated. The addition of a live second stage to the Saturn C-1 and the appearance of the much larger Saturn V dictated greater reliance on machines and computers. By the mid-1960s the Saturn checkout was well on the way to automation. Chapter 16 will address this subject in detail.

Fig. 16. Models of Jupiter, Juno, and Saturn I.

NASA had firmed up the Saturn C-1 program in late 1959 by adopting the Silverstein Committee's proposals (pages 13–14). Marshall Space Flight Center would start with the clustered booster (S-I) and dummy upper stages. A second block of missions would add a hydrogen-fueled second stage, and a third block would add a third stage to the stack. The Program Office listed the SA-10 launch, set for April 1964, as the Saturn C-1's debut as an operationally ready vehicle. Plans beyond the ten-vehicle research and development (R&D) schedule were indefinite. A 1960 NASA Long Range Program called for 50 Saturn C-1 and C-2 launches between 1965 and 1970. Twenty of these flights would launch Apollo spacecraft reentry tests, earth orbital missions, and circumlunar shots.[2]

These plans were altered in January 1961 when Wernher von Braun proposed to eliminate the third stage; a two-stage Saturn C-1 would meet the needs of the early Apollo missions. Following NASA Headquarters formal approval of von Braun's recommendation, the Saturn Office in Huntsville rearranged the ten-vehicle R&D program. Block I, beginning that fall, would consist of four S-I stage tests from LC-34 (mission numbers SA-1 through SA-4). Block II, the next six launches (SA-5 through SA-10), would add the second stage from the LC-37 launch pad, and from an upgraded LC-34.[3]

The Saturn C-1 test flights were to prove the design of the launch vehicle. The block I launches in particular would test the eight-engine propulsion system, the clustered tank structure, the first-stage control system's ability to cope with sloshing and nonrigid-body dynamics, and the compatibility of the vehicle and launch facility. During the block I series, Marshall engineers proposed a systematic buildup of tests to prepare the way for two-stage flights. Broadly stated, LOD's responsibilities were fourfold: assuring that transportation had not affected vehicle components, mating stages and ground equipment to verify the compatibility of the different stages, launching the rocket, and analyzing the performance of all vehicle systems immediately after launch to detect flight failures. Although the mission was referred to as "launch vehicle test and checkout," less than half of LOD's scheduled activities involved test performance. The balance of the total launch preparation effort included activities more properly described as assembly, installation, preparation for test, and evaluation of records.[4]

The Leadership

Entries in the LOD Director's daily journal during 1961 indicate that Debus kept a close eye on SA-1 operations. Other problems, however, occupied his time: a new launch facility for Saturn V—eventually the moonport for the moon rocket; Centaur facility development; and Mercury-Redstone, Pershing, and Ranger launches. On account of these duties, Debus did not deal with the details of the SA-1 checkout. That burden fell on his operations office chiefs, their deputies, and the veteran test engineers.

Dr. Hans Gruene headed the Electrical Engineering Guidance and Control office. A native of Braunschweig, Germany, he had earned his engineering degree at the technical university in his home town. Gruene had joined the Peenemünde operation in 1943 and emigrated to Fort Bliss after the war. Since 1951 he had served as the electrical networks chief for the launch team. Small in stature and unassuming, Gruene enjoyed great respect from his associates. Gruene's deputy, Robert Moser, had joined the von Braun team as an Army enlisted man in 1953, three years out of Vanderbilt University. He had reverted to civilian status in 1955, but stayed on in Huntsville as Gruene's right-hand man. Moser's launch countdowns resembled an orchestral performance and earned him high praise as test conductor for Explorer 1 and Alan Shepard's Mercury-Redstone flight. Gruene's office supervised the performance of all equipment affecting rocket guidance and control. This required a wire-by-wire knowledge of the electrical systems, both on board the vehicle and at the launch site. Gruene's men also evaluated preflight telemetry records relating to guidance, stabilization, control, and electrical networks of the vehicle.

Albert Zeiler's Mechanical, Structural, and Propulsion Office handled missile receipt and transfer, stage erection, and assembly. The team tested pressures, located leaks, and made necessary replacements, repairs, or modifications. One of the branch's sections was responsible for fueling the rocket, another for the firing. After the launch, the branch evaluated flight data to check on mechanical functions and make corrections for future flights. The Austrian-born Zeiler had served at Peenemünde throughout World War II, testing and launching V-2s. Following duty at White Sands, he had moved to Huntsville and worked with MFL. Robert Gorman, deputy in the Mechanical Office, had begun his engineering career in NACA's wind tunnels at Langley Field. A ready ear for subordinates' ideas contributed to his success. His calm manner balanced Zeiler's excitable nature, and the two provided the office with effective leadership.

Quiet and intense, Karl Sendler, chief of the Measuring and Tracking Office, seemed aloof to strangers, but to colleagues showed a warmth that sparked loyalty. He was Vienna-trained and reflected the traditions of the old Hapsburg capital in his manner and attire. At Peenemünde, Sendler had tracked the V-2s fired northward along the Baltic experimental range. He, too, had worked at White Sands before moving to Huntsville in 1950. His deputy, Grady Williams, had graduated from Auburn in 1949 and joined the von Braun team three years later. Associates considered him one of the friendliest members of the team. Like Sendler, Williams had a penchant for order. The two gave the Measuring and Tracking Office a reputation for being immaculate. During checkout, Sendler's systems engineers tested and calibrated the Saturn's measuring instruments—pressure gauges, thermometers, accelerometers, and the telemetry that relayed the measurements back to earth. At launch his office collected the flight data. Supporting ground radars tracked the flight for deviations in direction and range, which would reveal problems in the guidance and propulsion systems. Along with the other offices, Sendler's group prepared designs and established criteria for launch facilities. The unit's work brought frequent contact with other agencies investigating telemetry, high-frequency signals, and the measuring and tracking of launch vehicle flights. The branch's previous efforts had contributed to the development of three specialized tracking systems: DOVAP, "Beat-Beat," and UDOP.*[5]

The work of the three LOD operations offices involved close liaison with other Marshall divisions. Thus, Hans Gruene and his engineers spent more than half of 1960–1961 in Huntsville with the MSFC Guidance and Control Division. In turn, a dozen Guidance and Control engineers took part in the SA-1 checkout at the Cape. The launch team still considered itself an extension of Marshall. As one veteran recalled, "In the 1950s we looked at equipment when it came down here as not trusting a single thing in it. We were going to check everything from one end to the other."[6] Consequently, LOD's checkout was precise and exhaustive, "a laboratory type check on the pad."[7] Basic operating procedures were established and followed closely. Debus detailed some of these procedures in a letter to NASA Headquarters shortly after the first Saturn launch. LOD employed a test sequence that proceeded from components, through subsystems and systems, to overall tests. "If the preceding less complex tests are eliminated, as is tried frequently to shorten overall test schedules, any failure of one single component in an overall systems test necessitates activation of all other components whether

*DOVAP (doppler velocity and position) was a velocity-measuring system that used a ground transmitter, a transponder on the launch vehicle, and a number of ground receivers. The change of frequency between the signal transmitted from the ground and that later received on the ground, called the doppler shift, could be converted to the velocity of the rocket. Integrating the velocity with time provided distance, which applied to the known departure point indicated the rocket's position. The "Beat-Beat" system detected the deviation of a missile from a predetermined flight path. It derived its name from the use of two receivers that compared, or beat, two frequencies against each other. The system consisted of a pair of DOVAP receiver stations placed symmetrically about the flight path. When the missile deviated to the left or right, one receiver would detect an increasing frequency, the other a decreasing frequency. See W. R. McMurran, ed., "The Evolution of Electronic Tracking, Optical, Telemetry, and Command Systems at the Kennedy Space Center," 17 Apr. 1973, mimeographed paper. "Beat-Beat" could be used equally well with UDOP or telemetry signals. UDOP (ultra-high-frequency DOVAP), operating at 440 megahertz, offered certain advantages over DOVAP, including higher resolution and less loss of accuracy from ionospheric refraction.

critical to running time or not."[8] Debus insisted that his engineers conduct at least one systems test in its entirety to ensure a total working package. Other rules, established from long experience, included: calibrating sensors at the latest possible time, removing all connecting circuitry and components in a system when the cause for random irregularities could not be established, and disturbing a minimum of electrical and pneumatic connections after the final overall test. Some procedures concerned LOD's relations with other Marshall divisions. One provided for a speedy MSFC ruling on launch vehicle and ground support equipment modifications at the pad; another assured the availability of current Huntsville drawings.

The technical checkout of the various Saturn systems fell to LOD's test engineers. Debus considered these engineers "the backbone of LOD test activities"; they carried "full responsibility for preparing a launch vehicle to the point of launch readiness [and] merited equal status with . . . engineers in design, development, and assembly operations. While an error made in the design or development phase could be detected by a test engineer, a mistake by an LOD systems engineer would inevitably lead to mission failure."[9]

Conceding that launch site tests were part of a continuous program to assure reliability and quality, Debus stressed the test engineer's need for autonomy. "Since the systems engineer carries the full responsibility for the flight-readiness of his assigned system, this responsibility should not be attenuated by assigning a separate inspection or quality assurance team to check on the systems engineer for compliance to test procedures and test performance." Although limited manpower ruled out a two-shift operation at the Cape, Debus opposed it on principle: "A systems engineer had to be kept informed continuously of the status of his assigned system and all occurrences during the test period."[10] When problems arose, the launch team resorted to overtime. The work day during the SA-1 checkout varied from 8 to 16 hours.

The Test Catalog for SA-1

LOD began preparing for the first Saturn launch in mid-March of 1961 when Debus directed the Scheduling and Test Procedures Committee to review launch procedures. The Director did not want to "automatically transfer into the Saturn, things that may have been important in past operations."[11] The committee—composed of the operations office deputies Gorman, Moser, and Williams—agreed that Saturn required basic changes in launch procedures. For example, LOD personnel had conducted a detailed identification of component serial numbers on previous rockets. Since a serial inspection of Saturn components would require many man-hours, the committee proposed to rely on MSFC's detailed list instead. LOD would update the Marshall list when components were changed. The committee eliminated some redundant systems checkouts and recommended less component testing. During the Saturn C-1 launches, the emphasis would shift gradually from component testing to integrated systems testing. As the checkout for SA-1 was revised, other MSFC personnel undertook to coordinate all Saturn testing.[12]

The test catalog that emerged in May 1961 indicated the magnitude of the Saturn C-1 program. The catalog included 233 system tests, 102 of which were prepared by LOD. The tests were grouped in seven categories: electrical networks, measuring, telemetry, radio frequency and tracking, guidance and control systems, mechanical systems, and vehicle systems. The last category included overall tests, simulated flight tests, cooling systems tests, propellant loading tests, static firing, and fuel tank pressurization. Most of the tests ran from four to eight hours; a few required days. An example of LOD's contribution was 6-LOD-26, the fuel and LOX systems full-pressure tests: *6* indicated the category, mechanical systems; *LOD*, the responsible division; and *26* identified the particular test among 42 in that category. The test objectives were to "accomplish a pressure test of both propellant tanks to full working pressure, performed and monitored from the Blockhouse to determine if any major structural defects have occurred due to transporting, handling, erecting, etc. Pressure drop-off time, and pressure switch cycles will be recorded for system leakage analysis at full working pressures."[13]

While operations personnel were determining test requirements, construction at the launch complex progressed toward the 5 June 1961 dedication, when the Corps of Engineers would formally transfer LC-34 to NASA. LOD personnel began outfitting the service structure in early May. The propellants team used "live" fuel to run a "wet test" of the fuel system on the 19th. No serious leaks appeared in the LOX and RP-1 transfer lines, and the pumps worked satisfactorily. At the dedication ceremony the long cable mast and two short cable masts were the only major items missing. Redesign had slowed their development, but shipment from Huntsville was expected in mid-June.[14]

A new ground support requirement, however, threatened to delay the October launch date. On 11 May launch vehicle designers notified Maj. Rocco Petrone's Heavy Vehicle Systems Office that the high-pressure gas system would have to be modified. Model tests indicated that LOX sloshing in the Saturn tanks caused condensation of the gaseous nitrogen used for pressurizing the fuel, and this lowered the pressure to marginal limits. The solution was to pressurize the LOX tanks with helium. Petrone took immediate steps

Fig. 17. Assembling the long cable mast at LC-34.

to procure a helium facility through sole-source procedures—an emergency government purchase without competitive bidding. He transferred LC-37 funds to cover the expense and secured eight steelworkers, skilled in working on high-pressure tubes. Debus told von Braun the following day that the change to helium might hold up the launch. The Marshall director mentioned NASA Headquarters fear that a delay would have political repercussions, but assured Debus that Huntsville understood the problem. Modifications progressed rapidly, easing Debus's mind, and the helium facility was ready by mid-September.[15]

Up in Huntsville, the Fabrication and Assembly Engineering Division had fallen behind on its booster assembly schedule. Debus reluctantly agreed to have the work completed at the Cape. Albert Zeiler detailed a list of unfinished items in a letter to Debus on 14 July. Zeiler expressed particular concern about the scheduling problems posed by these requirements:

> 3 – Install hula hoops [rings that retained the heat shield] and coat uncoated portion on eight engines.

This would require 30 hours of unobstructed work in the tail section during the last 10 days before launch.

10 - Heat shield beams have to be coated, estimated time three days for application and up to ten days in addition where no work can be performed around the tail section because the coating discharges, during the curing time, burnable fumes.

Zeiler considered this a safety, as well as a scheduling, problem, but noted that the curing time could possibly be shortened.

11 - Four curtains for outboard engines will be prefitted, then coated, and then shipped.

The installation would require one day and should be done as late as possible to avoid any damage.[16]

Robert Moser was responsible for fitting the Fabrication Division's activities into the Saturn checkout. As SA-1 test conductor, he coordinated launch operations and ensured that proper procedures were used for the 102 formal tests. Moser's operations schedule, prepared in early August, included:

15 Aug. - Unloading barge and transporting S-I stage to pad 34.
17–21 - Erection of stages.
15 Sept. - Removing service structure for RF tests with the range.
20–25 - Overall systems tests.
2 Oct. - LOX loading test.
9 - Simulated flight test.
12 - Launch day.

Moser's schedule also listed much component testing and instrument calibration during the first half of the schedule; system and vehicle tests predominated in the second half.[17]

The Saturn Goes Sailing

Two years earlier Marshall Flight Center officials had decided to transport the Saturn booster (SA-1's only live stage) from Huntsville to Cape Canaveral by water. In April 1961, Test Division personnel loaded a water-ballasted tank, the approximate size and weight of the booster, and a dummy upper stage aboard the barge *Palaemon*. The barge, resembling a Quonset hut on a raft, made the first leg of its trial trip in five days, descending the Tennessee, Ohio, and Mississippi Rivers to New Orleans. There, a seagoing tug replaced the river tug. The *Palaemon* crossed the Gulf of Mexico to the

Florida Keys, sailed through the straits, and up the Atlantic coast via the Intracoastal Waterway. The LOD team on the Saturn dock, located at the south end of the Cape industrial area, witnessed a strange sight when the simulated booster emerged from the *Palaemon*'s hatch. The big spoked rings, 4.3 meters across, on each end of the 25 x 2.1-meter tank, looked like the wheels and axle of a gigantic vehicle. The simulation served its purpose, proving that both the *Palaemon* and the Cape's secondary roadways could carry the load.[18]

The *Palaemon* was undergoing modifications back at Huntsville in early June when the lock at Wheeler Dam, Tennessee, collapsed, stranding the barge upriver. Test Division and LOD personnel moved quickly to secure a reserve barge from the Navy's mothballed fleet at Green Cove Springs, Florida. Although there was not enough time to construct a cover for the second barge, the Avondale Shipyards at Harvey, Louisiana, made emergency modifications. Concurrently, the Tennessee Valley Authority enlisted the Corps of Engineers to build a bypass road and dock at Wheeler Dam. The Navy had identified its drab barge by a number, YFNB33. NASA rechristened the vessel *Compromise*, in hopes it would prove a workable one.[19]

The booster was ready for shipment in early August, following static firing and two months' further testing at Redstone Arsenal. To protect the booster during its voyage, the Test Division installed humidity and pressure regulating equipment within the LOX and RP-1 systems. Protective covers were placed on each end of the booster, as well as on the dummy upper stage and payload. After the assembled booster, with its support cradles, connecting trusses, and assembly rings, was jacked onto two axle-and-wheel units, an M-26 Army tank retriever towed the load to Redstone's dock. Marshall engineers had provided for the Tennessee River's three-meter fluctuation at the arsenal by building special ballasting characteristics into the *Palaemon*.[20]

The portage at Wheeler Dam, the reloading on the *Compromise*, and the journey to New Orleans went smoothly. Out in the Gulf of Mexico, however, the ten-man crew had rough sailing. Test Director Karl L. Heimburg attributed the handling problems to the *Compromise*'s insufficient ballast. Negotiating the Intracoastal Waterway proved even more difficult, and the *Compromise* went aground four times. Heimburg blamed this on unreliable channel depths due to the shifting of the loose, sandy bottom. Crosswinds were an additional hazard; besides threatening to blow the barge around, the wind caused several near-accidents at bridges. (The *Compromise* was to collide with a bridge on the return trip, causing minor damage.) Despite Heimburg's frustrations, the SA-1 arrived unscathed at the Cape's Saturn dock on the 15th.[21]

Fig. 18. The *Compromise* at Wheeler Dam, 5 August 1961, with SA-1 onboard.

Unloading the booster was relatively easy in the almost tideless Banana River. Henry Crunk's vehicle-handling unit towed the S-I transporter across the Cape at a majestic 6.5 kilometers per hour. Although the operation required little physical exertion, the ten-man team perspired freely on the treeless Cape. At pad 34 ocean breezes made the heat and glare more tolerable. Most visitors, associating Florida's beaches with leisure, would have found the mixed sounds of service structure cranes and pounding surf incongruous. The novelty for LOD veterans lay in the huge Saturn booster, which had at last arrived at its action station.[22]

The booster or S-I stage was erected on Sunday, 20 August. Crunk's unit had practiced maneuvering a dummy tank on the pad, but this was the first mating of the booster to the launch pedestal. With the service structure in place over the pedestal, an M-26 driver positioned the transporter parallel to the service structure base. The crew connected crane hooks to pickup points on the booster, a 60-ton hook to the forward sling and a 40-ton hook to the thrust frame sling. The crane operator raised the S-I stage vertically, brought it into the service structure, and lowered it onto four preleveled support arms. Removal of the transportation assembly rings proved the most time-consuming aspect of an uneventful operation. Early the following week, Crunk's unit hauled the dummy stages and payload from hangar D, where they had undergone inspection. The handling unit mated the dummy stages and the nose cone on the 23d. Cables and cable masts were installed, the four retractable support arms positioned, and network power applied on the 25th. Concurrently the Fabrication Division installed exhaust duct brackets, access doors, and the radio frequency shield.[23]

Fig. 19. Transporting SA-1 to the pad.

Beginning the Checkout

Andrew Pickett's Vehicle and Missile Systems Group (part of Zeiler's Mechanical Office) spent the next month installing the accessories of SA-1* and conducting a series of launch vehicle tests. In some, the purpose was to make sure that various components responded correctly to pressure stimuli. Others checked for leaks caused by the barge trip and the subsequent erection of the S-I stage. The first week the group performed pressure switch functional tests, verifying the pickup and dropout pressures for several hundred switches. The Saturn's 48 nitrogen bottles, which pressurized the RP-1 fuel tanks during flight, were then tested at one-half the operating pressure.

During the second week, the unit checked out the pressurizing and venting capability of the LOX tanks. Air pressure was applied to a switch in the tanks' electrical system. The switch, when functioning properly, would terminate pressurization at a certain level. If excessive pressure built up, a second switch would vent the hypothetical gaseous oxygen. LOX and RP-1 system leak checks followed; in both tests the team pressurized the tanks to about one-half the operating pressure, looking for seal leaks.

---

*Both the rocket and the mission carried the designation SA-1.

Concurrently Pickett's group conducted a series of engine tests. A nitrogen purge of the LOX dome, located at the top of the H-1 engine, served several purposes. A low-level purge, begun prior to propellant loading and continued until shortly before engine ignition, exceeded atmospheric pressure to prevent contaminants from entering the thrust chamber nozzle and flowing up to the injector plate and LOX dome. This also prevented moisture from condensing in the area. If a launch was cancelled, a full-flow nitrogen purge would quickly expel all LOX from the dome to avoid a possible explosion. Similar purges of the liquid-propellant gas generator, LOX-injector manifold, and the fuel-injector manifold of the thrust chamber prevented the entry of unwanted substances.

The full-tank pressurization test on 6 September ended the first phase of mechanical checkout. Allowing for the possibility of an explosion while bringing the launch vehicle to full pressure, LOD officials cleared the pad for the Wednesday morning test. The two-hour exercise went smoothly, and that afternoon engineers were back at the launch vehicle for further operations.[24]

Calibration of the measuring devices that were to report more than 500 flight measurements was a daily operation. Sensing devices such as transducers, potentiometers, thermocouples, and strain gauges measured pressures, propellant flows, temperatures, and vibrations. A signal from one of these sensors, measured in millivolts, was routed to a signal conditioner which amplified the reading until it could be read on a scale of 0–5 volts. The calibration of these signal conditioners, popularly referred to as black boxes, was a major concern of Reuben Wilkinson's Measuring Group (a unit of Sendler's Measuring and Tracking Office). The team sometimes stimulated a sensing device by tapping on a portion of the rocket to cause vibrations or by placing a hot soldering iron near a thermocouple. More often they simulated a signal with an electrical input through an "interrupt box" located between the sensor and the signal conditioner. While calibrating the black boxes, the launch team bypassed the telemetry system. The amplified signal went from the signal conditioner through a series of remote-controlled relays, and then over wires to a measuring station in the base of the service structure. The calibrating equipment in the station normally performed a five-step sequence, checking the reading of each instrument at 0, 25%, 50%, 75%, and 100% of maximum value. After the tests were completed, Wilkinson's team reconnected the measuring and telemetry systems for readings over the radio frequency (RF) links.* The Measuring Group removed faulty instruments

*According to the *Saturn SA-1 Vehicle Data Book*, the following types of measurements were made on the SA-1: "propulsion, expulsion, temperature, pressure, strain and vibration, flight mechanics, steering control, stabilized platform, guidance, RF and telemetering signals, voltage, current and frequency, and miscellaneous." Nearly 400 of SA-1's 510 telemetered readings concerned propulsion, temperature, or pressure. F. A. Speer, "Saturn I Flight Test Evaluation," 1st American Institute of Aeronautics and Astronautics Meeting, 29 June–2 July 1964, fig. 4.

Erecting SA-1

Figure 20

Figure 21

Figure 22

Fig. 20. Lifting the first stage from the transporter. Fig. 21. Hoisting the stage in vertical attitude. Fig. 22. Setting the first stage on the support arms at LC-34.

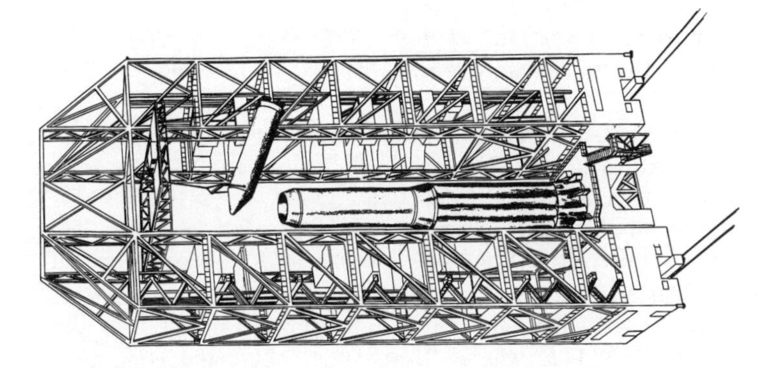

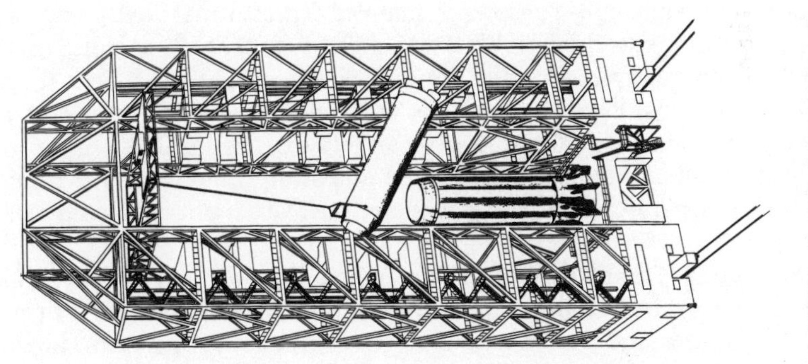

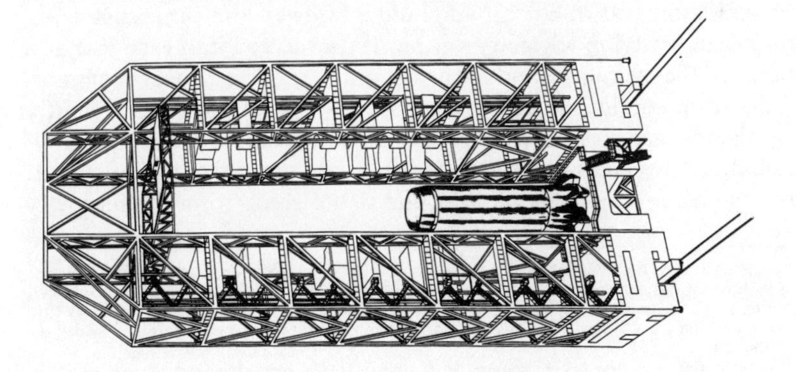

Fig. 23. Erecting the upper stages.

from the launch vehicle for further checks at calibration stands or in an instrument-calibration laboratory. The team was also responsible for the blockhouse measuring-station. Here LOD received 100 ground measurements on the rocket and ground support equipment, as well as telemetry data.[25]

Another of Sendler's units, Daniel McMath's telemetry team, checked out the booster's eight RF links. Seven of the links used the XO-4B package, a proved system from Jupiter flights. The XO-4B was a PAM–FM–FM (pulse amplitude modulated–frequency modulated–frequency modulated) system with 15 channels of continuous data and 54 multiplexed channels.*

The Guidance and Control Division in Huntsville had developed the eighth link to ensure sufficient data channels for the Saturn C-1. The central feature in the new XO-6B was a 216-channel electronic commutating system.† Sub-multiplexers sequentially sampled the same measurements for each of the eight engines. Sub-multiplex 1 might sample "temperature LOX pump bearing" while sub-multiplex 2 sampled "pressure at fuel pump inlet." The main transistorized multiplexer, in turn, sequentially sampled each of the 27 sub-multiplexers. The multiplexer's output was fed to a 70-kilohertz wide-band subcarrier. This frequency permitted the use of a commercially available oscillator that accurately carried the 3600-pulse-per-second wave train and utilized existing demodulation equipment. The result was that 216 separate Saturn measurements traveled on one radio frequency.[26]

McMath's Telemetry Group first tuned the two sets of antennas located at the forward end (top) of the S-I stage. The six-man team next performed transmitter and power amplifier checks. A third operation, alignment of the subcarrier channels, involved tuning each subcarrier oscillator to its center frequency and band edges. The test also ensured that signal output from the oscillators was of correct amplitude. Midway into the second week the team began verifying telemetry wiring. Data was fed into each line at a break between the measuring and telemetry systems. If range operations permitted, the team conducted an "open loop test," with the RF transmitter radiating the telemetry signal to receivers in the blockhouse and hangar D. But if radiating RF signals would interfere with any other activity in the area, the team operated "closed loop" with the signal going from the telemetry

*Each telemetry link employed one frequency, e.g., SA-1's link 3 used 248.6 megahertz. Oscillators within that system produced sub-carrier channels, referred to as straight channels because they carried continuous data from one sensor. Most measuring instruments, however, shared telemetry time by means of a multiplexer. On the XO-4B links, two 27-channel mechanical commutators provided the multiplex function.

†Commutation in telemetry is sequential sampling, on a repetitive time-sharing basis, of multiple-data sources for transmitting on a single channel.

link over wire to the telemetry ground stations. After all eight links were checked out, the team reconnected the measuring and telemetry systems for subsequent tests of the launch vehicle.[27]

During the first month of checkout, Jim White's Tracking Group worked on the tracking systems for the SA-1: cameras, UDOP and UDOP Beat-Beat, S-band radar, C-band radar, Azusa, Beat-Beat MKII Telemetry, and Telemetry ELSSE.* The two radar systems were controlled by the Air Force. The S-band provided position data by tracking the Saturn beacon. The C-band was a backup, should the Saturn beacon fail. LOD had eight UDOP stations in the Cape area, each connected by RF data links to a central recording station in hangar D. The Beat-Beat MKII Telemetry employed two baselines: one set of antennas located south of LC-34 determined whether the rocket made its proper turn out to sea; the other set, southwest of LC-34, ascertained flight path deviations downrange. The UDOP Beat-Beat system would fly on SA-1 as an experimental package.[28]

White's team employed a test transmitter to check out the UDOP stations. The test team simulated launch vehicle movement by varying the transmitted frequency. A drop in frequency simulated velocity away from the receiving station; conversely, a frequency increase represented rocket movement toward the receiver. These response tests checked the data-link equipment as well as the eight UDOP receiving sets. Preparation of the Beat-Beat systems included "walking the antenna," a basic test, but one which pointed up the importance of the tracking unit's work.† First, antenna connections were broken at one end of the baseline. Then a team member, equipped with a hand antenna and field telephone, walked a certain distance to set up a new baseline. Launch vehicle signals received at the new baseline indicated a theoretical rocket deviation from the previous flight path (read at the old baseline), the degree and direction of the deviation depending on the man's new location. By correlating the deviation and the new baseline, White's team determined whether the Beat-Beat system was functioning properly.[29]

---

*See footnote on p. 45 for descriptions of Beat-Beat and UDOP. Azusa dated back to the early 1950s and was named after the southern California town where the system was devised. The Azusa ground station determined the vehicle transponder's position by measuring range and two direction cosines with respect to the antenna baselines. ELSSE (Electronic Skyscreen Equipment) was used "to determine angular deviations of the missile from the flight line. The system consists of two ELSSE receivers placed behind the missile equidistant on either side of the backward extended flight line." W. R. McMurran, ed., "The Evolution of Electronic Tracking . . . at KSC," p. 3.

†According to LOD veterans, an incorrect performance of this test had cost the Air Force its first Thor shot several years earlier. After establishing its new baseline, an inexperienced contractor crew had picked up an LOD test transmitter frequency rather than the Thor's RF. Getting the opposite results from what they expected, the team had rewired the indicating device. When the Thor was launched, the range officer destroyed it unnecessarily, because the Beat-Beat system indicated a westward flight toward Orlando.

SA-1 required many modifications of equipment and procedures; as early as the second week the activities report listed among its major events, "engineering changes underway."[30] Characteristic of first launches, SA-1 was the most difficult and time-consuming of the Saturn block I launches. Robert Moser altered the schedule, when necessary, at the daily operations meeting in blockhouse 34.[31]

The scheduling committee planned an RF compatibility test for the midway point in the eight-week checkout (see table 3). The test was a major one for SA-1, marking the first time the vehicle stood alone (service structure removed from pad) for a complete check of the radio systems. Power was applied to the vehicle's RF systems to transmit signals to Cape receiving stations for telemetry, radar, and command and control. The launch team was particularly interested to see if the test would cause any interference in the command destruct system. Earlier launch programs had involved two to four telemetry links. SA-1's eight links increased the possibility of carrier and subcarrier frequencies beating against each other to produce harmonics that would feed back into receiving antennas. The effect might introduce spurious signals into the command destruct system.* The operations served both a validation and confidence function, proving each radio channel's performance and demonstrating that no serious interference would enter the destruct system. As an unexpected bonus, the test also demonstrated the launch vehicle's stability. Shortly after removal of the service structure, a sudden September squall subjected the rocket to 48-kilometer-per-hour winds without ill effect.[32]

LOD started integrated systems tests in the fifth week of checkout. Overall test (OAT) #1 (mechanical and network) was the first run of the launch vehicle's sequencing system, the relay logic that controlled the last minutes of countdown. OAT #2, a "plugs-drop test," put the vehicle on internal power with ground support disconnected. The key overall test, the guidance and control OAT #3, pulled all systems together in a check verifying the previous five weeks' work. The launch team began preparations for the test Saturday, 23 September. The advance work fell into seven categories: vehicle networks, ground networks, mechanical, electrical support, measuring, RF, and navigation. Vehicle network requirements included the connection and verification of telemeters, calibrators, radars, and 60 test cables, e.g., the Thrust OK Switch Engine #3 test cable. The checkout on Monday morning went well; MSFC officials were increasingly confident that SA-1 would fly.[33]

---

*In a subsequent Saturn I checkout, after additional telemetry links had been added and before LOD adopted a digital command receiver, the launch team had considerable trouble with interference in the command channel.

TABLE 3. RF INSTRUMENTATION TEST PROCEDURES, SA-1

Time	Step	Procedure
T-20	(20 minutes prior to launch).	
	1.M:	Telemeter 1, 2, 3, 4, 5, 6, 7, and 8 ON
	2.M:	Auxiliary Equipment ON
	3.M:	Azusa ON
	4.M:	UDOP ON
	5.M:	C-Band beacon to FILAMENT
	6.M:	S-Band to FILAMENT
	7.M:	Command Receiver +1 ON
	8.M:	Command Receiver +2 ON
	9.M:	Telemeter Calibration to PREFLIGHT
	10.M:	Telemeter Calibration Command to 50%
	11.RANGE:	Radars ON and away from pad
T-18	1.TM-D:	Telemeter Recording ON
	2.TM-B:	Telemeter Recording ON
T-17	1.M:	C-Band Beacon to B+
	2.M:	S-Band Beacon to B+
T-16	1.M:	Telemeter calibration command to 0% for 10 sec.
	2.M:	Telemeter calibration command to 100% for 10 sec.
	3.M:	Telemeter calibration command to 0%, 25%, 50%, 75%, 100%, 0% in 2 sec. increments
T-15	1.M:	Telemeter calibration to INFLIGHT
	2.M:	Telemeter calibration command ON & OFF
	3.RANGE:	Command Carrier ON
	4.RANGE:	Check Azusa and report verbal readout to Test Conductor
	5.RANGE:	Interrogate C- and S-Band Beacons and report verbal readout to Test Conductor
T-12	1.RANGE:	Cutoff command on request of Test Conductor
	2.RANGE:	Destruct command on request of Test Conductor
	3.RANGE:	Switch transmitters as required by Range and repeat functions
	4.RANGE:	Secure Command Carrier
	5.M:	Command Receiver #1 OFF
	6.	Command Receiver #2 OFF
T-10	1.M:	Telemeters 1, 2, 3, 4, 5, 6, 7, and 8 OFF
	2.M:	Auxiliary Equipment OFF
	3.TM-D:	elemeter Recording OFF
	4.TM-B:	Telemeter Recording OFF
T-5	1.M:	Azusa OFF (or sooner if RANGE readout is complete)
	2.M:	UDOP OFF
T-0	1.M:	C-Band Beacon OFF (or sooner if RANGE readout is complete)
	2.M:	S-Band Beacon OFF (or sooner if RANGE readout is complete)

Source: "Saturn Test Procedures, RF Instrumentation Test SA-1 (4-LOD-3)," Robert Moser papers. This test format is similar to, but briefer than, most of the several hundred other procedures prepared by LOD for SA-1.

Symbols:		
	M	Firing Room Measuring Panel
	TM-D	LOD Telemeter Station Hangar D
	TM-B	Blockhouse 34 Telemeter Station
	RANGE	Items for Test Conductor and Safety Officer

By early October the original launch date of the 12th had slipped eight days. On the 4th the launch team conducted the LOX loading test, a major exercise for SA-1 since it represented the first integration of the Cape's cryogenic support equipment with the Saturn vehicle. LOD followed this successful exercise with another plugs-drop test on the 10th. Engine-swivel checks were completed by the end of the week. The launch team began the ninth week of checkout with the simulated flight test, the last major preflight test. Robert Moser's 43-page procedure covered preparations for launch, the last 90 minutes of countdown, and activities for 5 hours after liftoff. The test went well, but MSFC delayed the launch another week while its Saturn Office debated the merits of adding more sensors near the base of the booster to provide additional information on the critical bending during the first 35 seconds of flight. It was finally decided that SA-1's instrumentation was adequate and the launch was set for 27 October. During the last week, LOD completed ordnance fitting (the command destruct system) and repeated the simulated flight test.[34]

The Launch of SA-1

Prelaunch preparation began at 7:00 a.m. on 26 October 1961. Mechanical Office tasks that morning included inspection of the high pressure gas panel, cable masts, and fuel masts; ordnance installation; and preparation of the holddown arms. At 12:30 p.m., Thomas Pantoliano's 12-man propellants section checked out the RP-1 fuel facility while Andrew Pickett's team pressurized the helium bottle. RP-1 loading began an hour later. The propellant team filled the launch vehicle's tanks to the 10% level, using a slow, manual procedure of approximately 750 liters per minute to check for leaks. A leak in the fuel mast vacuum breaker was easily repaired, and at 2:30 p.m. the launch team cleared the pad for the automatic "fast fill" operation. Fuel flowed into the launch vehicle at 7570 liters per minute, reaching the 97% level in about 35 minutes. The propellants team then reverted to the "slow fill" procedure. As the design of the Saturn included a fuel drainage system, Pantoliano's crew placed 103% of the required RP-1 aboard the Saturn. Just before launch, the propellants team would take a final density reading and drain sufficient kerosene to achieve the desired level.[35]

The ten-hour countdown started at 11:00 p.m. as LC-34 switched to the Cape's emergency generating plant. This facility supplied the launch team a current relatively free of the fluctuations common in commercial power. The Saturn's electrical circuits and components began warming up when vehicle power was applied at T – 570—570 minutes before launch time exclusive of holds. Five minutes later the measuring panel operator turned on the eight telemetry channels. A series of calibration checks followed. At T-510 range and launch officials initiated an hour of radar checks.[36]

Loading of liquid oxygen started after 3:00 a.m. on the 27th (T-350). The Saturn's LOX tanks were 10% filled to check for leaks in the launch vehicle or in the 229-meter transfer line, as well as to precool the line for the fast flow of super-cold LOX. While the automatic fast fill from the 473 000-liter LOX storage tank employed a centrifugal pump, the 10% precooling operation relied on the pressure in the reservoir. The 10% level in the Saturn's tanks was maintained for the next four hours by feeding LOX from the 49 000-liter replenishing tank.[37]

Testing of command and communication systems began at T-270. The flight control panel operator activated the guidance system's stabilized platform, the ST-90, to check pitch, roll, and yaw response. Ten minutes later the network panel operator placed the vehicle on internal power to ensure that the Saturn's batteries functioned properly. Meanwhile other engineers conducted Azusa, UDOP, radar, and telemetry checks. The operation was over by T-255, and the launch vehicle was returned to external power.[38]

Two hours from the 9:00 a.m. scheduled liftoff, an unfavorable weather report prompted launch officials to call a hold. When the count resumed at 7:34 a.m., the launch team rolled the service structure back to its parking area, 180 meters from the rocket. The propellants team set up the LOX facility for fast fill at T-100. The order to clear the pad came 20 minutes later; the blockhouse doors swung shut at T-65. One hour from launch the pad safety officer gave his clearance and the propellants team initiated a 6.5-minute precool sequence, a slow fill to recool the main LOX storage tank line, which had not been in use for four hours. When the "Precool Complete" light flashed on, the LOX facility's pump began moving 9500 liters per minute into the Saturn. In 30 minutes the tanks were 99% full. LOX loading changed over to the replenish system. An adjust-level drain* had already been made on the RP-1 tanks, bringing the fuel level down to 100%.[39]

Launch officials, concerned that a patch of clouds over the Cape might obscure tracking cameras, called a second hold at 9:14 a.m. A northeast breeze was soon clearing the skies, and within half an hour the countdown resumed. During the last 20 minutes, the launch team made final

*Establishing an exact ratio of RP-1 to LOX was important since simultaneous depletion of propellants at cutoff was desired. Flight data later indicated a 0.4% deviation in the RP-1 fuel density sensing system, 0.15% above design limits. Too much LOX (400 kilograms) and not enough RP-1 (410 kilograms) were therefore loaded. The error contributed to a premature cutoff 1.6 seconds ahead of schedule.

checks of telemetry, radar, and the command network. Automatic countdown operations commenced at T – 364 seconds. A sequencer or central timing device controlled a series of electrical circuits by means of relay logic; i.e., if event A occurred (e.g., opening a valve), the sequencer triggered event B, and so on through the required functions to liftoff. The sequencer monitored tank, hydraulic, and pump pressures; ordered a nitrogen purge of the engine compartment; and closed the LOX tank vents to pressurize the liquid oxygen. The Saturn vehicle switched to internal power at T – 35 seconds. Ten seconds later the sequencer ejected the long cable mast. The pad flush command at T – 5 seconds began a flow of water around the launcher base. At that time, a number of possible malfunctions (a premature commit signal, insufficient thrust in one or more engines, rough combustion, short mast failure, detection of fire, or voltage failure) could still cause the automatic programmer to terminate the countdown.[40]

Away from launch complex 34, Cape watchers gazed uncertainly at the Saturn rocket as the countdown neared completion. No previous maiden launch had gone flawlessly, and the Saturn C-1 was considerably more complicated than earlier rockets. LOD officials gave the rocket a 75% chance of getting off the ground, a 30% chance of completing the eight-minute flight. Although odds on a pad catastrophe were not quoted, launch officials acknowledged their vulnerability. With the construction of LC-37 barely begun, a pad explosion could delay the Saturn program a year. Critics had questioned the wisdom of the clustered booster design. Propellant pumps were supposedly reaching design limits and the Saturn C-1 had 16 pumps in eight engines. Local wags derisively referred to the SA-1 launch as "Cluster's Last Stand."[41]

Saturn backers, while expressing confidence in the rocket, were concerned about its launch effects. During test firings at Redstone Arsenal, residents 12 kilometers away had reported shattered windows and earth tremors. The launch team had set up panels and microphones at the Cape to register the Saturn's shock and sound waves. At the press site, 3 kilometers from pad 34, reporters were issued ear plugs as a precautionary measure. LOD officials had assured local residents that fears of the rocket were exaggerated. Still, everyone wondered what it would be like. The moment of truth came at 10:06 a.m. Contrary to popular belief, no one pushed a firing button to send SA-1 on its way. Launch came when the sequencer ordered the firing of a solid propellant charge. The gases from the ignition accelerated a turbine that in turn drove fuel and LOX pumps. Hydraulic valves opened, allowing RP-1 and LOX into the combustion chambers, along with a hypergolic fluid that ignited the mixture. The engines fired in pairs, developing full thrust in 1.4 seconds. A final rough combustion check was

Fig. 24. Liftoff of Saturn I. Note the long cable mast falling away on the right.

followed by ejection of the LOX and RP-1 fill masts from the booster base. The four hold-down arms released the rocket 3.97 seconds after first ignition. SA-1 was airborne.

Spectators saw a lake of flame, felt the rush of a shock wave, and then heard the roar of the eight engines. Trailer windows at the viewing site shook in response to the Saturn's power. Yet for many of the thousands watching the launch, the roar was a letdown. Reporters thought the sound equaled an Atlas launch viewed at half the distance.* The Miami *Herald* headline the next morning read: "Saturn Blast 'Quieter' Than Expected."[42]

Although the Saturn's roar failed to meet expectations, the human noise at LC-34's control center was impressive. Bart Slattery, a NASA information officer, told reporters that when the rocket passed maximum Q (point of greatest aerodynamic pressure) at about 60 seconds into the flight, "all hell broke loose in the blockhouse." Kurt Debus's face reflected the happy sense of accomplishment hours later when he informed the press that it had been a nearly perfect launch.[43]

---

*Marshall Center scientists, after studying readings taken in nearby communities during launch, explained that weather conditions were such that sound was absorbed by the atmosphere. As a result, sound levels were less than those experienced during static firings at Huntsville.

The success was particularly welcome to the Kennedy administration, coming at a time of high tension between the United States and the Soviet Union. The raising of the Berlin Wall had stunned the Western world in August 1961. President Kennedy had responded with a partial mobilization of U.S. reserve forces, but most political analysts considered the events a Russian victory. In late October, as the Soviet Union prepared to test a 50-megaton H-bomb, the President had proposed a massive fallout shelter program. On the day of the SA-1 launch, Russian tanks moved into East Berlin for the first time in several years.

The space race was an important element in a Cold War that threatened to turn hot. With the success of the Saturn booster, the United States had achieved a launch capability of 5.8 million newtons (1.3 million pounds of thrust). Space reporters were quick to point out the limits of the American success. The Soviet Union already had workable upper stages for their first stage. Furthermore, the current Russian tests in the Pacific would likely result in sizable booster advances. Despite these caveats, commentators agreed that SA-1 was an important step toward a lunar landing.[44]

4

ORIGINS OF THE MOBILE MOONPORT

Ambitious Plans and Limited Space

The original commitment of the Saturn program to a Cape Canaveral launching site was for the research and development launches only.* A launch site for operational missions remained an open question long after construction started on LC-34. Four major questions were involved: Would blast and acoustic hazards require an isolated—perhaps offshore—launch pad for larger Saturn rockets? If not, could the pads be safely located on the coast of Florida or elsewhere—Cumberland Island, Georgia, perhaps? Would the Saturn become America's prototype space rocket? If so, how many Saturn launches per year would be required? In the midst of these questions was one stern reality: Cape Canaveral was running out of launching room.

By early 1960 the Cape resembled a Gulf Coast oil field. Launch towers crowded the 16 kilometers of sandy coastline with less than a kilometer of palmetto scrub separating most of the pads. The busy landscape testified to the recent advances in America's space program, but the density of the launch pads posed a problem for NASA and Air Force officials. Launch programs were under way for Titan, Polaris, Pershing, and Mercury; plans for Minuteman and Saturn were well along. A Department of Defense management study, prepared in April 1960, reported that the Atlantic Missile Range was "substantially saturated with missile launching facilities and flight test instrumentation."[1] This seconded a 1959 congressional study that criticized the range's severe shortage of support facilities.[2] With the siting of the second Saturn launch complex (complex 37) near the northern boundary of the range, launch officials were running out of real estate.

The lack of room at the Cape did not deter Marshall Space Flight Center personnel from preparing plans for 20, 50, even 100 Saturn flights a

---

*In mid-1960, 10 R&D launches were scheduled. LC-34 was to launch the first four Saturn C-1 shots (testing the booster). Six subsequent C-1 R&D missions with upper stages would be launched from a modified LC-34 and from LC-37. The latter complex would also be used for an undetermined number of C-2 R&D shots. Operational launches were still very tentative; a NASA Headquarters schedule in late 1960 called for 50 C-1 and C-2 launches between 1965 and 1970, 20 of them concerned with the Apollo program (reentry tests, earth orbital missions, and circumlunar missions).

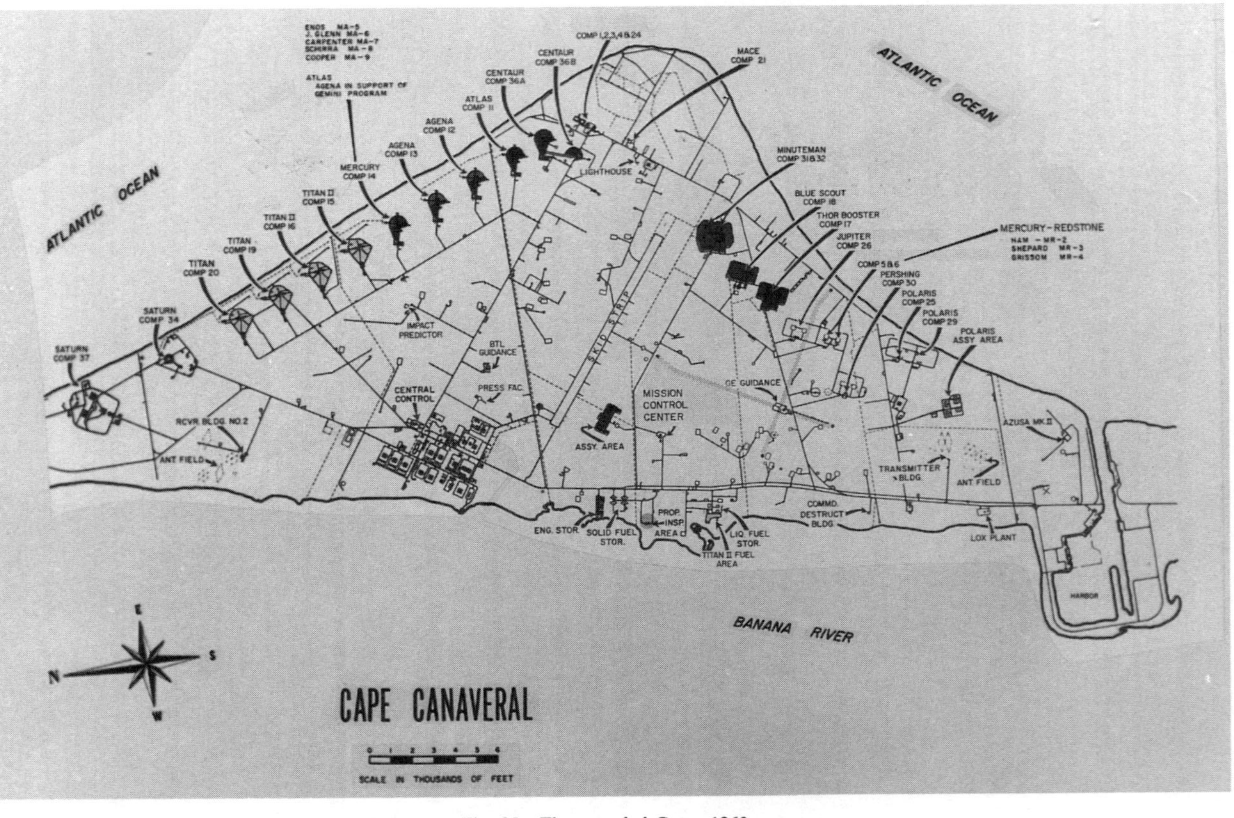

Fig. 25. The crowded Cape, 1963.

year. The Army's failure to carry out Project Horizon and put a squad of men on the moon had not dulled Hermann Koelle's enthusiasm (page 11). Now under NASA, his Future Projects Office was investigating earth-orbital space stations, a permanent scientific facility on the moon, a "switchboard in the sky" to serve communications satellites, and manned exploration of Mars. The last project would extend into the 1980s and involve sending several spaceships to that planet.[3]

NASA's ability to implement Koelle's plans depended upon the development of the launch vehicle in Huntsville. With the Saturn C-1 off the drawing boards, Huntsville planners were working on Saturn C-2. This three-stage rocket was to use the two stages of the C-1 configuration and insert a new second stage incorporating Rocketdyne's J-2 engine. A cluster of four J-2s, fueled by liquid hydrogen and liquid oxygen, could produce 3 520 000 newtons (800 000 pounds of thrust), giving the C-2 a total of 10 428 000 newtons (2 370 000 pounds of thrust). The C-2 could carry a payload 2.5 times that of the C-1; large enough to send a 3630-kilogram manned spacecraft to the vicinity of the moon, that payload would still be far short of what was needed for a direct ascent lunar landing (flying one spacecraft to the moon, landing, and returning to earth). An alternative to direct ascent was the use of earth-orbital rendezvous. This scheme involved launching a number of rockets into earth orbit, assembling a moon rocket there, and then firing it to the moon. NASA officials estimated that an earth-orbital rendezvous would take six or seven C-2 launches to place a 3630-kilogram spacecraft on the moon, nine or ten launches for a 5445-kilogram spacecraft. With this in mind, Koelle warned Debus at a 15 June 1960 meeting that such programs might require as many as 100 C-2 launches annually.[4]

Debus considered Koelle's projections plausible. Future Projects Office charts indicated that the cost per launch vehicle might drop as low as $10 million at the higher launch rate. If the space program received 3% of the annual gross national product for the next two decades, the American launch program could reach 100 vehicles per year.[5] A launch rate of such magnitude seemed unrealistic to other Launch Operations Directorate (LOD) members in light of their experience with the Redstone and Jupiter missiles—programs that had not exceeded 15 launches per year. Some doubted the Atlantic Missile Range's capability to sustain so large an operation, as well as the nation's willingness to fund it. Aware of the impact his program would have on LOD, Koelle asked Debus to determine the highest possible firing capability for Saturn from the Atlantic Missile Range.[6]

There was general agreement within LOD that launch procedures at complex 34 could not satisfy the Future Projects Office plans. Debus and his associates estimated that LC-34 could launch four or five vehicles per year, depending upon the degree to which checkout was automated. This allowed two months for vehicle assembly and checkout on the pad and a month for rehabilitation after the launch. With its two pads, LC-37 could handle six to eight launches annually.[7] The two complexes together barely satisfied Koelle's lowest projection for the C-2 study (12 launches annually); 48 Saturn launches per year would require at least 10 launch pads. Since the protection of rockets on adjacent pads might entail a safety zone of nearly 5 kilometers, a Saturn launch row could extend 48 kilometers up the Atlantic Coast. Purchase of this much land would be a considerable expense, and the price of maintaining operational crews for 10 pads would eventually prove even more costly. Limited space, larger launch vehicles with new blast and acoustic hazards, a steeply stepped-up launch schedule—all combined to set up a study of new launch sites for the Saturn. How and where to launch the big rocket?

Offshore Launch Facilities

As early as 1958, Livingston Wever, a member of the Army Test Office's Facilities Branch, had proposed the use of a modified Texas Tower* as an offshore launching platform for big rockets. Concerned about the Saturn's noise-making potential, Wever renewed his proposals in March 1960. Preliminary calculations, extrapolated from the noise levels measured during Atlas booster tests, indicated the Saturn C-1 would generate acoustical levels as high as 205 decibels at a distance of 305 meters from the launch pad. Peaks of 140 decibels, the threshold of pain, could be expected more than 3000 meters from the pad. Wever was particularly concerned that the Saturn vehicle might emit a shock wave in the early stages of its trajectory (at heights from 600 to 900 meters) that would cause serious damage in nearby towns. He proposed to solve the acoustical problem by moving the launch platform to a structure 169 kilometers southeast of Cape Canaveral and 56 kilometers north of Grand Bahama Island. Wever noted that "because of the shallow waters and slight tide actions in the proposed area, it would not be unfeasible to construct a rugged, but unadorned, steel platform as large as 500 feet [150 meters] square, not only for immediate static tests of the Saturn, but also for actual launchings of the Saturn and large boosters of the future." Venting the rocket's exhaust into ocean water would save the cost of an expensive

---

*Named for their similarity to offshore oil rigs in the Gulf of Mexico, Texas Towers were skeletal steel platforms built in the mid-1950s by the Air Force. The structure's massive triangular platform, supported by three 94-meter stilt-like legs, provided space for three large radars and a 73-man crew. Three of these towers were placed about 128 kilometers off the northeast coast of the U.S. to provide early warning of air attack.

flame deflector. Wever also anticipated savings on the construction cost of the firing room (blockhouse).[8]

Wever's proposal met with mixed reactions at the Army Test Office's Facilities Branch. Although Nelson M. Parry, assistant branch chief, approved Wever's effort to circumvent blast and acoustical problems, Parry disagreed with the solution. Parry himself had been working on plans to develop artificial islands for several years. In a study completed December 1958, entitled "Land Development for Missile Range Installations," Parry proposed an artificial island large enough to contain a blockhouse, instrumentation, camera mounts, fuel storage, and launch pad and tower. His process involved pumping sand from the shallow waters just off the Cape. Parry estimated that an artificial island 1.6 kilometers square, with a mean elevation of 1.8 meters above high water, could be constructed for $9 million. This compared favorably with the $11 million cost of one Texas Tower in the early warning defense system. More important, the island would be a fixed platform; the Texas Towers swayed in moderate winds. Parry also objected to Wever's proposal to remove the launching site from the Cape to the Bahamas. This would introduce problems of telemetry, coordination, tracking, and camera coverage.[9] Although supporting Parry's landfill procedures, Facilities Branch Chief Arthur Porcher considered the Banana River a better site for an island than the ocean floor off the Cape. He thought that any attempt to build up islands in the Atlantic would run into construction difficulties.[10]

In the Launch Operations Directorate, the job of evaluating offshore launch facilities fell to Georg von Tiesenhausen's Future Launch Systems Study Office. Tall, thin, and scholarly in appearance, von Tiesenhausen's looks befitted his "think-tank" role. His interest in offshore launch facilities dated back to World War II. Following the Allied bombing of Peenemünde in August 1944, von Tiesenhausen had recommended construction of floating pads to permit the dispersion of V-2 static firings. His plan had employed two barges, with the missile emplaced on cross bars.[11] At the Cape, von Tiesenhausen assigned direct responsibility for studying offshore facilities to Owen Sparks, a former U.S. Army colonel and the team's unofficial technical writer. Sparks's first task was to prepare a preliminary survey for Debus.

Sparks's May 1960 report listed a number of launch problems for the Saturn program. These included the shortage of space at the Cape, safety hazards, and the problem of constructing an adequate flame deflector. The noise factor merited attention but was secondary. He suggested locating an offshore launch complex downrange in the nearest ocean area with a depth of 15 meters of water. He believed such a site would satisfy the requirements

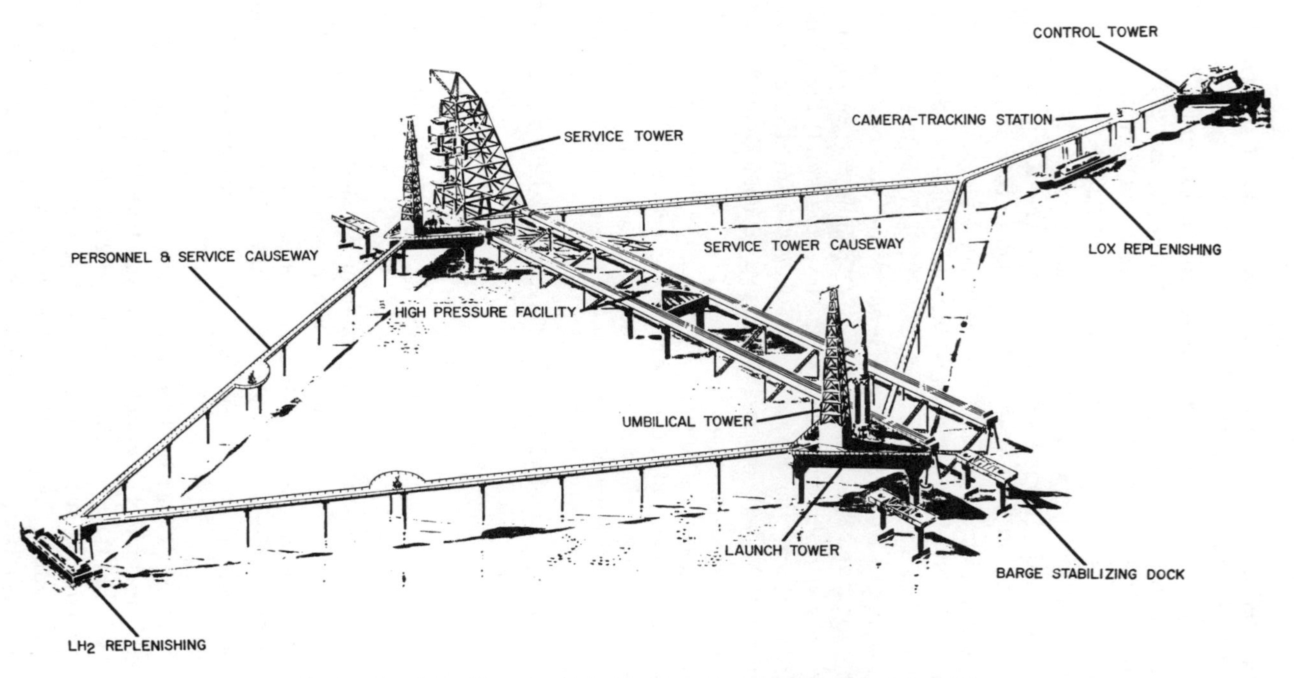

Fig. 26. Possible offshore launch facility, from a study by Owen Sparks in 1961.

of blast absorption without unduly complicating range support. Since marine construction involved a great many problems, the design should be as simple as possible. Sparks recommended the use of a stiff-leg derrick combined with the umbilical tower to reduce gantry requirements, and the employment of a knock-down mobile service structure. Beyond provision for both static firings and launches, any offshore facility should, he said, be expansible into a multipad complex.[12]

Sparks followed his first estimate with a preliminary feasibility study in late July 1960. His rationale for an offshore launching site had not changed. An evaluation of a half-dozen facilities favored the Texas Tower. This kind of facility, Sparks noted, could be placed in deep water where blast and sound posed no problems. Among other advantages, the offshore location would provide unlimited room for expansion, and fuel supplies could be kept on barges at a savings, compared to storage facilities on land. Sparks was no longer certain that the exhaust should be vented into the ocean—the resulting waves might damage the pad. Major disadvantages of a Texas Tower included the high cost of marine construction, the logistical problems of waterborne support for the facility, and the difficulty of providing a stable platform for handling vehicle stages and propellants. Sparks suggested further investigation of oceanographic conditions and their effects on launch structures, platform stability, and space vehicle requirements.[13]

Texas Tower vs. Landfill

Under increasing pressure to develop a greater launching capacity, LOD spent early 1961 examining the merits of offshore facilities and landfill proposals. In February the Office of Launch Vehicle Programs at NASA Headquarters asked LOD to step up its planning. Samuel Snyder, assistant director for Launch Operations, feared a pad explosion might shut down both LC-37A and LC-37B, and this in the face of a possible demand for nearly simultaneous C-2 firings on rendezvous missions. With space at the Cape already in short supply, he predicted it might be further limited if the Air Force stepped up its Dyna-Soar (glider-bomber) program. He asked LOD to plan a third fixed complex for FY 1963. Although Debus objected that the Saturn schedule did not at that time warrant an additional launch complex, LOD continued studies to find additional space.[14]

Debus then asked Col. Asa Gibbs in the NASA Test Support Office to obtain information on the cost of land reclamation, in either the Atlantic Ocean or the Banana River. Debus said he needed space for three additional dual-pad complexes and wanted to compare the expense of this operation with offshore Texas Tower facilities.[15] Gibbs's office responded on 9 March with two proposals for land development in the Banana River using hydraulic fill. A "maximum" concept involved filling approximately 2.5 square kilometers of Banana River tideland. The pad and support areas would rest on compacted earth about five meters above mean low water. Two of the proposed launch complexes could be built in this area, with the third pad on existing land north of LC-37. The total cost was $25 200 000. A "minimum" concept provided for two islands in the Banana River, each 610 meters in diameter, with 15-meter-wide causeways to link each island with the Cape, and a cost of $5 830 000.[16] Debus asked Gibbs in early April to secure Atlantic Missile Range approval for the tentative siting of the larger plan.[17]

At the same time, the survey of offshore facilities was accelerated. Concerned by a recent report on the blast hazards of the liquid hydrogen engine, Debus established an ad hoc committee under von Tiesenhausen's direction to select contractors who would conduct the offshore study. Early in February, Debus set the scope of the study. It should include expansion of the Cape northward by reclaiming and pumping up land; semi-offshore sites using Texas Towers or manmade islands; an offshore launch complex at some distance from the Cape; and a floating pad capable of location anywhere on the oceans.[18] Plans to solicit proposals moved ahead in February and March, but the offshore launching sites encountered heavy going. Sparks's study, submitted to Debus on 4 April, failed to satisfy the Director. He thought that transferring present launch methods to a Texas Tower would not suffice.[19]

Offshore facilities received a further setback in May with the presentation of Nelson Parry's land development scheme. Parry's list of drawbacks, two pages long, reflected the results of his interviews with Launch Operations personnel. Disadvantages included higher construction and maintenance costs, increased problems of communications and logistics, and a morale problem. While Parry's report did not give specific costs for remote offshore facilities, he was certain that land development would be cheaper than Texas Towers. His cost estimate sheets, prepared by James Deese of the Facilities Design Group, further indicated that building islands on the Atlantic shelf would be much more expensive than reclaiming land in the Banana River. A 2.3-square-kilometer island, 16 kilometers off the Cape, would cost $12.7 million; an island of 15 square kilometers, $59.9 million. He contrasted these figures with price tags of $18.7 million for dredging 7 square kilometers in the Banana River and $16 million for buying 750 square kilometers on Merritt Island.[20] Working independently, Rocco Petrone's Heavy Launch Vehicle Systems Office reached similar conclusions. The construction costs for causeways in the Florida Keys convinced them that the expense of building facilities in the ocean east of the Cape would be prohibitive.[21]

The ad hoc committee finally selected two study contractors on 15 May, but events rendered the C-2 offshore launch study moot. Marshall planners dropped the proposed rocket and started planning for a larger C-3 model. An even more decisive vote was cast by the Air Force–NASA Hazards Analysis Board (below, pp. 87–88), which found that "operational hazards for liquid and solid boosters did not dictate going to offshore launch sites."[22] Large vehicles could be launched from the coastline if Merritt Island was purchased as a safety zone. On 24 May, Debus told von Braun the contracts would not be let as the studies were no longer required.[23] Perhaps the biggest reason for the verdict against offshore facilities was seldom mentioned. In January 1961, a Texas Tower, part of the U.S. Air Force early warning system, had disappeared in a heavy storm with a loss of 28 lives.[24] Despite assurances from engineers that a similar catastrophe could be avoided, LOD leaders did not want the task of convincing Congress and the American public that an offshore facility would be safe against storm hazards.

The Mobile Launch Concept

During the early months of 1961, LOD took under consideration a third launch alternative, one that would eventually place men on the moon—the mobile launch concept.* The great advantage of a mobile launch concept lay in its promise of faster launch operations. With the fixed launch operation, e.g., SA-1 at LC-34, all rocket systems were mated and went through a thorough checkout at the pad. In the new scheme, LOD proposed to mate the vehicle and conduct these checks in an assembly building some distance from the pad. Only a brief prelaunch checkout at the pad would be needed to verify the rocket systems. Two digital computers, one in the launch control center and one on the transporter, would accelerate the checkout program and detect any change in rocket systems that might occur during the transfer to the pad. The computers were part of an automatic checkout system under development at Huntsville. By combining a mobile concept and automation, LOD leaders expected to reduce time on the pad from two months to no more than ten days.

There were other advantages to a mobile concept. Cape weather had corroded earlier rockets and might affect an exposed Saturn. An assembly building would provide cover for both the launch vehicle and the launch

***Concept* vied with *interface* for first place in Cape Canaveral jargon. Meaning of *concept* ranged from the first "batting around" of an idea to its fruition in a multi-million-dollar building or procedure. While the authors have tried to limit their use of the term, they confess to ill success especially in the early days when LOD planners were dealing with many contingencies and termed each tentative plan a concept.

team. Having worked on rockets in the open, LOD leaders knew how difficult it could be for technicians laboring in wind, rain, and lightning at the upper levels of the space vehicle. Finally the mobile concept offered considerable savings in labor costs. Concentrating the work force in one assembly building, rather than on the ten pads projected for 48 launches per year, would reduce personnel requirements substantially.

The idea of assembling a rocket in a location remote from the pad and then moving it to the launch area dated back to World War II. At Peenemünde the German rocket team had transported V-2s in a horizontal attitude to a hangar where they were erected in checkout stalls. Following transfer to a rail-mounted static-firing tower, each V-2 was rolled out in a vertical attitude—sitting on its tail—for an engine calibration test and static firing. The missile received a final checkout in the hangar before being placed horizontally on a *Meillerwagen* for the ride to the launch site.[25] Both the Redstone and Jupiter programs had employed a mobile launch concept with the rockets traveling from assembly building to pad in a horizontal attitude. LOD officials had hoped to use the same principle at LC-34, but time and money dictated otherwise. The Saturn C-1 test series permitted at least four months between launches, which was enough time to assemble and check out each vehicle on the pad.

Space planners outside NASA appreciated the merits of the mobile concept. The Air Force in 1960 had commissioned the Space Technology Laboratory to determine an optimum vehicle system for military use from 1965 to 1975. Entitled "The Phoenix Study Program," the work was subsequently completed by Aerospace Corporation and the Rand Corporation in June 1961. One of the recommendations of the study was an integration building where assembly and checkout could be completed before the vehicle was moved—sitting on its tail—to the firing area. It was estimated that pad time for the Atlas-Agena could be reduced from 28 days under the current operation to 5 days with a mobile system.[26] In similar fashion, two Saturn C-2 launch studies, conducted by the Martin Company and Douglas Aircraft, concluded that Marshall's high launch rates would require a mobile complex.

The Mobile Concept—Initial Studies

Although LOD officials had appreciated the advantages of a mobile launch system for years, a Russian space achievement provided the impetus for the study that culminated in launch complex 39. Reports in early 1961 indicated a Russian capability of launching rockets from the same complex within a few days' time. LOD leaders saw a need to reassess American launch methods. Appropriately, considering the thousands of hours of overtime put into the future moonport, the initial plans were laid after duty hours. On the first weekend in February 1961, Debus discussed a new Saturn launch concept with Theodor Poppel and Georg von Tiesenhausen. At the end of the meeting, von Tiesenhausen was given the task of preparing several mobile launch alternatives.[27]

After von Tiesenhausen's Future Launch Systems Study Office began work in mid-February 1961, time clocks were ignored. One team member wryly recalls the two weeks compensatory time he enjoyed later in the year as scant repayment for the many hours of overtime devoted to the study. The survey considered moving the rocket from assembly area to pad in either a horizontal or vertical attitude and by barge or rail.[28]

While the Study Office examined the new proposal's impact on launch facilities, other LOD officials considered operational aspects. At a 21 March staff meeting, Debus challenged his subordinates to point up the concept's weakness. There was opposition, mostly on the grounds of cost. After a second day of debate, Debus appointed a formal committee under Albert Zeiler to consider the operational aspects. Any major problem area was to be brought to his attention before 31 March, at which time Debus intended to introduce the concept to the Marshall Space Flight Center Board. On 30 March, Rocco Petrone described the new plan to Abraham Hyatt, director of NASA's Office of Program Planning and Evaluation. The following day Debus made his presentation before the Marshall Board. Von Braun and other MSFC officials reacted favorably and asked for a comparison of vertical versus horizontal transfer costs. Debus promised to provide the results of an in-house survey in four weeks. The Board also considered hiring Connell and Associates to conduct a more detailed investigation. On 10 April, LOD officials briefed Gen. Don R. Ostrander, director of the Office of Launch Vehicle Programs in NASA Headquarters, who exercised general management over Marshall and LOD. Although receptive to the new launch concept, Ostrander strongly opposed any idea of trying to incorporate it into LC-37. Budgetary planning was too far along to permit extensive changes. He cautioned Debus that any launch concept had to be compatible with the launch vehicle. Reliability, rather than high launch rates, should serve as the guiding principle.[29]

The Future Launch Systems Office was ready by mid-April to submit its findings to Debus. Included among numerous charts and drawings prepared for the briefing was an analysis of the new proposals (table 4), from which von Tiesenhausen's group concluded that a mobile concept based on a horizontal barge transfer was most economical.[30] The projected cost advantages of the mobile proposals were good news, especially at a rate of 48

TABLE 4. COMPARISON OF PROPOSED LAUNCH COMPLEXES

Mobile complex		Fixed pad (similar to LC-37)
Vertical transfer	Horizontal transfer	
Expensive assembly building	Economical assembly building	No environmental protection for vehicle during assembly
Minimum loss by catastrophe	Minimum loss by catastrophe	Maximum loss by catastrophe
Hurricane protection	Hurricane protection	No hurricane protection
Maximum vehicle handling	Maximum vehicle handling	Minimum handling of assembled vehicle
Maximum R&D	Maximum R&D	Minimum R&D
Reconnect cables and retest on pad	Reconnect cables and retest on pad	No electrical or pneumatic disconnections required after checkout
Wind loads critical at transfer	Wind loads not critical at transfer	Wind loads critical during erection

Operational costs using barge or rail transport from assembly area to pad, and using a fixed pad (in millions of dollars)

Vertical transfer	Horizontal transfer	Fixed pad
At a launch rate of 8 per year:		
barge $125	barge $60	$60 (LC-37)
rail 105	rail 70	
At a launch rate of 48 per year:		
barge $210	barge $130	$370 (6 LC-37s)
rail 180	rail 145	

Source: O. K. Duren, *Interim Report on Future Saturn Launch Facility Study,* Future Launch Systems Study Office, MSFC, MIN-LOD-DL-1-61, 10 May 1961.

launches a year. Other questions remained unanswered. The offshore studies might still affect the choice of a launch concept. There was some question about the delivery dates for automated checkout equipment. The latest word from Haeussermann's Guidance and Control Division placed complete automation three to four years away.

Despite these uncertainties, Debus was anxious to secure approval from NASA's top management for further studies. A meeting with Robert Seamans, NASA Associate Administrator, was set on 25 April 1961 for this purpose. Debus met with von Braun one week earlier to review Marshall's position on launch facilities. The two men agreed that work on LC-37 should continue as planned. January 1964 was set as a tentative date for establishing the LC-39 criteria, allowing LOD nearly three years to investigate the mobile concept. The Seamans briefing went well, and feasibility studies for the new concept were authorized. The Associate Administrator told Debus to base the planning for LC-39 on technical considerations; cost was not to be the overriding factor.[31]

Martin and Douglas Aircraft Companies, at work on the C-2 operational-modes study since November 1960, were logical choices to conduct a feasibility study of the mobile launch concept. Both Martin and Douglas engineers believed the present facilities would be satisfactory for a rate of 12 Saturns per year. For higher launch rates, a mobile concept was recommended "because of more efficient utilization of personnel and equipment, and reduced land requirements by virtue of its centralized assembly and checkout procedures."[32] Douglas recommended transporting the mated booster stages from an assembly building in an upright position and adding the payload at the pad. Martin employed a rail-mounted vertical transporter or A-frame and called for mating the spacecraft in the assembly area with only propellant loading and countdown left for the pad. Both companies agreed that a mobile concept would provide more flexibility "because a greater latitude of launch rates is realized for any given expenditure." However, a Martin group working on Titan at the Cape recommended that LOD continue to assemble the rocket on the pad.[33]

NASA Plans for a Lunar Landing

The task of extending the Martin and Douglas study contracts to include the mobile concept was complicated by an unanswered question: what rocket would be launched from LC-39? Since the fall of 1960, NASA officials had given much thought to ways of accomplishing a lunar landing. A meeting in early January 1961 revealed the divisions within NASA as to the best means to accomplish this goal. The Space Task Group, responsible for Project Mercury, and the Headquarters Office of Launch Vehicle Programs favored using the Nova rocket for a direct flight from earth to the moon.* Marshall Space Flight Center advocated the use of several smaller Saturn launch vehicles to rendezvous in earth orbit, refueling one vehicle for the flight to the moon. A group at Langley Research Center supported a third mode—a lunar-orbital rendezvous. This involved placing a spacecraft into lunar orbit where it would detach a portion of the ship for the short trip to and from the moon. During the month of January 1961, a committee headed

---

*Nova was the name used by NASA during 1959-62 to describe a very large booster in the range of 44-88 million newtons (10-20 million pounds of thrust). The rocket never advanced beyond the conceptual stage, as was also true of the Saturn C-2 and C-3.

by George Low, Program Chief for Manned Space Flight, examined the manned lunar landing program. The committee concluded in its 7 February report that both direct ascent and earth-orbital-rendezvous methods were feasible. Using the Saturn C-2, the latter could be achieved at an earlier date (1968–69), but posed a high launch rate in a short period of time (six or seven C-2s for a 3630-kilogram spacecraft) and a mastery of rendezvous techniques. The direct ascent mode would take two years longer, depending on the development of the Nova rocket.[34]

Doubts about the adaptability of the Saturn C-2 to lunar landing missions appeared in March. Testifying before the House Committee on Science and Astronautics, Abraham Hyatt said that the Saturn C-1 would be used for an earth-orbiting laboratory and the C-2 for orbiting the moon. For missions beyond this such as a lunar landing, "payload capabilities greater than that of the Saturn C-2 appear to be necessary."[35] NASA officials had in mind a Saturn C-3 employing the new F-1 engine. Under development by Rocketdyne Corporation since January 1959, the F-1 burned the same fuel as the H-1 engine in the Saturn C-1's first stage. The F-1, however, dwarfed the H-1 in size and thrust: two F-1s in the proposed Saturn C-3 would produce 13 344 000 newtons (3 000 000 pounds of thrust), nearly double the lift of the Saturn C-2's proposed first stage.[36]

NASA's revised budget request of 25 March sought and obtained additional funds for the Saturn C-2 launch vehicle and the F-1 engine. Plans to accelerate C-2 development were announced 31 March, but the program was shortlived. Marshall engineers concluded in May that a Saturn vehicle more powerful than the C-2 was needed for circumlunar missions. Von Braun announced the demise of the C-2 the following month, at the same time stating that NASA's effort would be directed toward a clarification of Saturn C-3 and Nova concepts.[37]

May 1961 found LOD personnel grappling with a changing launch vehicle, the dangers of blast and sound from the large vehicles, and the demand for new launch facilities. The Director's daily journal reflected the frequent changes in the organization's planning:

26 April - Marshall's Future Projects Office initiated with LOD help an extension of the C-2 operational modes study (Martin and Douglas).

1 May - Debus informed NASA Headquarters that he would probably reorient launch study from offshore to mobile concept.

9 May - Von Tiesenhausen directed to proceed immediately with preparation and issuance of following studies: 1. C-2 offshore facilities with high firing-rate capability; 2. facility

for a solid booster of 44–88 million newtons (10–20 million pounds of thrust) from offshore, semi-offshore, and land; 3. add $100 000 to the C-2 operational modes study contracts to permit consideration of liquid-fueled vehicles of 22–44 million newtons (5–10 million pounds of thrust).

12 May - Von Braun requested a consideration of modifying LC-37 to accept a booster with either two F-1 engines or a 20-million-newton (4.5-million-pound thrust) solid motor.

15 May - Two contractors selected for offshore launch facilities study.

23 May - Cancellation of offshore study as designed.

26 May - C-3 launch facility contract with Martin initiated.

29 May - Nova offshore contract initiated.

5 June - NASA Headquarters notified LOD that C-3 and Nova studies were disapproved. Ostrander rescinded that disapproval at a Cape meeting.[38]

The Fleming Committee

In Washington, President Kennedy's announcement on 25 May spurred NASA's examination of the requirements for a lunar landing. An ad hoc committee chaired by William Fleming (Office of Space Flight Programs, NASA Headquarters) was conducting a six weeks' study of the requirements for a lunar landing. The Fleming Committee, judging the direct ascent approach most feasible, concentrated their attention accordingly. They devised a launch schedule employing Saturn C-1s for manned orbital flights in late 1964, a Saturn C-3 for circumlunar flights in late 1965, and a Nova, powered by 8 F-1 engines, for lunar landing flights in 1967. Seamans was unwilling to adopt the Fleming recommendations without a quick look at the rendezvous thesis. In early June, Bruce Lundin, deputy director of the Lewis Research Center, led a week-long study of six different rendezvous possibilities. The alternatives included earth-orbital rendezvous, lunar-orbital rendezvous, earth and lunar rendezvous, and rendezvous on the lunar surface, employing Saturn C-1s, C-3s, and Novas. His committee concluded that rendezvous enjoyed distinct advantages over direct ascent and recommended an earth-orbital rendezvous using two or three Saturn C-3s. NASA officials were sufficiently impressed to postpone a decision pending further studies.[39]

The Fleming Report's flight schedule caused some anxiety at the Cape. During his 5 June visit, General Ostrander suggested that the committee's recommendations might force a reevaluation of the new mobile launch proposals. In fact, the report indicated that the Saturn C-3 launch rate would not exceed 13 per year. This was a far cry from the Future Projects Office's revised projection of 30 to 40 annual Saturn C-3 launches. Debus called von Braun to point out the significance of the Fleming schedule. LOD's estimates of the economic crossover point between fixed and mobile launch facilities placed the figure around 15 launches per year. If NASA Headquarters adopted the Fleming recommendations, conventional launch facilities would probably be more appropriate. After checking into the matter, Marshall officials informed Debus that the 13 annual launches represented only a part of the future Saturn C-3 launch rate. Earth-orbital flights and interplanetary missions would keep the rate well above the economic break-even point for a mobile launch facility.[40]

Another troublesome matter stemming from the report had to do with NASA's possible use of solid-fueled rockets. The Fleming Committee's proposed launch vehicles included solid-liquid versions. In the C-3 configuration three solid-propellant motors would take the place of the two F-1 engines in the first stage. NASA Headquarters officials wanted the C-3 and Nova launch study contractors to design a facility that could service solid as well as liquid rockets. Debus objected, insisting that a "dual use" facility would penalize the liquid program. Solid motors, because of their greater weight and blast, would require expensive modifications to either conventional facilities or the new mobile concept. Furthermore, Debus was anxious to get the C-3 launch facilities study started and detailed criteria for solid rockets were not yet available. The difference of opinion took several weeks to resolve, but LOD's position prevailed. When LOD received data for the solid motors, additional studies might be done. In late June, Martin started work on the C-3 (liquid version) launch facilities study.[41]

Debus-Davis Study

The Fleming Committee's final report, 16 June 1961, listed construction of the launch complex as a "crucial item" and recommended that a "contractor immediately be brought aboard to begin design."[42] One week later Robert Seamans initiated a joint NASA–Air Force study of "launch requirements, methods, and procedures" for the Fleming Committee's flight program. LOD would concentrate on establishing mission facility criteria; Maj. Gen. Leighton I. Davis's Air Force Missile Test Center would determine support facility criteria.[43] In a second letter Seamans stated the study's objectives more precisely. The LOD-AFMTC team was to examine launch site locations, land acquisition requirements, spacecraft and launch vehicle preparation facilities, launch facilities, and launch support facilities.[44] The ensuing four-week study produced the *Joint Report on Facilities and Resources*

Required at Launch Site to Support NASA Manned Lunar Landing Program (the Debus-Davis Report). Because of its major recommendation that Merritt Island be the launch site for the Apollo program, the report will be discussed at some length in the next chapter. But the study advanced LOD thinking in regard to the mobile launch concept and must therefore be taken up at this point.

Two of the ground rules governing the Fleming Committee complicated LOD's work on the subsequent Debus-Davis study. One was that intermediate major space missions, such as manned circumlunar flights, were desirable at the earliest possible date to aid in the development of the manned lunar landing program. This envisioned a flight program using two radically different launch vehicles, the C-3 and the Nova, and consequently two distinct launch procedures. The second involved NASA's intention to develop liquid- and solid-propellant rockets on parallel lines. LOD planners would have to calculate costs and requirements for a liquid Saturn C-3, a solid-liquid C-3, a liquid Nova, and a solid-liquid Nova (table 5). The study was further complicated by NASA's decision to examine eight possible launch sites (see p. 91). The launch team faced the plight of a dressmaker, called on to outfit a beauty queen a month before she is selected from 50 contestants.[45]

The men who developed the Apollo launch facilities recall this study as one of the more hectic periods in the program's history. Some planning sessions extended into the early hours of the morning. Onc participant recalls arriving at his Cocoa Beach motel on a Saturday evening with the Miss Universe contest on TV. To his wife's amazement, his interest in feminine pulchritude gave way to fatigue and he was asleep before the final selection. Work on the study continued right up to the 31 July deadline, and the report was collated on the flight to Washington. Despite some embarrassing errors on the charts prepared for the NASA–Defense Department briefing, the 460-page survey was a real achievement.[46]

TABLE 5. DIMENSIONS AND WEIGHTS OF PROPOSED LAUNCH VEHICLES

	1st stage diameter (meters)	Total length (meters)	Weight at liftoff (kilograms)
Saturn C-3, July 1961			
liquid fuel	8.2	70.1	1 254 000
solid/liquid fuel	10.3	65.5	1 881 000
Nova, July 1961			
liquid fuel	13.4	102.1	4 336 000
solid/liquid fuel	13.7	97.5	5 561 000
Saturn V, Dec. 1961	10.0	84.9	2 860 000

A spirit of competition with the Air Force Missile Test Center spurred on the LOD effort. Air Force personnel caused some friction by offering unsolicited assistance in LOD areas. One such incident involved an Air Force recommendation to build a liquid-hydrogen plant at Cape Canaveral. There was uncertainty at this time as to how long liquid hydrogen could be stored at 20 kelvins (−253 °C) and therefore a question as to how much production capacity was needed. LOD officials considered the Air Force proposal technically infeasible; the proposed plant's electrical power needs would far exceed what the central Florida area could reasonably provide. Instead LOD wanted to purchase liquid hydrogen commercially, and the final report clearly stated that view. Working relations during the study were generally good, but some LOD officials believed that their Air Force counterparts wanted to assume a larger role in the manned lunar landing program.[47]

Debus appointed Rocco Petrone, Heavy Vehicle Systems Office, to represent LOD on the study's Executive Planning Committee. As a young ordnance officer, Petrone had helped the Director launch the first Redstone in 1953. Impressed by his work, Debus welcomed Petrone's reassignment to the launch team in July 1960. The joint study began Petrone's rise to prominence in the Apollo program. In various positions during the next nine years he would direct the Saturn program, first the facilities planning and construction, later the launch operations. He would acquire influence at the launch center second only to Debus. Tenacity, intellectual honesty, aggressiveness, and ambition were the basic ingredients in Petrone's advancement. A native of Amsterdam, New York, Petrone had been a tackle on the Blanchard-Davis teams at West Point. A determined pursuit of knowledge characterized his tour with the Missile Firing Laboratory in the 1950s. Associates recall that he devoured every piece of Redstone literature. His knowledge of launch operations made him a logical choice for Saturn program management. Petrone could get along well with people and even be charming. He demanded honesty, however, and did not hesitate to brand poor work for what it was. Consequently, some controversy accompanied his success. Described by intimates as basically shy and sensitive, Petrone displayed an aggressive exterior. His drive made workdays of 12–14 hours typical. Perhaps most important, Petrone's high ambition matched the Apollo program's lofty goals.[48]

Debus-Davis Report—Launch Concept

Although the mobile launch concept would not reach fruition for another year, by July 1961 its four major features were clear:

- Vertical assembly and basic checkout of the space vehicle on a mobile launcher-umbilical tower, located within an industrial and environmentally controlled building;
- Transfer of the assembled space vehicle and mobile launcher to the pad for final checkout, fueling, and launching;
- Control of operations from a remote launch control center; and
- Automation of vehicle checkout and launch.

The Debus-Davis Report represented considerable progress since the Study Office's May report. All aspects of the Saturn concept were described in greater detail, particularly the automated checkout. The flexibility that would characterize LC-39 was evident. The basic concept assumed a launch rate of 26 Saturns per year, but LOD plans allowed for additional pads and assembly bays to accommodate higher launch rates and special missions involving the launch of several vehicles in a brief period. Expediency dictated that rail be the only form of transfer considered. There was not enough time to prepare good cost estimates for canal and road. Further, LOD officials were confident from their LC-34 experience that a rail system would work.[49]

One of the initial mobile concepts, the horizontal transfer, had been eliminated by mid-1961 and was not mentioned in the Debus-Davis study. In its May report the Study Office had noted "certain operational limits of the horizontal transfer which might prohibit good reliability."[50] The statement reflected Albert Zeiler's concern that inspectors would damage wires and tubing during checkout of a horizontal vehicle. (During a vertical checkout workers would stand on platforms extending around the rocket. With the vehicle in a horizontal position, it would be difficult to keep workers from damaging the rocket's thin skin.) Maintenance of umbilical connections during a horizontal transfer was another problem. Fear of the stresses generated in lifting a large launch vehicle from a horizontal to a vertical position was the third and decisive consideration leading to the concept's demise. Huntsville engineers were aware of the strain placed on the 21-meter Redstone's joints and outer skin during this operation. The stress on the 70-meter Saturn might well be excessive.[51]

The Saturn C-3 (liquid) launch complex plan comprised a vertical assembly building (VAB), a launcher-transporter, an arming area, and launch pad. The VAB would consist of assembly bay areas for each of the stages, with a high bay unit approximately 110 meters in height for final assembly and checkout of the vehicle. Buildings adjacent to the VAB would house the Apollo spacecraft and the launch control center. The launcher-transporter would incorporate three major facilities: a pedestal for the space vehicle, an umbilical tower to service the upper reaches of the space vehicle, and a rail transporter. An arming tower would stand about midway between the

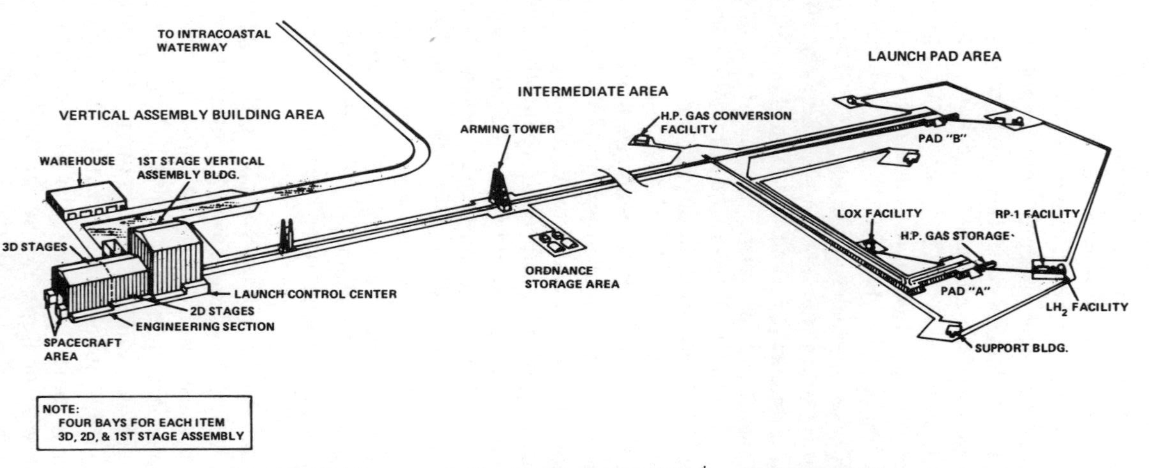

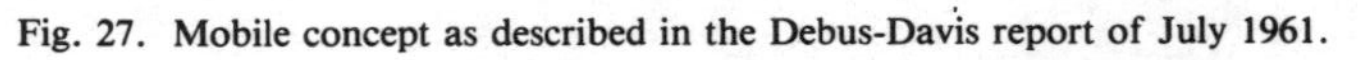

Fig. 27. Mobile concept as described in the Debus-Davis report of July 1961.

assembly building and the pads. The Apollo Saturn would carry a number of hazardous explosives: the launch escape system (the tower on top of the vehicle that lifted the spacecraft away from the launch vehicle in case of an emergency), retrorockets to separate the stages, ullage rockets to force fuel to the bottom of tanks, and the launch vehicle's destruct system. Launch officials wanted to install these solid-propellant items in an area apart from the rest of the operation.[52]

By July 1961 LOD engineers had fixed the requirements for the mobile launch concept's electrical checkout. These were fourfold: first, the electrical ground support equipment was to be designed so that checkouts could be conducted simultaneously on vehicles in the VAB and on the pad; second, the electrical systems of the vehicle and launcher-transporter would remain intact after checkout in the VAB; third, the launch control center would be able to launch rockets at a distant pad and check vehicles in the nearby VAB; and fourth, there would be a minimum of connecting cables between the launch pads and the control center because of the distances involved. The plan required the use of two digital computers, one located on the launcher-transporter and the other in the launch control center. The former would be used for checkout of the launch vehicle both at the VAB and on the pad. The performance of the computer on the launcher-transporter would be remotely controlled by the computer in the launch control center. Two firing rooms were necessary—one for control of checkout procedures in the VAB and the other for launch pad operations.[53]

The significance of the initial mobile launch studies lay more in the timing than in the content. LOD officials would not agree on a final concept for another year. By mid-1961, however, they were confident that some form of vertical transfer would work. Debus's initiative in February 1961 provided LOD time to examine the concept and make some reasonable judgments. When the Kennedy administration announced the lunar landing program in May 1961, LOD officials had a suitable launch concept in mind. Without the three months gained by the February decision, it is doubtful that LOD would have ventured on a new launch concept. The Apollo facilities might well have resembled a larger LC-37.[54]

5

Acquiring a Launch Site

Hazards Board Recommends Merritt Island

While Dr. Debus took the occasion of the top-level meeting at Huntsville on 25 April 1961 to brief Robert Seamans, NASA Associate Administrator, on the mobile launch concept, the conferees discussed other questions, especially the lack of space at Cape Canaveral. Gen. Donald Ostrander, Dr. Wernher von Braun, and William Fleming, soon to be head of the Project Review Division, participated in the discussion. At its conclusion, Debus was directed to meet with Maj. Gen. Leighton I. Davis, commander of the Air Force Missile Test Center, to discuss NASA's need for additional land.[1] The presidential challenge (a man on the moon by 1970) lent urgency to Debus's inquiry. Very likely, the launching of a moon rocket, Saturn or Nova, would create blast hazards requiring a large safety zone around the pad. Acquisition of many acres of real estate was the next step in building the moonport and the question facing the Launch Operations Directorate (LOD) was, Where? The answer would prove twofold: NASA would build the moonport on land (Merritt Island) within the Air Force sphere of influence at Cape Canaveral, but in the process would work out an understanding with the Air Force that would secure freedom of action in NASA's launch area.

Before recommending any land purchase, NASA and the Air Force had to determine the dangers involved in testing and launching a moon vehicle. In the last week of May 1961, the two groups set up a Joint Air Force–NASA Hazards Analysis Board to study the effects of blast, noise, fire, fragmentation, radiation, and toxicity. It would also prepare preliminary design data as a basis for safety perimeters for personnel and facilities within government-controlled areas, as well as for people and property in areas adjacent to the launch site.[2] Since NASA had reached no decision on the vehicle for the moon landing, the analysts considered the use of a Saturn C-3 booster of 13 million newtons (3 million pounds of thrust) and Nova boosters of 53, 98, and 164 million newtons (12, 22, and 37 million pounds of thrust). These were further classified as to fuels: liquid propellants, solid propellant for the booster and liquid propellants for the upper stages, and liquid propellant for the booster with a nuclear-powered upper stage.

On 1 June 1961, the board published a preliminary report of its findings and recommendations. The hazards analysis indicated that the minimum distance required for overall safety between the launch pad and uncontrolled areas varied from 5270 meters for the Saturn C-3 to 15 240 meters for the 164-million-newton (37-million-pound-thrust) Nova booster. The minimum safe distance for nuclear stages reached 16 kilometers. The board concluded that if the government acquired additional land on Merritt Island, vehicles without nuclear upper stages could be launched from onshore facilities along False Cape north of Cape Canaveral. Further, since persons working within government-controlled areas could be given adequate protection, Merritt Island provided suitable land for industrial and technical support areas.[3]

A New Home in Georgia?

When the news spread that NASA was investigating launch sites, a group of Georgia businessmen suggested the coastal islands of their state. A survey team of NASA, Air Force, and Pan American personnel found many advantages at Cumberland Island: undeveloped land, railroad facilities, a coastal waterway, and port facilities. The team concluded that Cumberland Island merited further investigation as a site for launching large rockets.[4]

Beginning near the Florida state line, Cumberland Island extends north for 32 kilometers. It varies in width, being some 5 kilometers at the widest point. Extensive tidal flats, saltwater marshes, and the Intracoastal Waterway separate it from the Georgia mainland. Deepwater docks along the Intracoastal Waterway provided access to cheap water transportation. King's Bay Ammunition Facility was close at hand, owned by the government, with readily accessible railroad sidings. Anticipated real estate costs were relatively low; to the north, however, were expensive island resorts.

In the meantime, the Air Force went ahead with proposals to purchase 93 square kilometers adjacent to Cape Canaveral at an estimated cost of $10 million.[5] One month after the Mercury flight of Alan Shepard, General Ostrander and Samuel Snyder from NASA Headquarters, Eberhard Rees from Huntsville, and Debus met with a group at Cape Canaveral on 5 June 1961. The conferees agreed that the NASA program would require more than the 93 square kilometers. A few days later, General Ostrander suggested that the group give greater consideration to Cumberland Island.[6] An ad hoc committee under the chairmanship of William A. Fleming began work on 8 May. Its report, turned in on 16 June, spoke favorably of the southeast Georgia site. "There are alternate possibilities, besides AMR. . . . One of the most promising . . . is the King's Bay area along the Georgia coast. . . ."[7]

The advantages of the Canaveral area were nevertheless overwhelming. It lay at the head of the Atlantic Missile Range, a series of tracking stations that reached southeastward almost 9000 kilometers to Ascension Island (with further extensions under way for the Mercury program). Its trained personnel had launched many missiles. No big cities stood in danger from accidental explosions or wandering missiles. The noise would not disturb a large civilian population. Finally, while the Cape itself was filled up, there was room for expansion on Merritt Island and along the coastline north of False Cape.

The Canaveral area and Cumberland Island shared one advantage over other possible sites. Barges from Huntsville could sail down the Tennessee, Ohio, and Mississippi Rivers, through the Gulf, and up the east coast of Florida. In view of the mammoth proportions of the Saturn and Nova boosters under consideration as moon vehicles, this access to barge transport was an important consideration.

Organizing for the Debus-Davis Study

High-level agencies in Washington took a hand in the matter. On 16 June 1961 Roswell L. Gilpatric, Deputy Secretary of Defense, alerted the Secretaries of the Army, Navy, and Air Force to the joint planning by NASA and the Department of Defense concerning all elements of the space program, "including the extension of ground facilities."[8] He directed them to instruct commanders of national ranges and other officers in charge of space resources to lend their full support. At the Cape this responsibility fell to Gen. Leighton Davis. Montana-born, Davis had excelled at West Point as a student and instructor. After the entry of the United States into World War II, he had expressed dismay at the quality of the sighting equipment on the planes in his bomber command. The Army transferred him to research and development of gun and bomb sights at Wright Field in Ohio. Other R&D assignments followed, prior to his taking command of the Air Force Missile Test Center in May 1960.[9]

On 23 June, Robert Seamans formally requested Debus and Davis to study all major factors concerning launch requirements and procedures for direct or orbital flights to the moon (page 80). NASA was to set up criteria for mission facilities, and AFMTC was to arrange for support facilities; both were to suggest guidelines for management structure and division of authority. On the 30th Seamans asked the two men to study all possible sites—mainland, offshore, and island locations. Their responsibility extended to the facilities and the acquisition of land, but not to worldwide tracking and command stations.[10]

On 6 July, Petrone for the Director of the Launch Operations Directorate, Col. Leonard Shapiro for the Air Force, and Col. Asa Gibbs, NASA Test Support Office Chief, drew up a detailed outline for the Debus-Davis study of the facilities and resources required at the launch site to support NASA's manned lunar landing program.* Petrone was responsible for operational plans and concepts and mission functions, launch facilities, operations control, and support requirements. Shapiro would develop plans for range support to be provided by the Department of Defense, including support facilities, utilities, and instrumentation in the launch area and downrange. Gibbs was responsible for analyzing and recommending appropriate management relationships at the range, including flight control and ground safety.[11] Since Shapiro had limited experience in this field of work, Col. Verne Creighton took his place for a time. Later, Shapiro returned to finish the report.[12]

While the opening section of the Debus-Davis Report explicitly set out a "NASA Manned Lunar Landing Program," the section on funding revealed a different point of view on the part of the Air Force representatives. NASA's proposal called for NASA to provide funds for construction of range support facilities, all mission facilities, and all instrumentation required for the Manned Lunar Landing Program. The Department of Defense would budget and fund for operation and maintenance costs of the range in support of this program, and NASA agreed to assist the Department of Defense in justifying these costs.[13]

The Air Force, on the other hand, saw the program as "national," combining civilian and military control, rather than as a strictly civilian (NASA) enterprise. In the same Debus-Davis Report, the Air Force recommended that NASA and the Department of Defense budget their needs separately. NASA would budget and fund mission requirements. The Department of Defense would budget and fund range support requirements. The Defense proposal then spelled out its viewpoint: "Budgetary requirements for the Manned Lunar Landing Program will be submitted and justified as a 'Joint Package,' segregated by agency and department," and "funds apportioned to the respective organizations will be administered according to policies and procedures internal to the agency or department."[14] Several years would elapse before the two organizations would clarify this delicate matter.

---

*As Chief of the NASA Test Support Office since its inauguration in April 1960, Gibbs had served as liaison officer between the Launch Operations Directorate and the Air Force Missile Test Center.

Recommending a Launch Site

During the month of July, the NASA–Air Force team considered eight sites:

- Cape Canaveral
- Offshore from Cape Canaveral
- Mayaguana Island in the Bahamas
- Cumberland Island, Georgia
- A mainland site near Brownsville, Texas
- White Sands Missile Range in New Mexico
- Christmas Island in the mid-Pacific south of Hawaii
- South Point on the island of Hawaii

Of the eight, the Debus-Davis Report estimated that White Sands would cost the least to develop and operate. These advantages were offset by its land-locked location; the lack of water transport would virtually dictate construction of the space vehicle assembly plant and test-firing stands near White Sands. Cost alone eliminated the island sites of Mayaguana, Christmas, and Hawaii, where construction and operation costs would be more than twice the estimates for White Sands or Cape Canaveral. The islands also posed severe problems of logistics. Although Brownsville costs were reasonable, launches from the Texas coast entailed a serious over-flight hazard for populated areas in the southeastern United States. Construction costs for an offshore complex at Cape Canaveral ran about 10% more than the costs of land purchase and development on shore, and maintenance estimates for the offshore sites were much higher.[15]

Cumberland Island enjoyed some of Cape Canaveral's advantages: accessibility to deep water transport and railroads and no problem with over-flight or booster impact. However, the Air Force listed a number of problems at Cumberland:

- Interference with the Intracoastal Waterway.
- Expensive launch area instrumentation would have to be duplicated.
- Land-based instrumentation for the early portion of flight would not be available.
- Extensive communications tie-ins with Cape Canaveral and down-range stations would be necessary.
- Towns in the area were small. The local economy might not support the large influx of people.
- The land area involved was primarily marshland.[16]

The Air Force listed only two disadvantages for Cape Canaveral: comparatively expensive land acquisition and higher-than-average cost for electrical power and water. Among the advantages for the Cape, the Air Force noted that "The Titusville-Cocoa-Melbourne area of Florida is a dynamic area which has been continuously growing with Cape Canaveral since the Cape's inception. Therefore, we expect a minimum of problems in the further area expansion which will be necessary for this program." Since "practically the entire local area population is missile oriented," the Air Force foresaw a "minimum of public relations type problems due to missile hazards and inconveniences."[17]

The NASA portion of the report cited two disadvantages at the Cape: labor conditions and the possibility of hurricanes. Local lore assured Canaveral newcomers that the eye of a hurricane had never passed over the area. Hurricanes had indeed passed near Merritt Island in 1885, 1893, 1926, and 1960—one year before.[18] As for its labor problem, Florida had never been an industrial state. Skilled workers in most categories were scarce, nonexistent in others. This meant that NASA and its contractors would not only have to call in engineers, scientists, and other experts from all parts of the country, but would have to attract craftsmen or train local men on the job for a wide variety of skills. Along with the men, manufactured goods would have to pour in from elsewhere, "such as copper wire, power and instrumentation cable, transformers, oil circuit breakers, generators"—to list but a few.[19]

Some shortcomings of the Cape went unmentioned in the report. Debus subsequently stated: "The chief drawback with this particular site was the danger of being swallowed up by the existing organization."[20] This concern perhaps underlay the interest in Cumberland Island. There were also doubts as to the area's ability to support the Apollo program. Remoteness—a positive factor in the matter of safety—had its disadvantages in the lack of housing, stores, schools, and recreational facilities for new residents. The fastest growing county in the nation, Brevard had scarcely been able to keep up with the needs of pre-NASA expansion. Debus was keenly aware of the impact of a NASA-engendered boom on the people of Brevard County, an interest that later took such forms as a Community Impact Committee set up by Debus, Davis, and Governor Farris Bryant of Florida.[21]

The Questions Begin

Even before submission of the report, Debus had misgivings about NASA's grip on the purse strings in the event the moonport was located within the Air Force sphere of influence at Cape Canaveral. Someone in the Department of Defense, it appeared, had already initiated plans to take over funds for LOD instrumentation and facilities. During a conference with Eberhard Rees, Associate Director of the Marshall Space Flight Center, Debus emphasized that the Launch Operations Directorate should control these funds at the Atlantic Missile Range and gave several instances of past problems to substantiate his position. Since Rees would be in Washington when Petrone was to deliver the report, it was agreed that Petrone would furnish Rees with arguments supporting NASA's retention of funding control.[22]

On 31 July 1961, scarcely a month after starting its work, the committee presented the Debus-Davis Report to Seamans in Washington. Two days later NASA Headquarters announced a worldwide study of launching sites for lunar spacecraft. Reflecting the concern of many inside and outside NASA, a *Washington Post* article stated that the size, power, noise, and possible hazards of Saturn or Nova rockets would require greater isolation for public safety than current NASA launch sites offered.[23]

At this juncture, Milton W. Rosen, Acting Director of Launch Vehicle Programs, submitted a report to Webb and Dryden that called for a more complete study of Cumberland Island before a final decision in favor of the Canaveral area. Rosen wrote:

> At Cumberland, however, there is an opportunity, one which we should not lose, to operate in a much simpler and more effective and less time-consuming manner. At Cumberland there could be at the beginning, at least, essentially one project directed toward a single major objective. The newness of Cumberland would be an asset. Both White Sands and Canaveral had simpler and more direct and less time-consuming procedures in their early days, when they did not have to cope with their present volumes of traffic.

Rosen noted that personnel living in the northern suburbs of Jacksonville could drive to work at Cumberland through less traffic than employees faced at Cape Canaveral. The cost of duplicating instrumentation was minor in contrast to the total investment at either site.[24]

On the same day, however, the highly respected scientist-administrator Dr. Hugh Dryden sent in his conclusion: "In my judgement, the nation's interests would best be served by expanding the existing range rather than developing an entirely new and separate installation at this time."[25] NASA Headquarters announced plans six days later (24 August) to acquire approximately 324 square kilometers north and west of the Cape Canaveral launch area, largely on Merritt Island, for manned lunar flights.[26] While most observers felt that the deciding factor was financial, Gibbs believed that "the Hazard Report [of June 1961, pp. 87–88] was the whole basis on which the selection was really made."[27] Petrone thought the decision had a wider base:

the low cost, the proximity to available range resources, and compatibility with program requirements. In response to a direct question on the weight given in the Debus-Davis study to Merritt Island's proximity to the tracking system, Petrone placed it "very high." He also noted that when the decision was made, complex 34 was ready for operation and complex 37 was under construction on Cape Canaveral. With NASA making preliminary Saturn launches from these pads, locating the moonport hundreds of miles from the Cape would have created severe dislocations.[28] Whatever the decisive factors, NASA was committed to launching its manned lunar flights from the Florida facility. Working out of the same geographical area, NASA and the Air Force would have to face the magnitude of the man-in-space program, and the Air Force would have to recognize that NASA was not simply another range-user, waiting in line for its turn. New policies and procedures were called for.

The Webb-Gilpatric Agreement

On the same day that NASA announced its intention of obtaining land on Merritt Island, it signed an agreement with the Department of Defense that set guidelines for managing and funding the Manned Lunar Landing Program. This agreement, which took its name from James E. Webb, whom President Kennedy had just appointed to head NASA, and Deputy Secretary of Defense Roswell Gilpatric, set down three preliminary considerations: the Department of Defense and NASA recognized the great impact of the Manned Lunar Landing Program on the Atlantic Missile Range; in the national interest, the two should pool their resources to make the most effective use of the facilities and services; and the traditional relationship between range-user and range-operator would continue.[29]

The agreement contained 11 provisions that gave NASA ultimate responsibility for acquiring the new land, improving it, constructing necessary buildings, and operating the Manned Lunar Landing Program facilities on the new site and elsewhere. The seventh provision read:

> (7) As agent for NASA, the Department of the Air Force will: a) prepare and maintain a master plan of all facilities on the new site, to include the selection of sites for mission and range support facilities (NASA will be represented on the Master Planning Board); b) prepare design criteria for all land improvements and range support facilities subject to NASA approval, and arrange for the construction thereof; c) design,

develop, and procure all communications, range instrumentation, and range support equipment required in support of NASA at or near the launch area.[30]

Unfortunately, this hastily drafted document neither defined some critical terms nor included interpretative guidelines. The two parties resolved some simple disputes easily; others they found harder to overcome.[31]

Disagreement centered on the definition of the word *agent* in paragraph (7). According to NASA, an agent was one who acts for or in the place of another by authority from the principal. In the NASA view, the intent of the Webb-Gilpatric Agreement was not to give authority to the Air Force for master planning on Merritt Island; rather, the Air Force was to exercise the master planning functions by authority of NASA and subject to its approval. The Air Force, in contrast, stated that since range users never had the right to locate their launch facilities at the Atlantic Missile Range, it was the range commander's responsibility to site all facilities in accordance with needs of all users. The Air Force, however, had no intention of assuming responsibility for design planning of any NASA mission facilities, such as launch pads.[32]

The Air Force quite simply viewed the new area as an extension of the Cape Canaveral Missile Test Annex. To avoid unnecessary duplication of facilities and personnel, it seemed best that a single manager should control the operation. Responsible for development of the Eastern Test Range since October 1949, the Air Force had supported other agencies, including NASA, with manifold facilities in the areas of range safety, logistics, and tracking. From November 1958 to August 1961, first as the Atlantic Missile Range Operations Office, then after 1 July 1960 as the Launch Operations Directorate, NASA had funded the construction of blockhouses, launch pads, and assembly buildings for its specific programs on the Cape. The Air Force Missile Test Center had purchased and improved land and incorporated the new facilities into its real property accountability system. "Only certain specified services and functions," the *History of the Air Force Missile Test Center* pointed out, "were provided NASA on a reimbursable basis."[33]

Now there was to be an important departure from the Air Force policy of retaining control of all real property at Canaveral. The Department of Defense could not provide money for an immediate purchase of Merritt Island. NASA would have to buy the land. During deliberations in the Office of the Secretary of Defense, preliminary to the Webb-Gilpatric Agreement, the Department of Defense Research and Engineering representatives had inserted a clause in the draft agreement to the effect that all land acquired in behalf of NASA should be transferred to the Department of Defense and incorporated into the Atlantic Missile Range. Gilpatric had questioned the need for such a clause and transfer, saying that the land belonged to the government. Gilpatric's attitude would prove an unfavorable harbinger to Air Force enthusiasts who viewed Merritt Island as an extension of their Cape.[34] NASA eventually took the position that the Air Force, as the agent for NASA in relation to the new land, had assumed a completely new management position, and that NASA had the authority to control the management actions of its agent in these new and separate areas.[35]

For the time, a reading of the Webb-Gilpatric Agreement, especially the controversial seventh provision, along with an understanding of the traditional Air Force viewpoint, might lead one to wonder how the Director of the Launch Operations Directorate had presumed he would have sufficient freedom of movement. Sometime later Debus recalled his reasons for agreeing to these arrangements. He stated,

> Although it may appear that this agreement was to the advantage of the Air Force, you must remember that the Air Force did everything—everyone else was a customer. All their efforts were space oriented and anyone encroaching on this area was considered a challenge by the Air Force. During this period we had to continually make an effort to understand the Air Force's position.[36]

At the time, Debus discussed the tenancy aspects of the Webb-Gilpatric Agreement with Samuel Snyder, Associate Director of Launch Operations at NASA Headquarters, and General Ostrander. While he had suggestions for improving several points, Snyder had urged that "if we could live with it," NASA should sign.[37] Debus and the Commander of the Missile Test Center hoped that they could avoid referring most issues to Washington, preferring to settle them locally.

The Launch Operations Director came to feel during the ensuing months that he needed a stronger hand in site selection and approval of facilities and could not live with the Air Force assumption that Merritt Island was simply an extension of Cape Canaveral. Even a casual observer could see that the two groups would not always be working in harmony and that their areas of operation overlapped at certain points. A new arrangement would eventually have to succeed the Webb-Gilpatric Agreement of August 1961.

Merritt Island Purchase

On 1 September, NASA asked Congress to authorize the purchase of 324 square kilometers of land on Merritt Island, immediately north and west of the existing missile launching area at Cape Canaveral. In support of the proposal, Senator Robert Kerr of Oklahoma, Chairman of the Senate Committee on Aeronautical and Space Sciences, stressed several factors. Stringent time schedules for the lunar program made the area ideal. NASA could reduce costs by use of existing resources, facilities, and personnel. The tracking network stretched almost 14 500 kilometers into the Indian Ocean. If NASA tried to start from scratch in another area, this one aspect of the program would be prohibitive. NASA could plan efficiently for future expansion in the new complex. And lastly, Senator Kerr insisted that this facility would be used for many years to come. Congress was favorable.[38]

On 21 September, Seamans requested the Army Corps of Engineers to undertake the land acquisition.[39] Congress adjourned before authorizing the purchase. Without such authorization, NASA could not ask for the appropriation; but the agency's reprogramming authority made it possible to start purchasing land before the end of 1961. NASA transferred funds from its Research and Development account to its Construction of Facilities Account, and advanced the money to the Army Corps of Engineers, its agent in purchasing the land, and balanced the books the following year.[40]

The use of the Corps of Engineers in this way followed an established pattern of cooperation between NASA and the Corps.*[41] Morris A. Spooner, Chief, Real Estate Division, Jacksonville District Engineers, supervised the buying of the land. After notifying the public of NASA's plans and the exact boundary of the area involved, the Corps opened an office in Titusville, the county seat, before the end of September. When all owners had listed their holdings, 440 tracts were involved. Three-fourths of the owners were absentee; three-fifths lived outside of Florida. The Corps hired experienced land appraisers from firms in Lakeland, Miami, Jacksonville Beach, and Melbourne and issued a booklet to explain the procedures to property owners.[42] First, the Corps would identify the owner, map the land, and describe it legally. Then the appraiser would evaluate each tract. Finally, the Corps would negotiate with the owner. If negotiations proved successful, the direct purchase representative closed the deal; sometimes negotiations broke down and the government had to begin condemnation proceedings.[43]

According to the NASA plan, one group of owners had to vacate their property by the end of February 1962. Many complained to the Titusville *Star-Advocate* that the Corps had not gotten in touch with them and offered a fair price. An editorial on 17 February 1962 maintained that the

---

*Relations between LOD and the Corps did not always run smoothly. After a March 1962 visit to the Jacksonville office of the Corps, an LOD finance officer noted that the Corps was anxious to "dump" administration charges on NASA. In interviews, NASA officials have commented that Corps support did not come cheaply.

Corps had not moved as fast as it should have. It insisted that the agents of the federal government should have placed an equitable price on each piece of property and mailed the offer to the owner with a self-addressed return envelope. If the homeowner agreed, he could have notified the Engineers. If he did not, the Engineers could proceed with the suit in court.

> It is common knowledge [the editorial went on] the Corps of Engineers is making offers for property subject to negotiation. Is this proper? Should the federal government agents go into the horse-trading business? . . . To send in negotiators is nothing less than high-pressure tactics to get the most for the least.

It urged the owners not to allow the Engineers to high-pressure them. If any delay occurred, the editorial concluded, "the Corps of Engineers should carry this delaying responsibility."[44] In spite of this and other complaints, most land acquisitions moved ahead without too much delay. Many individuals took the Corps and NASA into court, but in almost every instance the jury verdict was in the government's favor or close to the figure the government had offered.[45]

While not involving a great number of people, the exodus had its poignant elements—as do all such transfers. This was home for many people, and a lovely home. One family had come down from Savannah, Georgia, a few years before and purchased a small estate near Happy Lagoon, about three kilometers north of where the assembly building was to rise. Husband and wife had come to cherish their new location. The Corps of Engineers assured them that if they purchased similar land north of Haulover Canal, they need never worry about moving again. They took the advice, only to have NASA subsequently reassess its needs and decide to expand farther north. The couple moved to Orlando.[46] The government retained 60 homes for interim use by NASA, the Corps of Engineers, or the Air Force.[47] Some individuals moved their houses to the mainland or to the south end of Merritt Island.

The Titan III Problem

During the fall of 1961, the Air Force was faced with the problem of finding a launch area for its new Titan III. This 39-meter missile consisted of a liquid-fueled central rocket flanked by two solid boosters of great power. Launch sites on Cape Canaveral, including pad 18, pad 20, and the tip of the Cape, were deemed unusable on account of blast and toxicity factors. Events

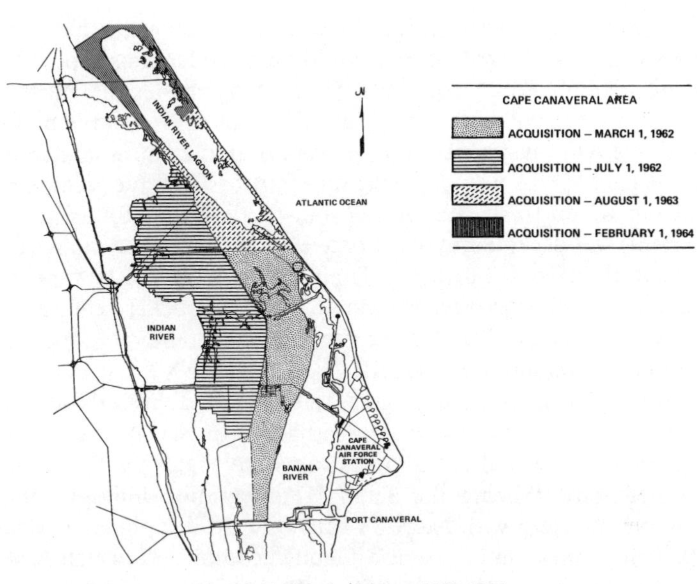

Fig. 28. Land acquisition, 1962–1964.

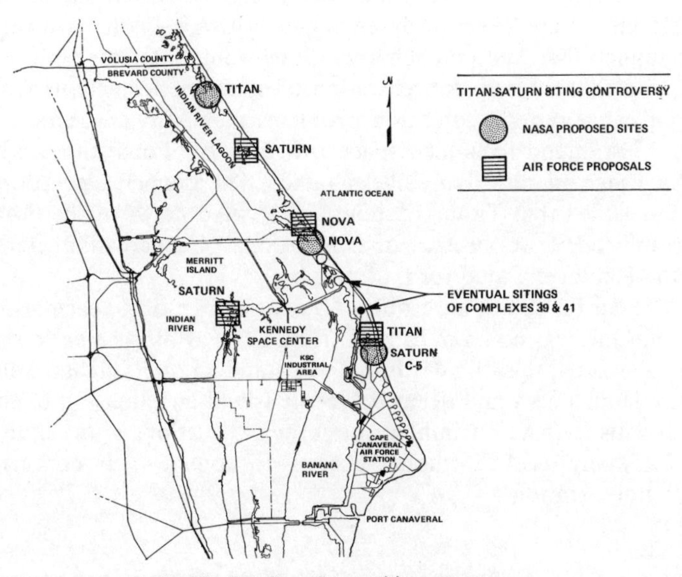

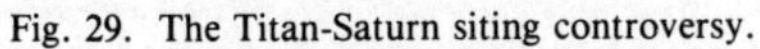
Fig. 29. The Titan-Saturn siting controversy.

took a collision course when Missile Test Center administrators decided the new NASA land on Merritt Island could be considered as a possible Titan launch site. The Air Force would place the Titan complex just north of complex 37, spacing the pads for use of class IX* explosives. On the premise that the Air Force had master planning powers over the entire launch area, including the land NASA was acquiring on Merritt Island, the recommendation to site Titan III north of complex 37 and partly on NASA land (and submerged land) was accepted by Air Force Headquarters and approved by the Department of Defense. Further, a Titan overflight of LC-37 appeared to be no problem, and the corrosive effects of the Titan rocket exhaust would be negligible. The Missile Test Center proposed that NASA move its launch pads north to accommodate Titan III.[48]

To this, Debus could not agree. LOD believed the corrosive effects of the Titan exhaust would pose a serious hazard to NASA space vehicles on launch complexes 34 and 37, and that any overflight would create serious safety restrictions. Placing the Titan III integration building on Merritt Island would interfere with NASA's canal and bridge plans. The proposed Titan III firing rate would close down launch complex 37 or pad A of launch complex 39 once every ten days. Moving LC-39 farther north would double the distance from assembly building to launch area, increase the cost of communications lines by $1 million, and force NASA personnel to detour around the Titan III area in going from the Cape to LC-39. In sum, LOD believed a Titan III failure could seriously endanger NASA's flight hardware, pads, and personnel; that Titan launch operations would interfere with NASA activities; and that a heavy concentration of escaping propellants from Titan III might cause serious corrosion problems in NASA spacecraft. Finally, LOD did not intend to launch spacecraft over Air Force sites and did not want Air Force missiles flying over its pads. The Launch Operations Directorate concluded that Titan III should be located north of the NASA area and recommended the purchase of an additional 60 square kilometers of land above the Haulover Canal for that purpose.[49]

The Air Force was agreeable to buying this land and earmarking it for NASA, but this was no balm to LOD. The Titan affair seemed to say, if not in so many words, that the Air Force was standing on its rights as master of the entire launch area and deemed Merritt Island an extension of the Cape. Debus and his staff were troubled about the implications of the situation and tried—for many weeks without success—to convey their concern to the NASA administration.

---

*In the U.S. military forces, this designation identifies high explosives such as dynamite, materials that are very susceptible to ignition by spark or friction and burn with explosive violence.

In an effort to work out some of the problems, General Shriever, General Davis, and their staffs met on 19 February 1962 with a NASA team of D. Brainerd Holmes, Debus, and others. The conference produced a lamentable communications gap. Shriever understood that Holmes and Debus were agreeable to siting Titan III in the south where the Air Force wanted it, deferring selection of a moonport site, and purchasing additional land north of Haulover Canal. The actual NASA position, as set out at a Management Council meeting under Holmes's chairmanship on 27 February, was that "the preferable solution to Cape siting problems is immediate acquisition of additional land to the north and siting the Titan III at the north, Nova in the center, and [Saturn] C-5 to the south."[50] Much of the following month was devoted to the solution of this impasse, a process complicated by misunderstandings within the NASA command. Much to Debus's disappointment, NASA agreed tentatively to the southern sitings.[51] On 27 March a statement of the acceptability of overflight was signed by L. L. Kavanaugh, for the Department of Defense, and Robert Seamans. A still unreconciled Debus told the Management Council that NASA should retain control over NASA-purchased lands and seek an amendment to the Webb-Gilpatric Agreement providing for joint master planning.[52]

Congress Says NASA

The LOD director was to have his turn sooner than he knew, but for the time being NASA Headquarters appeared loath to cross swords with the Air Force. Nor was the Air Force ready to relinquish any of its perquisites. On 29 March, John H. Rubel, Assistant Secretary of Defense, presented a statement to the Subcommittee on Manned Space Flight of the House Committee on Science and Astronautics. Rubel reviewed the procedures at the Atlantic Missile Range during 11 years. He gave no indication that the Air Force saw any noticeable difference between the manned lunar landing program and the other programs that had used the Cape during that time. The range commander had to have authority to make decisions in the common interest of all range users. At the same hearing, Seamans pointed out that some of the launch pads for Titan III would be on lands funded by NASA, just as some of the Saturn C-1 facilities were located on land originally funded by DoD. He stated: "It is not a question of our land or their land. It is the country's land. It is a national range."[53]

The subcommittee chairman, Olin E. Teague (D., Tex.), told Rubel that the dispute confused him:

> The main thing that troubles the committee is, we go to the Cape, for example, we talk with some of your responsible people there, we talk with some of Dr. Seamans' responsible people and we come away confused, frustrated, disturbed, and they don't agree on this overflight matter, and they don't agree to a Titan siting next to a Saturn. . . . We have some questions we are going to submit to you, Mr. Rubel and Dr. Seamans, which we want answered for the record.

As a result the Teague committee sent 28 questions to the Department of Defense and NASA. Typical of the questions were:

- What is the management arrangement between DoD and NASA regarding NASA's utilization of the national missile ranges?
- From a management and technology standpoint, does NASA lack the necessary capability to do their own siting and preliminary design for the new area?
- Would it be considered undesirable duplication to allow NASA to develop their own support unit for activities at AMR?[54]

In a generally conciliatory set of answers, NASA recognized the contribution of the Air Force on the Cape, but strongly insisted that, "since NASA has MLLP [Manned Lunar Landing Program] responsibility to the Congress, it must exercise management and funding control of all its aspects."[55] The Air Force felt that in answering Teague's 28 questions, NASA had shifted its position and was interpreting the terms of the Webb-Gilpatric Agreement in a manner that would place the Air Force Missile Test Center in a subordinate role to NASA in every range matter concerned with the manned lunar landing program. The Air Force felt it could not give up its traditional role in the management and operation of the range, including the new area, "without a deterioration of its services to all agents collectively."[56]

Having stated its case, the Air Force did something about it. When the question of custody of the title to the land on Merritt Island arose, General Davis requested the District Chief of the Corps of Engineers to transfer the title of all property on Merritt Island to the Air Force.[57] Moreover, despite indications that land ownership was going to give NASA special status as more than an Air Force tenant, NASA Headquarters at Washington seemed ready to concede the point with the proposed new purchase. It would let the Air Force buy and hold title to the 60 square kilometers at the north end of the range which, it was agreed, should be acquired for NASA's use in lieu of the land lost to Titan III.

In looking back at the issues, Rocco Petrone stated flatly in an interview some years later: "The ownership of the land . . . was a key one." "In those days NASA was a pretty small customer," Petrone admitted, "and tackling DoD was a tough game. . . . Webb knew that at all costs he had to have peace in a federal family, the two agencies that could go into space, NASA and DoD." Further Webb had to face one of the most prestigious men in the new administration, Secretary of Defense Robert S. McNamara; and Webb had to recognize that the Air Force had long considered space its province. Petrone felt that only the presidential decision had given NASA priority in the lunar program.[58]

For the time being NASA Headquarters was cooperating with the Air Force to enable the latter to purchase the land earmarked for NASA in compensation for the Titan sites. Seamans wrote Webb on 13 April that "although the Debus-Holmes recommendation is that NASA seek to acquire the additional acreage, it is my feeling that since the Titan III program forms the basis for this need, it is more desirable for DoD to seek this additional land."[59] Webb agreed and notified McNamara of NASA's acquiescence in the Air Force siting of the Titan pads, and the Air Force purchase of compensatory acreage.[60] An article in *Missiles and Rockets* for 30 April 1962 reported that the Air Force wanted to put its Titan pads at the south end of the coastal area of the expansion tract (NASA's Merritt Island purchase), and that this would force NASA to relocate its pads. "The NASA position is that this is fine as long as the Air Force provides the funds."[61] The Bureau of the Budget approved the Department of Defense request.[62]

By this time it appeared to NASA people at Canaveral that Headquarters in Washington had given in and agreed that the lunar team was only one of many tenants using Air Force facilities at the discretion of the Air Force. But help came from another quarter. Robert Seamans and Dr. Brockway McMillan, the Assistant Secretary of the Air Force for Research and Development, appeared before the Military Construction Subcommittee of the Senate on 8 May to testify in favor of DoD's acquiring the additional land. Their testimony backfired. Henry Jackson (D., Wash.), Chairman of the Subcommittee, saw the wisdom in the purchase of the new land. But the testimony showed that the additional acreage would support NASA development. Since NASA was a civilian agency, he would not honor the request and so wrote McNamara on 21 May.[63] In a reply three days later, McNamara explained the Air Force's position, but conceded the Senator's point that it could well be a NASA purchase "provided the use of this and all other land at the Cape is subject to the joint use policy under a single manager."[64] McNamara concluded his letter with the assurance that NASA was in the process of presenting the request through the proper congressional committee. NASA then took over the task of pushing the matter with Congress.

On 14 June, Debus notified Davis of word received from Washington. NASA and the Department of Defense had agreed that NASA would buy the additional 60 square kilometers of land and was submitting the recommendation to Congress for the FY 1963 authorization bill. He understood that the concerned congressional committees had not opposed the purchase.[65] James Webb appeared before the Senate Subcommittee on Appropriations on 10 August and explained in full the need for additional land. Chairman Warren Magnuson (D., Wash.) and Senator Leverett Saltonstall (R., Mass.) did most of the questioning as Webb went beyond the simple request for more funds to a wide statement on the whole program.[66] NASA's 1963 Authorization Act, passed four days later, included funds for the additional land north of Haulover Canal and included a key statement as to jurisdiction: "All real estate heretofore or hereafter acquired by the United States for the use of the National Aeronautics and Space Administration shall remain under the control and jurisdiction of that Administration, unless it is disposed of in accordance with the Federal Property and Administrative Services Act of 1949 (63 Stat. 377), as amended."[67]

At this point the bureaucratic infighting reached a draw. The Air Force had placed its Titan III facilities on part of NASA's Merritt Island land, but NASA retained jurisdiction over the land, nailed down by its further acquisition of the last 60 square kilometers at the northern limits of the Florida launch area. NASA had established its status as more than a tenant of the Air Force. It would be a mistake to make too much of the disagreement. At the Cape, NASA and Air Force personnel were working together on a day-to-day basis, and the Launch Operations Center was always quick to acknowledge its debt to the Missile Test Center. There is some force to an Air Force suggestion that it was creating issues to get clear-cut decisions from Washington on the powers and responsibilities of the two agencies. The decision finally came down—NASA, and not NASA and the Air Force, would put a man on the moon. During the negotiations John Glenn and Scott Carpenter had orbited the earth, and the American public was cheering for its new space agency.

A New Agreement

It was time for a review of the Webb-Gilpatric Agreement. NASA and the Department of Defense had distinctive programs. The Department of Defense agencies that used the range were primarily research and development users of a test facility for the development of weapon systems. NASA, in addition to doing R&D, was an operational user of launch facilities for the exploration of space. Congress indicated its intent that the land on Merritt Island remain under the control of NASA by the way funds were appropriated for its purchase, but NASA did not propose to disturb in any significant way the arrangements at Cape Canaveral or the downrange facilities and intended to pay a prorated share of the operating expenses of the Atlantic Missile Range. Under these circumstances, Webb wrote to Gilpatric on 14 August 1962, with a draft that he hoped would replace their earlier agreement.[68] During the fall and early winter of 1962, NASA and the Department of Defense engaged in a series of conferences that led to a clarification of relationships at Cape Canaveral and Merritt Island. General Davis wrote to the Secretary of Defense, for instance, pointing out the duplication of support activities that might be required—such things as guard services, printing plants, fuel analysis laboratories, instrument repair shops, fire protection, and weather forecasting. He admitted that a division on a geographical basis was possible, but advised that NASA be prepared to accept the responsibility for the necessary duplication.[69] On 17 January 1963 NASA Administrator Webb and Secretary of Defense McNamara signed a new agreement.

NASA gained two points. Paragraph B of the General Concept stated: "In recognition of the acquisition by NASA of MILA (Merritt Island Launch Area) and its anticipated use predominantly in support of the Manned Lunar Landing Program and in order to provide more direct control by NASA of MILA development and operation, the Merritt Island Launch Area is considered a NASA installation separate and distinct from the Atlantic Missile Range."[70] In the area of master planning, NASA also had more liberty. Further agreements and additions during the spring and summer of 1963 settled many of the minor problems that remained.[71]

Land, Lots of Land—Much of It Marshy

While NASA and the Air Force pursued their own battle for beachheads, the Corps of Engineers continued its less spectacular efforts to stake out NASA's new land holdings on Merritt Island and at the north end of the range. Within two and a half years of its initial commission (that is, by 1 February 1964), the Corps had acquired the bulk of the needed land. Out of the more than 1500 ownerships involved initially, a few were to remain unsettled for several years more.[72] Not unexpectedly, the absentee owners of large tracts who could delay and negotiate came off better than the small owners who sometimes found their awards inadequate to purchase similar property in the neighborhood. At least one person owning cultivated land on Merritt Island sold the tract for $244 an acre. But one year later, when she wanted to purchase a similar plot on a non-NASA section of the island, she found the price to be $3000 an acre.[73]

The buildings did not prove as simple an acquisition as the land. The Corps sold some for salvage, transferred 44 to the Brevard County School System for use as temporary classrooms, and turned one old building into a museum. The Air Force, among others, used the Standard Oil station to service official vehicles, the Roberts residence as a first-aid station, and several homes as security patrol offices. The purchase included a considerable number of trailers that eventually served in a variety of capacities.[74]

The disposition of more than 12 square kilometers of citrus trees proved one of the most difficult problems. NASA at first proposed to lease the land to the growers for five years. A representative of the Merritt Island citrus growers stated that they were willing to vacate their dwellings and farm the groves in accordance with NASA regulations, but desired to retain title to the property. They were afraid that a lease system would not guarantee them any right of repurchase if NASA no longer needed the tracts. The growers rightfully pointed out that it was difficult to spray, fertilize, and cultivate the groves without a guarantee that they could gather the fruit. Debus met with J. Hardin Peterson, a lawyer representing the Florida Citrus Mutual, as early as December 1961 and assured him and representatives of the citrus growers on North Merritt Island of NASA's good will.[75]

A group of citrus growers carried their complaints to Senator Spessard L. Holland (D., Fla.), Chairman of the Subcommittee on Appropriations, who asked some sharp-edged questions of NASA Associate Administrator Robert Seamans in the April 1962 hearings on the Second Supplemental Appropriation Bill. Seamans referred the queries to Ralph E. Ulmer, Director of Facilities Coordination. Ulmer tried to dismiss Senator Holland's question with the statement: "We have received no recent complaints from landowners on that score." Senator Holland answered flatly: "You have received them, because I passed them on myself directly to Mr. Webb and to others in NASA, going back to last fall." The Senator insisted that the heart of the matter was NASA's attitude toward the production of citrus on eight square kilometers of valuable groves. Ulmer promised to give the matter careful attention. Late the following year, the Corps of Engineers announced a lease plan for the Merritt Island citrus groves that seemed much more satisfactory than earlier arrangements. In place of the original five-year lease plan, the Corps offered the original grove owners a lease until 30 June 1968, with an option to renew the lease for an additional five years. Two factors tended to make this option essential for the growers: young trees required more than five years to develop and the high cost of equipment could not be recovered in five years.[76]

The space agency finally took 340 square kilometers by purchase and negotiated with the State of Florida for the use of an additional 225 square kilometers of submerged lands. Much of the latter lay within the Mosquito Lagoon, separated from the ocean by a narrow strip of beach on the east. The property cost $72 171 487.[77] The Space Center invited Brevard County to maintain a public beach north of the launching facilities, to be used whenever activities on the pads did not create a hazard. In 1963 NASA empowered the National Wildlife Service to administer those areas of the Space Center not immediately involved in space launch operations. At the time this covered about 230 square kilometers and formed a safety belt between the launch area and the population centers to the west and northwest. A few years later the manager of the Merritt Island National Wildlife Refuge was to report the identification of more than 150 species of birds. During the winter season the waterfowl population exceeded 400 000. Animals included alligators, wild pigs, and bobcats.[78]

NASA and the Department of the Interior were to finalize the arrangements between KSC and the Refuge some years later. NASA added lands, submerged lands, and waters, increasing the total under the control of the Refuge to 508 square kilometers. By this agreement, the Refuge would administer the citrus groves and lease fishing camps, previously handled by the Corps of Engineers; operate Playalinda Beach at the north end of the Cape; and cooperate with the Brevard Mosquito Control District. NASA provided fire protection and would continue to maintain all major highways, bridges, and traffic signals required for employee and public access to the spaceport and adjacent facilities. NASA could make use of these areas at any time in conjunction with the space program. NASA could terminate the agreement when the space program demanded it, or if the Bureau of Sport Fisheries and Wildlife failed to use the premises according to the terms of the agreement. The Bureau, on its part, could withdraw if the nature of the space activities rendered the area unsuitable for wildlife purposes.[79] NASA would have all the land it needed for the foreseeable future, as well as a safety belt that served a second purpose as a wildlife refuge.

6

LC-39 Plans Take Shape

Rapidly Evolving Hardware

In the year following the Debus-Davis study, Huntsville planners kept coming up with a larger Saturn, only to discard it for a still bigger one. Their bigger-rocket designs, coupled with lunar-orbital rendezvous, could drop the Apollo launch rate from 13 Saturns a year to 6, well below what Debus had warned was an economic use for the mobile concept. Critics in and out of NASA began to question the wisdom of the mobile concept, but it rolled on. For one thing, the plan was under way and time and money had been invested in its development. For another, Debus and Petrone were proving effective advocates, stressing the concept's flexibility when declining launch rates undercut its major premise. Finally Congress and the country wanted NASA "to travel first class" if it meant beating Russia to the moon. The Launch Operations Directorate (LOD) men believed their proposals promised first-class travel to the moon and beyond.

Although acceptance of the Debus-Davis Report was a more-or-less green light for the mobile concept, several major questions remained about moving a gigantic rocket over Merritt Island's marshes from assembly building to launch pad. Cost remained a primary consideration. But during the last six months of 1961, LOD's great concern lay in the plethora of rocket designs and rendezvous studies that kept pouring out of Huntsville and Washington. An orderly account of events belies the tentative manner in which the Debus team had to plan launch facilities for problematical rockets flying on undetermined flight paths to the moon.

The Lundin Committee had taken a "quick look" (one week) at the rendezvous mode of accomplishing the manned lunar landing (see page 79). In late June 1961 Associate Administrator Seamans directed Air Force Col. Donald H. Heaton of NASA Headquarters to conduct a more detailed study. Heaton's committee supported the Lundin finding that an earth-orbital rendezvous promised the earliest lunar landing and at less cost than a direct ascent. Its August report recommended the use of a Saturn C-4 with four F-1 engines. The C-4's bigger payload would reduce the number of rendezvous vehicles, with "a higher probability of an earlier successful manned lunar landing than the C-3."[1]

Despite the Heaton Committee's recommendation, General Ostrander's Office of Launch Vehicle Programs urged an early start for the Saturn C-3 program. Seamans was not ready to commit himself, having agreed in July to a NASA-DoD launch vehicle study. Nicholas Golovin, a mathematician who had previously worked on the Mercury project, directed the joint study. Although the group failed to establish a national launch vehicle program, it outlined alternative programs (including developmental flights) for a manned lunar landing:

- ***Lunar-orbit rendezvous.*** 28 Saturn C-1 flights and 38 C-4 flights. First landing possible in October 1967. Cost of program, $7.33 billion.
- ***Earth-orbit rendezvous.*** 32 Saturn C-1 flights and 53 C-4 flights. First landing possible in July 1968. Cost of program, $8.16 billion.
- ***Direct ascent.*** 22 Saturn C-1 flights and 38 flights of a Nova configuration with eight F-1 engines in the first stage, eight J-2 engines in the second stage, and two J-2 engines in the third stage. First landing possible in October 1968. Cost of program, $6.39 billion.[2]

Contemporary with the changing studies in Washington, the Saturn launch vehicle evolved rapidly in Huntsville, going from a C-3 version in June to a C-5 in December. Plans for the C-3 were barely under way when Marshall Space Flight Center initiated studies of a larger C-4. The C-4, incorporating four F-1 engines in the booster and five J-2 engines in the second stage, at first seemed large enough to power a lunar landing mission via either lunar-orbital or earth-orbital rendezvous. As spacecraft weight estimates continued upward, Marshall officials began to question this assumption. Von Braun's proposal to add a fifth F-1 engine, making the C-4 a C-5, was approved in November when Milton Rosen, NASA Director of Launch Vehicles and Propulsion, made another launch vehicle study. Rosen's team spent two weeks in Huntsville matching potential launch vehicles with lunar landing missions. The group's findings reinforced von Braun's argument for a C-5; the C-4's capability for a rendezvous mission was marginal. Since the clustering of the four F-1 engines left a large open space in the C-4's first stage, a fifth engine would strengthen the Saturn design. Rosen pointed out that a fifth engine could be mounted at the junction of two very strong crossbeams that supported the other four engines. This eliminated a potential trouble spot since the junction would have been exposed to excessive exhaust backwash and a serious overheating problem. Marshall engineers estimated that the C-5 would place 108900 kilograms in earth orbit or lift 40200 kilograms to escape velocity. Still short of a direct ascent capability (68000 kilograms to escape velocity), the C-5 provided ample power for a rendezvous mission.[3]

Decisions came rapidly during the next four weeks. On 4 December 1961, Seamans agreed to the Rosen Committee's recommendations. NASA selected the Boeing Company as a possible prime contractor for the first stage on the 15th. The frame (10-meter diameter, 42.7 meters in length) would be manufactured at NASA's Michoud plant just east of New Orleans. At its first meeting on the 21st, the Manned Space Flight Management Council* approved the C-5 configuration of five F-1 engines in the first S-IC stage, five J-2 engines in the second S-II stage, and one J-2 in the third S-IVB stage. The same day NASA Headquarters began negotiations with Douglas Aircraft Company to modify the C-1's S-IV stage for use as the S-IVB. As NASA had indicated in September that North American Aviation would build the S-II stage, the Douglas selection rounded out the team of contractors for the Saturn C-5. Formal announcement that Marshall Space Flight Center would direct C-5 development came in January 1962.[4]

The Space Task Group, NASA's spacecraft organization, went through an equally hectic six months after the lunar-landing decision. STG and McDonnell Aircraft Corporation had been considering advanced Mercury projects since September 1959; proposals included a maneuverable Mercury capsule, extended missions of 14 days, a two-man vehicle, and a rendezvous attempt. In May 1961, Martin Company spokesmen approached NASA officials about the use of the Titan II missile in a post-Mercury program. Further presentations convinced Robert Gilruth, Space Task Group chief, of the Titan II's merits. Engineers prepared a project development plan calling for the two-man Mercury spacecraft and a modified Titan II booster. As a rendezvous capability seemed very important for Apollo, the project included an Agena rendezvous target, boosted into earth orbit by an Atlas launch vehicle. The project won approval in December and was formally christened the Gemini program† the following month.[5]

Work on the Apollo spacecraft also moved forward. NASA Headquarters announced on 9 September 1961 the establishment of a Manned Spacecraft Center at Houston. The center would design, develop, evaluate, and test Apollo spacecraft and train astronauts for space missions. Robert

---

*NASA Headquarters underwent a major reorganization during the fall of 1961. An Office of Manned Space Flight was set up to supervise the Apollo program. Field center directors no longer reported to Headquarters program offices but directly to the Associate Administrator, giving the directors additional power. D. Brainerd Holmes came from RCA to head the Office of Manned Space Flight. One of his first actions was to establish a Management Council to provide overall direction for the Apollo program. MSFC, MSC, and LOD (Debus) were represented, as well as key members of the Manned Space Flight Office. The Council played an important decision-making role in 1962–63. Robert L. Rosholt, *An Administrative History of NASA, 1958–1963*, NASA SP-4101 (Washington, 1966), pp. 274–75.

†See Barton C. Hacker and James M. Grimwood, *On the Shoulders of Titans: A History of Project Gemini,* NASA SP-4203 (Washington: 1977).

Gilruth would head the new organization with his Space Task Group as its nucleus.[6]

The home and organization were new, but not the mission. The Gilruth team had prepared the preliminary guidelines for an advanced manned spacecraft in March 1960. In subsequent months the group had enlisted research assistance from other NASA centers, briefed American industry, and awarded contracts for spacecraft feasibility studies. By mid-1961 Gilruth was ready to invite bids on the prime Apollo spacecraft. The 28 July work statement described three phases of the Apollo program. Manned earth-orbital flights and unmanned reentry flights comprised phase one missions. NASA would qualify spacecraft systems and the heat shield, study human reactions to extended periods in space, conduct experiments related to the lunar mission, and work on flight and ground operational techniques. The second phase involved circumlunar flights to develop the Apollo spacecraft and conduct lunar reconnaissance. Manned lunar landings would come in phase three.[7]

The work statement called for the design and manufacture of a command module and associated ground support equipment. The contractor would also provide test spacecraft for Saturn C-1 developmental vehicles and mockups. A second major assignment involved the integration of the spacecraft modules with each other, with the launch vehicle, and with ground support equipment. During operations the contractor would prepare the spacecraft for flight and monitor its systems. Description of the command and service modules ran more than 20 pages. Major systems of the two modules included guidance and control, vernier propulsion for longitudinal velocity and thrust-vector control, mission propulsion, reaction control, provisions for escape during launch, environmental control, electrical power, communications and instrumentation, and a number of crew-related systems. Although NASA had not decided on the mission mode, the Space Task Group nevertheless included some general plans of a lunar landing module for direct ascent or an earth-orbital rendezvous mission. Twelve companies bid on the contract that would eventually cost NASA over 2.2 billion dollars. In November, NASA announced the selection of North American Aviation for the task.[8] Mission, rocket, and spacecraft were taking form.

The Mobile Launch Plan Comes under Fire

While rocket and spacecraft plans were proceeding, the Martin Marietta Corporation of Baltimore began work on a two-part launch facility study. In part one Martin was to recommend an "optimum concept for facilities to launch Saturn C-3 vehicles at specified rates"; part two involved design of a launch complex based upon the selected concept.[9] The Martin team reported its part one findings orally at Huntsville on 27 September 1961. As in its earlier C-2 study, the Martin Company found the fixed concept superior for a launch rate below 12 Saturns a year and the mobile concept clearly preferable at annual launch rates above 24. The team recommended moving the rocket by canal. The 3350-meter safety distance between assembly building and pad (almost twice that for the C-2) and the C-3's greater weight had multiplied rail costs. Martin placed the cost of one barge launcher-transporter and pad at $8.152 million, while estimating the cost of comparable rail facilities at $21.965 million. Other advantages of the canal system included more room for bigger cargoes (growth potential for the Nova), a turning basin that compared favorably with complicated switching arrangements by rail, and best use of the Cape's marshy terrain. Although acknowledging a lack of data, the team discounted the wind effect on a barge transporter.[10]

At the end of the presentation, von Braun asked the Martin team to interrupt their C-3 study and conduct a quick investigation of launch requirements for a Saturn C-4. Martin's mid-October report contained no major changes. A Launch Facilities and Support Equipment Office (LFSEO) study, completed in late October, reached similar conclusions. Assuming an annual launch rate of 30 Saturn C-4s, LFSEO placed the cost of fixed facilities at $350.5 million, of rail $278.2 million, and of barge $259.1 million. The barge savings came entirely from the canal's lower cost. The study noted that "movement of a transporter launcher with vehicle by barge will present some difficult engineering problems [but] preliminary investigation has shown that it is feasible and within current 'state of the art' capability."[11]

As LOD moved ahead with LC-39 planning, some of its members began to have second thoughts. Georg von Tiesenhausen noted in October that "after an initial period of general acceptance, various segments of LOD are now reluctant to go ahead to develop this [mobile] concept." The size of the C-4, the boldness of the concept, and uncertainty about future launch rates contributed to the uneasiness. Von Tiesenhausen did not agree with the critics: "There is no insurmountable problem involved, engineering-wise or operationally, which appears, that cannot readily be solved . . . . This concept is highly flexible, readily expandable, and most economical for launch rates to be expected in the future."[12]

Connell & Associates, engineering consultants on LC-34 and LC-37, did not share this optimism and volunteered criticism in November. Harvey Pierce's eight-page letter to Debus acknowledged certain advantages of the

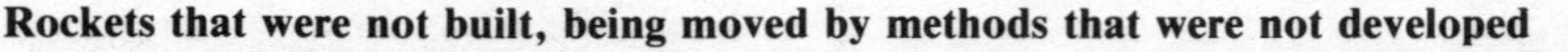

Rockets that were not built, being moved by methods that were not developed

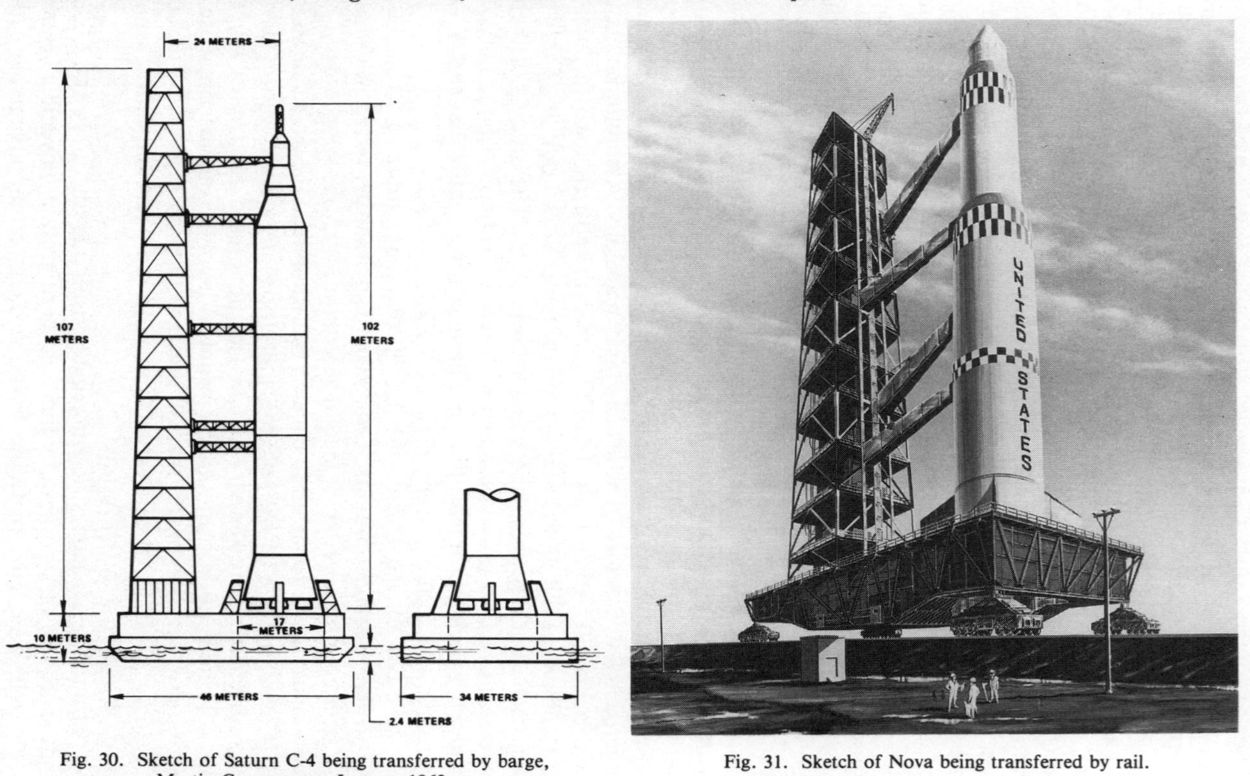

Fig. 30. Sketch of Saturn C-4 being transferred by barge, Martin Co. concept, January 1962.

Fig. 31. Sketch of Nova being transferred by rail.

mobile concept: more efficient use of land and personnel; only one launch control center; assembly and erection inside a building; and a brief checkout period (one week on the firing pad). The disadvantages, however, were more significant:

- Pad stay time is estimated at one week. During this entire period the vehicle is unprotected and subject to the elements. Since the weather cannot be predicted accurately for such a period, the vehicle must be designed for stability in line squall winds up to 70 knots . . . . This may comprise a severe penalty in vehicle design.
- Transporting the erected vehicle over a considerable distance must subject it to vibration which has not previously been encountered.
- A bending moment due to tilting the very tall vehicle away from true vertical will result from a wheeled transporter traveling up a slope or from a water-borne transporter under high wind loading . . . . The bending load must be considered additive to the wind load, and will add structural weight to the flight vehicle.
- Tests with cryogenic fluids must be made at the launch pad. If leaks are detected, repairs probably cannot be made without withdrawal to the remote area . . . . There is no reason to assume a lower incidence of these leaks in the future than in the past.
- This concept places the maximum emphasis on correct first guesses, and the maximum penalty on a wrong guess. The remote assembly-checkout facility, the transporter device, and the route development for the transporter must have the ability to handle all future vehicles, and will soon limit the vehicle design to fit their capabilities. This is an extreme limitation to accept this early in any program.
- In addition to some immediate decisions on some very difficult criteria predictions, the chances of having a usable facility in the near future are minimized by the difficult problems which are anticipated but unsolved . . . . Considering all factors it appears that the vehicle could easily be ready and available many months, perhaps years, in advance of available launch facilities.

The letter called for a thorough examination, with model studies and wind tunnel tests, of design and construction requirements for the remote assembly building, stability of the rocket in transit, shock and sound over-pressure effects on launcher-transporter equipment, placement of launcher-transporter and flame deflector at pad, transporter propulsion, barge stability, and rail switching. Although the Connell engineers agreed that all technical problems could be resolved with sufficient time and money, they recommended the use of fixed launch facilities for LC-39.[13]

A Trip by Barge or a Trip by Rail?

The Connell letter pointed up the crucial role of the launcher-transporter in LC-39 planning. Its characteristics determined the design criteria of other facilities. The success of the mobile concept rested on the transfer system; the system's development involved some of LOD's most difficult engineering problems. Understandably, the selection of a transporter became a major event in the LC-39 story.

The launcher-transporter fell within the purview of Theodore Poppel's Launch Facilities and Support Equipment Office (LFSEO). A Poppel directive on the October C-4 study indicates that the item, while crucial to LC-39, was a small part of the office's workload:

- Mr. [Chester] Wasileski will start on the propellant systems immediately.
- Mr. [Donald] Buchanan will start on the launch transporter and the fixed launch sites as soon as possible.
- Mr. [Robert] Moore's office will supply certain paragraphs and photographs that are generally applicable in this study.
- Mr. [Julian] Hamilton's outfit will come up with a light coverage of transportation with an illustration or two. Mr. [Georg] von Tiesenhausen will start with some overall layouts.
- Mr. [R. P.] Dodd will start on the assembly building immediately as to cost and arrangement. Mr. [Lester] Owens will determine blast distances. Mr. [O. K.] Duren will be in charge of the overall coordination and the written material . . . .[14]

Everyone in LFSEO was busy, but perhaps the heaviest workload fell to Donald Buchanan. After four years of Air Force duty in World War II, Buchanan had earned a degree in mechanical engineering at the University of Virginia. He had joined the National Advisory Committee for Aeronautics at Langley Field, Virginia, in 1949, moving on to Redstone Arsenal in 1956. Buchanan's responsibilities as Launcher Systems and Umbilical Tower Design Section Chief included pad arrangement and deflector design. Although Poppel and Lester Owens, Deputy Chief of LFSEO, intentionally left the launcher-transporter selection open to the entire office, Buchanan took the lead in the barge investigations. In April 1962 he assumed responsibility for transporter development.[15]

Cost estimates on a canal system were favorable, but the use of a barge as the launcher-transporter raised a number of engineering questions: How to position the barge and flame deflector at the launch site? What means of propulsion and steering to use? How to ensure a stable platform for the launch vehicle? While Martin Marietta examined these matters in the second part of its C-3 study, LOD stepped up its own inquiry. On 2 November an LOD team inspected the elevating mechanism of a Gulf Coast offshore oil rig. A possible solution to the positioning problem at the launch site involved the use of Texas Tower legs on the barge-transporter. The long tubular legs, actuated by a hydraulic jacking system, would be located at each corner of the barge. While the barge was under way, the legs would be raised until flush with the bottom of the barge. At the launch position, the legs would be lowered to rest firmly on a concrete basin. Then the hydraulic system would raise the barge on the legs to provide sufficient clearance for the flame deflector to float beneath it. However, a Launch Facilities and Support Equipment report opposed the hydraulic jacking system since it would place the launch platform at least 18 meters above ground level. In its place, the report recommended a deeper concrete launch basin with the barge positioned on supports extending outward from the basin walls. A lift-gate (lock) would allow sufficient water to be drained to permit passage of the deflector beneath the launcher. This plan offered a low profile (the launch platform would be only 2.4 meters above ground level), but this advantage would be offset by the increased costs of the lift-gate and deeper basin.[16]

Lacking expertise in barge propulsion and stability, LOD hired a Baltimore naval architect, M. Mack Earle, "to review the static and dynamic stability programs . . . and prepare a model test program." Earle's preliminary report warned that LOD would likely encounter problems with the propulsion system in restricted canals. Early in the new year Earle began arranging for a test program at the David Taylor Model Basin in Washington, D.C.[17]

Martin Marietta Corporation submitted the second part of its C-3 launch facility study on 11 January 1962. The report recommended use of a barge 55 × 41 meters, with 1.8 meters draft. Thirteen kilometers of canal, 61 meters wide and 4.6 meters deep, would service the three-pad complex. Four to six Murray and Tregurtha Harbormaster motors would propel the barge. Rated at 530 horsepower, this large outboard motor was capable of achieving nearly 900 horsepower for limited periods. Estimating 45 pounds of thrust per horsepower, Martin calculated that six Harbormaster units would overcome the drag of a 60-knot wind. Fixed legs, designed by DeLong Corporation and R. G. LeTourneau, Inc. (specialists in offshore oil drilling platforms), would elevate the barge out of the water at the vertical assembly building, the arming tower, and the launch pad.[18]

After NASA chose to develop the Saturn C-5 for the moon mission, little time remained to select a transfer mode. On 23 January, American Machine & Foundry Company presented the results of a comprehensive survey that included railway wheels, pneumatic tires, crawler treads, barge, and special ground effects, and recommended a rail-barge combination possibly using mechanical mules.[19] Debus agreed with their report; he informed Petrone a week later that he tentatively supported a plan "to let the barge weight be carried by water, but use for stabilization and propulsion a rail which carries only partial weight." The LOD Director reviewed transfer modes with Zeiler, Poppel, and O. K. Duren on 30 January, discussion centering on the merits of another launch vehicle transfer study. Although the group postponed an award in hopes that additional suggestions might appear, Debus did not intend to wait long. Summarizing the meeting for Petrone, Debus wrote: "It appears urgent that we have a program for the crucial engineering studies and possibly cost estimates for these studies early next week because a decision to proceed on 39 is imminent."[20]

In this atmosphere, a chance meeting at Huntsville introduced a new transporter to the LC-39 competition. Duren, an Auburn University graduate, had been with von Braun since 1951, most recently as Deputy Chief of the Future Launch Systems Study Office. On 2 February, Duren received a call from Barry Schlenk, a Bucyrus-Erie Company representative. While discussing Titan silo overhead cranes with Thiokol Corporation, Schlenk had overheard a remark about LOD's transport problem. The two men spent the afternoon examining some pictures of Bucyrus-Erie's steam-shovel crawler used in the Kentucky coal fields. The vehicle seemed suited to LOD's needs; its characteristics included a leveling capability to balance a load on uneven terrain. Caught up in Schlenk's enthusiasm, Duren called Albert Zeiler about his find. Zeiler was skeptical, but agreed to look into the matter.[21]

Four days later, LOD laid plans for barge, rail, and crawler studies. The staff concurred in a three-month barge study at David Taylor Model Basin, employing a 1:10 scale model of the barge. Additional tests would be run in a wind tunnel with a 1:60 scale model. A consulting engineer, William G. Griffith, would assist the Launch Facilities and Support Equipment Office on another rail study, this one concentrating on dynamic loads and foundation costs. Poppel's group (LFSEO) would follow up the Bucyrus-Erie lead with an inspection of the crawler shovel.[22]

When Donald Buchanan and George Walter arrived in Washington on 20 February, David Taylor Model Basin officials brought some uncomfortable facts to light. LOD's proposed canals were too narrow and would cause serious propulsion and steering problems. The steering problem resulted from the venturi effect. The relative motion of water to barge in the 3-meter space between the canal bank and the barge decreased the pressure on the side of the barge, causing a suction effect. The David Taylor officials recommended a wider canal—and that would raise costs considerably. Then wind-tunnel tests indicated that the drag effect in a 60-knot wind might be

Fig. 32. Bucyrus-Erie's steam-shovel crawler, used for surface mining of coal in Kentucky.

three times the estimated value. Basin tests also revealed that the arrangement of the six Harbormaster motors, three across the bow and three across the stern, reduced motor efficiency. There were several possible solutions: tugboats fore and aft of the barge, air jets placed below the waterline, and spuds (vertical steel pipes) to anchor the barge in heavy winds. These involved new tests and cost projections.[23]

In his rail study William Griffith concentrated on ways to reduce the cost of the roadbed. The continuous concrete beam (2.4 meters deep and 3.5 meters wide) supporting the service structure runway at LC-34 cost more than $3000 per meter—a prohibitive amount for LC-39's proposed 19 kilometers of rail foundation. Griffith proposed, instead, concrete ties supported by rock ballast on vibro-compacted soil. In a 3 April report, George Walter criticized Griffith's suggestion, arguing that the concrete ties and ballast would not stabilize the track horizontally. Walter opposed Griffith's recommendation of curved tracks. In rounding a curve the transporter's outside trucks would each follow a different route (the transporter would ride on four rails rather than two) and would require a complicated switching arrangement. Negotiating rail curves would also pose a serious problem in synchronizing the transporter's drive units and maintaining a balanced load.[24]

Presented with contradictory reports, LOD asked Connell & Associates to conduct a more detailed study. The findings of the Miami firm supported Walter's position. Curved tracks were judged unacceptable because "the switches required would be fantastically complex . . . . The matter of maintenance of track alignment of the curves is another difficult aspect of this system to which an economical solution is not apparent."[25] The Connell engineers recommended a perpendicular set of railbeds for north-south and east-west travel with switching from one line to another accomplished by one of the Connell team's own inventions: hydraulic equalizer jacks to raise the truck assemblies and a worm or pinion drive sector gear to rotate them. The Connell report questioned the feasibility of Griffith's foundation. Ballast deflection would occur under the heavy horizontal wheel loads, causing track misalignment. Connell recommended a three-layer foundation: compacted fill, a soil-cement subbase, and a reinforced concrete pavement on top. Concrete ties would be keyed transversely to the reinforced pavement. The Connell proposal would reduce the expense of the foundation by over 50%, but even so LC-39's roadbed figured to cost more than $28 million.[26]

The Crawler Makes Its Debut

On Lincoln's birthday, 1962, an LOD team visited Paradise, Kentucky, to watch a Bucyrus-Erie 2700-metric-ton crawler-shovel in action.

Albert Zeiler's report compared the crawler favorably to LC-34's service structure. The work platform, stabilized by hydraulic cylinders at the four corners, varied no more than one-half degree from level. Nearby, Bucyrus-Erie was constructing for the Peabody Coal Company a larger crawler-shovel which would have a load-bearing capacity in excess of the expected weight of the Saturn C-5 and its support equipment. Although maximum speed for the existing crawler was only 6.1 meters per minute, more speed could be built into the new model. Impressed with the crawler's potential, the LOD representatives asked their hosts to propose a study program for LC-39.[27]

Bucyrus-Erie began such a study one month later. An LOD phone call on 23 March requested preliminary information for Petrone's congressional briefing that afternoon. Thomas Learmont, Bucyrus-Erie's chief design engineer, provided tentative estimates: the crawler, jacks, hydraulic system, and steering mechanisms would cost $3650000, the umbilical tower $1500000, the box structure (launch platform) $800000. The crawler figure reflected the cost of Bucyrus-Erie's new model with few changes. Later Bucyrus-Erie incorporated a redundant power system and a more sensitive leveling mechanism, raising estimates an additional million dollars. Although the crawler's reliability and flexibility were attractive, the cost was a major disadvantage. LC-39 plans called for five launcher-transporters, putting the price of the crawler units at nearly $25 million. In early April, Buchanan suggested separating the launcher from its transporter and building only two crawlers. The proposal would increase total launcher-transporter weight (the separate crawler would require a heavy platform), but the cost savings more than compensated. After Buchanan's idea won approval, LOD supplemented Bucyrus-Erie's contract to include a "separate crawler" investigation.[28]

By May the crawler was scoring the highest marks of the three transfer proposals. On the 10th Poppel, Buchanan, and Duren inspected barge tests at the model basin and reviewed the adverse findings from the wind tunnel. The following day Bucyrus-Erie's final presentation was well received by NASA personnel. The crawler would go 1.6 kilometers per hour under load. Its turning radius was 152 meters. The hydraulic leveling system would keep the platform within 25 centimeters of the horizontal when moving on a 5% grade. The Jacksonville engineering firm of Reynolds, Smith, and Hills reported crawlerway costs per mile of $447000 on high ground and $1200000 across marsh. The latter figure included the cost of removing 6 meters of silt so that a firm roadway could be constructed. The estimate was close to the eventual cost of $7.5 million for ten kilometers of crawlerway. On 15 May, Harvey Pierce summarized Connell's rail study. Although the new railbed appeared sound, it was unproven and twice the cost of a crawlerway. Perhaps more important, the switching arrangements looked like trouble to operations personnel.[29]

The crawler received a further boost from a 1 June Corps of Engineers report. During a three-week study, the Jacksonville office focused on Merritt Island's ability to support the different transporters. Rail fared the worst.

> As a result of the nonhomogeneity of the foundation materials, differential settlement is inevitable along any long embankment. The effect of such settlement would be most detrimental to any system using rails or concrete slabs. Flexible pavements would be less affected and the effect on canal design would be negligible.[30]

A barge transporter would entail high construction costs for a launch basin and docking facilities at the vertical assembly building; the Corps of Engineers estimated $20000000 for the launch basin alone. The crawler presented no serious problems.

The decision to use the crawler came at an LC-39 conference on 12–13 June. Representatives from NASA Headquarters, the Manned Spacecraft Center, Marshall divisions, and private industry joined LOD at the Cape meeting. The launcher-transporter's crucial role placed it first on the agenda. After reviewing LOD's search, Donald Buchanan compared the three major contenders. Although the barge concept offered the best growth potential, there were unresolved design problems with propulsion, steering, platform stability, and placement at the launch pad. Buchanan noted, "If meeting a tight schedule has any bearing on the choice of modes, it would be difficult to assign a low enough value to the barge to illustrate the situation as it now stands."[31] The barge's operational shortcomings included a vulnerability to blast and a slow reaction time (evacuating the rocket in an emergency from the launch pad). While both the rail and crawler systems were within the state of the art, the latter enjoyed advantages of cost and flexibility. Buchanan's crawler recommendation met no serious objections.[32]

Plans for a VAB

The complexity of LC-39 planning dictated formal program management. Debus moved to provide this in the summer of 1961 with the establishment of the Heavy Space Vehicle Systems Office. Rocco Petrone and two assistants constituted the primary working force at the outset. J. P. Claybourne, a Minnesota native and New York University graduate, had handled program management with Petrone in the Saturn Systems Office the previous year. William Clearman, raised in Georgia and educated at Georgia Tech, had served with naval aviation during and after World War II. By early 1962 Petrone's office was providing other LOD offices with program criteria: details such as hook height, service platform levels, umbilical tower service arm heights, and weight loads for the transporter. This involved frequent liaison with MSFC, Houston's Manned Spacecraft Center, and NASA Headquarters.[33]

The vertical assembly building received much of the Heavy Vehicle Office's attention. As Petrone noted in a March 1962 congressional briefing, "the building is our most expensive item. On this item we put forth greatest study."[34] At the time Petrone estimated the VAB would cost $129.5 million of a total of $432 million for the entire complex. The earliest plans for the VAB envisioned a circular assembly building with a turntable to position the transporter. An alternate scheme resembled Martin Marietta's Titan II assembly building design with high bays in line. LOD's October 1961 study placed the high bays back-to-back with the transporter routed down the middle of the VAB. Martin's C-3 study proposed a box-shaped VAB in which six high bays enclosed water channels—transportation by barge was still being considered. There were two unattractive features. An extensive canal system within the VAB would hamper operations and raise the humidity. Negotiating right angle turns into the high bays with the barge would require a floor plan of 204 × 303 meters, nearly 50% larger than the eventual VAB. LOD vetoed the design in January 1962.[35]

At the LC-39 conference 6 February 1962, the Launch Facilities and Support Equipment Office agreed to compare open and enclosed VAB designs. Much of the subsequent study was performed by Brown Engineering Company of Huntsville. Ernest Briel directed 20 men investigating two VAB concepts with a barge transfer: one, a fully enclosed box structure with outward-opening bays; the second, an open, in-line structure with silo vehicle enclosures for the launch vehicle. R. P. Dodd supervised the Brown effort; James Reese performed liaison. Brown's reports on 2 April rated the enclosed VAB good for operating characteristics but poor for expansion potential because of canals on three sides and a low bay on the fourth. With the in-line version, the canal would run along the front side, permitting expansion. Low cost was a second advantage; Brown engineers placed a $65 million price tag on the open VAB, $10 million less than the enclosed version. Since a major reason for the remote assembly building was protection from the weather, operations personnel opposed the open concept.[36]

The operations group carried the day at the 13 June LC-39 conference. Gruene led the attack against the open design, arguing that environmental control would be a problem because of the umbilical openings; lightning would be a hazard in an open VAB, particularly if a rocket returned

Drawings of possible assembly buildings for C-5

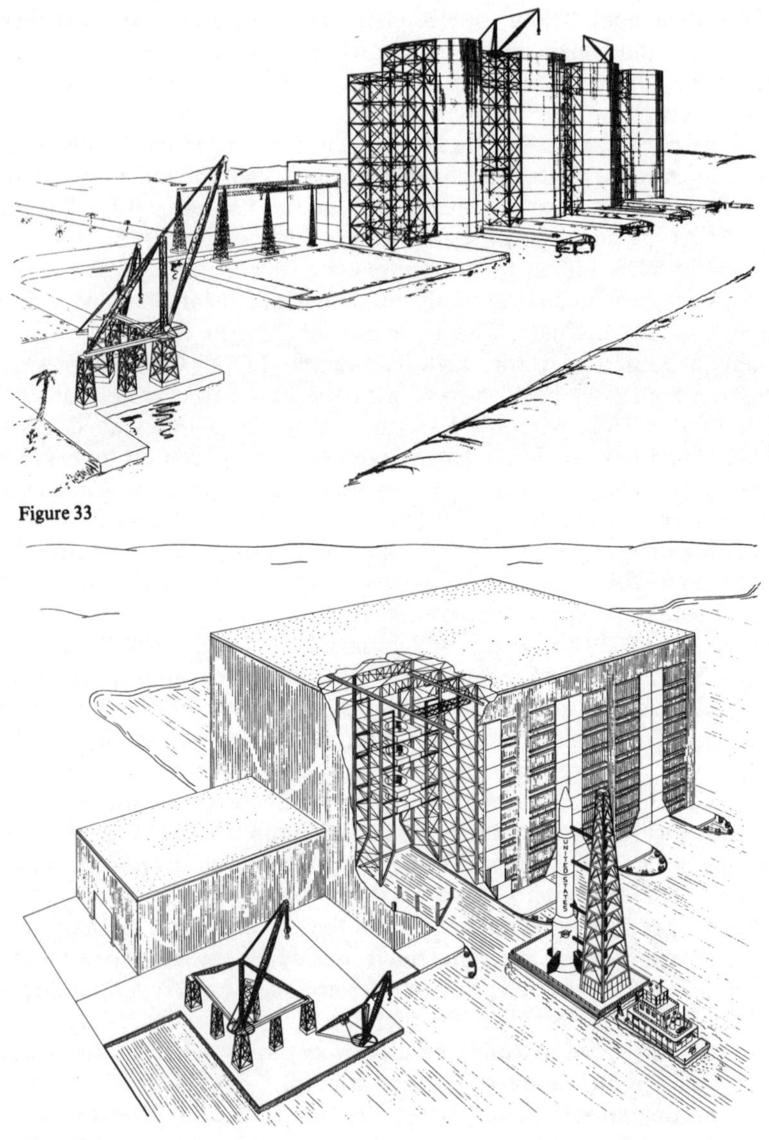

Figure 33

Figure 34

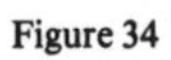

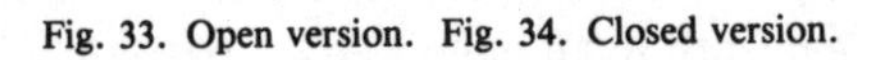

Fig. 33. Open version. Fig. 34. Closed version.

from the pad with ordnance aboard; with the silo enclosure open during assembly, high winds could curtail operations; and work at umbilical arm heights would be difficult. The conference agreed to a closed VAB, but no choice was made between an in-line and a box design.[37]

While selection of the crawler simplified VAB planning, the design remained tentative the rest of the summer. At an 18 June meeting, Deese presented a design of six high bays in line and a low bay to the rear, the high bay areas to be constructed in three increments. The low bay, completely air conditioned, would provide checkout areas and aisle space for the upper stages and spacecraft. After erection of the first stage on a launcher-umbilical unit (accomplished by a 250-ton crane at the barge unloading dock), the crawler would carry it into a high bay through a 43-meter-wide door and position the launcher on a set of concrete piers. Mating of the remaining stages would take place in the high bay where five retractable platforms provided access to the rocket. The launch control center and the central instrumentation facility would probably be housed within the VAB, using the roof as an antenna platform. Deese stated that an early definition of requirements was needed for both facilities.[38]

VAB design was again discussed at a 31 July meeting convened by Petrone. Hook height for a 60-ton crane to mate the upper stages was set at 139 meters; the door would extend 3 meters higher. The first of four high bays would be ready for use in January 1965. The launch control center would go either on top of the low bay roof or between the transfer portals that opened to the high bays. Matters were still unsettled at a mid-August briefing for the center director. When LOC engineering presented a VAB plan with four enclosed high bays in line, Debus expressed reservations about the number of bays and the in-line design.[39]

The architectural-engineering consortium URSAM won the contract for detailed VAB criteria in late August 1962 and quickly went to work (see pp. 222–25). On the 30th, URSAM received a set of documents from the Cape that included: "An Evaluation of an Enclosed-in-Line Concept of a C-5 Vertical Assembly Building," prepared by Brown Engineering Company; an evaluation of an open concept for the VAB, also prepared by Brown; NASA organizational charts and schedules; a general site plan of the Cape Canaveral missile test area; a "Geology and Soil Report" made by the Corps of Engineers the previous June; configurations of the C-5; plans of the retraction mechanism for the umbilical tower arms; general instructions; and discussions of the function of the VAB.[40]

By September a Facilities Vertical Assembly Task Group consisting of Arthur J. Carraway, Jack Bing, and Norman Gerstenzang of NASA, and Wesley Allen and Ernest M. Briel of Brown Engineering, was busy defining requirements for URSAM—the general layout of the VAB, the needed shops, general support engineering, and work areas. Some 600 people were expected to work in the VAB, including 100 Pan American maintenance people. A variety of things had to be resolved, from the requirements for a cafeteria to the umbilical arms in the low bays. On 6 September the group worked out methods of obtaining critical and emergency power; the cable requirements from the pad to the VAB, from the launch control center to each high bay, and within each high bay; the power requirements for the launcher umbilical tower; and the launch control center layout.[41]

Four days later an URSAM team arrived at the Cape and, in its first meeting, reached a major decision. It proposed that NASA place the bays in the VAB back-to-back rather than in-line, to gain the following advantages:

- Availability of all four high bays for vehicle erection and assembly without any restrictions.
- Reduction in the number of cranes required from seven to three.
- Elimination of extensive handling of the upper stages on rail-mounted dollies, thus avoiding complex turntable installations and differential settlement problems.
- Simplification of booster and upper-stage transfer and erection procedures.
- Greater adaptability for expansion.[42]

Another consideration, the paramount one for many LOD engineers, was the wind load factor. The huge assembly building would be subjected to tremendous wind pressures and a back-to-back design promised more stability.[43]

The Mobile Launch Concept—Debate and Approval

Debus had little trouble with critics of the mobile concept within LOD. It was a different story outside the launch team. At NASA Headquarters, Milton Rosen questioned both cost and feasibility. In early January 1962, he commissioned a launch facility study by three engineers of the Office of Manned Space Flight. Drawing their information from NASA and aerospace corporation studies, the team concluded that fixed pads were preferable to the mobile concept. The judgment rested on three grounds: the automated checkout equipment and increased reliability of space vehicles would reduce the minimum interval between launches from a fixed pad to one month; the high launch rates, for which the mobile concept was designed, were increasingly unlikely; and the mobile launch concept involved too many risks and engineering uncertainties.[44]

The mobile concept came under more fire in March. On the 6th von Braun notified Debus that an adverse Air Force report had triggered further doubts at NASA Headquarters. Debus stuck to his guns and was supported by Seamans and Holmes. During congressional testimony in early April, Holmes responded to an inquiry regarding the VAB's importance:

> This is an absolute necessity. It is a basic element in our lunar program. If we don't go to this type of vertical assembly, protected from weather, where assembly can take place with integrated checkout equipment for our lunar program, I really think we will end up with the same kind of rather crude facilities we now have for launching, where we assemble them on the pad for 2 or 3 months, where we do not have spares, and it would probably be impossible to use Earth orbital rendezvous.[45]

LOD's opportunity to defend LC-39 came on 23 March when Representative Olin Teague's Manned Space Flight Subcommittee visited the Cape. After describing the mobile concept's advantages in general terms of flexibility and high launch potential, Debus listed seven specific advantages:

- Pad staytime reduced to a week.
- A minimum of equipment exposed to launch area hazards.
- Repetitious testing eliminated by automation.
- Pad unaffected by different vehicle stage arrangements since the transporter-launcher carried the checkout equipment.
- Considerable savings in land costs.
- Minimum construction costs for high launch rates.
- Economic utilization of personnel.

Petrone stressed the last two points. LC-37's $432-million price tag was a bargain compared with the $900-million cost of nine fixed pads for 36 annual launches. If LOD planned facilities for a maximum launch rate of 24 per year, LC-39 still represented a saving of $168 million. One congressman considered Petrone's manpower savings estimates the best argument for LC-39. The complex would employ 2200 men, 1500 fewer than the requirement for nine fixed pads. The annual savings in salaries would amount to $18 million; comparing LC-39 to six fixed pads, Petrone estimated savings of $8 million per year.[46]

The committee questioned the VAB's availability for Nova. Petrone pointed out that Nova dimensions were not firm and postponing LC-39 plans would delay the Saturn C-5 program. The VAB design would allow modification at a later date. Col. Clarence Bidgood, Facilities Chief, stated that flexibility was desirable at three points in the complex: the assembly building, the transporter, and the launch pad. Although LOD was attempting to provide growth potential and a capability for handling solids or liquids, "you might build so much expense into it to get flexibility that it would be very, very uneconomical in the first place." The congressmen were silent on two important matters affecting LC-39: the likelihood of high launch rates and the technical problems of the mobile concept. Perhaps they were unaware of the engineering difficulties that bothered Harvey Pierce and Milton Rosen. They may have feared delay in a pacing item* of the Apollo program. As Teague said, the committee was well disposed toward LOD's project. Their main concern was defending LC-39 before the House Appropriations Committee.[47]

By late May planning on LC-39 was well along; preliminary schedules called for design criteria contracts within three months. Debus moved to secure approval of the mobile concept at the Office of Manned Space Flight Management Council meeting on 29 May 1962. He acknowledged that launch rates were at a break-even point and cost savings no longer a major factor. LC-39, however, offered distinct technical advantages. Milton Rosen accepted Debus's arguments, but thought there should be further study of the disadvantages. Robert Gilruth expressed MSC's concern that LC-39 would not provide servicing of the spacecraft at the pad. Von Braun then interjected a telling point. The fundamental question, the Huntsville director stated, was whether they believed "a space program is here to stay, and will continue to grow." The Council responded with approval of Debus's plan.[48]

Despite the vote of confidence, the issue reappeared at the 22 June Management Council meeting. Rosen warned that LC-39 would be three years in the making and any slippage would delay the launch program. He recommended modifying the complex to allow for on-pad assembly. As a compromise Debus suggested transporting the arming tower to the pad for assembly purposes or spacecraft checkout.† Although Holmes requested more information pending a final decision, the mobile concept was a virtual certainty. Rosen had told Debus on the 15th not to worry about further questioning; Headquarters was going along with LC-39.[49]

---

*The term *pacing item* refers to a facility or equipment that is essential to a program, with little or no margin for delay. During the Apollo program different items earned this distinction. In the spring of 1962, the Mississippi Test Facility (where the C-5's first stage would be test-fired) and LC-39 were pacing items.

†Most members of LOD wanted a stationary arming tower midway between the assembly building and the pad. Ernest Briel's 31 July notes from a Petrone meeting include the statement, "an AT arming tower *NOT* to be used as service structure." Because of weight constraints, the service arms on the launcher-transporter could not provide 360° of access to the spacecraft. MSC's insistence on this capability eventually forced LOD to accept a mobile service structure (see pp. 130, 163).

June 1962 brought other Apollo decisions, including selection of lunar-orbital rendezvous (LOR) for the mission mode. NASA had studied the issue since the late 1960s. At first, either direct flight with a Nova or earth-orbital rendezvous (EOR) with Saturns seemed likely choices; but by May 1962, debate had narrowed to EOR versus LOR. Lunar-orbital enthusiasts at Langley, Houston, and Headquarters stressed the advantage of landing on the moon with a light vehicle specially designed for the mission. MSFC engineers continued to support EOR for practical as well as technical reasons: much of their workload would disappear if EOR was dropped. An impasse seemed likely, until von Braun announced his support for the lunar-orbital mode on 7 June. The decision was brought on by the influence of LOR's technical advantages, assurances that Headquarters would compensate MSFC with new tasks, and concern for the Apollo program. In explaining the about-face to his Huntsville team, von Braun stated: "If we do not make a clear-cut decision on the mode very soon, our chances of accomplishing the first lunar expedition in this decade will fade rapidly."[50] With Houston and Huntsville in agreement, the matter was pretty well settled. The Management Council and Administrator Webb approved LOR within a month. At its 22 June meeting the Management Council also endorsed immediate development of a lunar excursion module and an intermediate rocket, the Saturn IB. The new member of the Saturn family would use an uprated S-I stage (first stage of the Saturn C-1) and the new S-IVB stage for testing the Apollo spacecraft in earth orbit.[51]

The summer's weekly staff reports to Debus reveal the breadth of LC-39 activities. On 5 July Karl Sendler reported on the telemetry studies of the Manned Lunar Landing Program (MLLP) Instrumentation Planning Group. Two weeks later the group organized an eight-man task force to determine LC-39's requirements for weather data. The continuing dispute over LC-39 siting was a frequent topic of Colonel Bidgood's Facilities Office reports. On 5 July Bidgood notified Debus that a site proposal was ready for the MLLP Joint Facilities Planning Group; it called for placing the complex near the ocean. Although the Air Force no longer insisted that NASA place LC-39 north along the Mosquito Lagoon, it wanted the complex 4.5 kilometers inland. Air Force officials believed that location would provide space for additional launch complexes at a later date. The matter dragged on for six more weeks before the Air Force Missile Test Center yielded. Bidgood reported two major achievements on 23 August: Air Force concurrence on siting and initiation of criteria work for LC-39.[52]

The Launch Support Equipment Office began a study of the mobile arming tower in June, following Debus's offer to investigate the matter for the Management Council. Poppel announced the study's completion in his

16 August report: "it is not only feasible but highly recommended since this added flexibility to the C-5 complex can be achieved with little increase in cost." The flexibility concerned the use of the mobile arming tower to erect upper stages at the pad if necessary. The study rejected using the 116-meter tower to erect the booster, since the addition of a huge crane would impose severe structural problems.[53]

LC-39 was the sole topic at a meeting of the Launch Operations Working Group on 18–19 July that brought together 113 representatives from LOD, MSFC, and the launch vehicle contractors: Boeing, North American, Douglas, and General Electric. In Petrone's absence, Phillip Claybourne and William Clearman chaired the sessions. Claybourne's welcoming remarks described the role of the working group panels, teams that were to be organized later in the day to exchange information and accomplish specific tasks. Clearman followed with a general description of LC-39.

Following Donald Buchanan's report on the crawler and launcher-umbilical tower, Chester Wasileski briefed the meeting on propellant systems. Although LC-39 would involve no new propellants, loading requirements would dwarf LC-34 operations. Each pad would need storage for approximately 3 407 000 liters of LOX, 946 000 liters of RP-1, 2 460 000 liters of LH_2, and 946 000 liters of LN_2. Propellant loading rates would be:

S-IC	38 000 liters per minute of	LOX
	7 600	RP-1
S-II	19 000	LOX
	38 000	LH_2
S-IVB	3 800	LOX
	15 200	LH_2

LOC planned to automate propellant loading on all Saturn launch sites; controls in the launch control center would operate through the data link on the launcher. A compression-converter facility near .the VAB would provide gases to charge high-pressure spheres on the launch vehicle and to keep certain ground support equipment free of moisture and dust. Wasileski proposed redundant sensors in the loading system and asked the panels for further comment.

Robert Moore and Bradley Downs of the Firing-Equipment Design Group (Launch Support Equipment Office) described the seven arms of the launcher-umbilical tower that would provide personnel access and support electrical cables, propellant lines, and pneumatic lines to the launch vehicle. Prior to the rocket's first motion, five arms would disconnect and begin withdrawal. Arms 4 and 6, providing hydrogen vent ducting and services to the S-II stage and the instrumentation unit, would retract at liftoff. Moore asked the groups responsible for individual stage operations to reexamine their service needs. Lengthy but inconclusive debate followed on a remote reconnect capability for aborted missions.[54]

With this meeting, LC-39 was just about ready to go. After it won final approval, Marvin Redfield, co-author of the NASA Headquarters report that had criticized the mobile concept, congratulated his friend, Rocco Petrone, but insisted the price would far exceed the launch team's estimates. Petrone accepted the challenge, wagering a case of Scotch that costs would not run over $500 million. The bill eventually came to about $500 million despite a significant reduction in LC-39 components, e.g., four high bays instead of six in the VAB. When Petrone insisted he had won the bet, Redfield grudgingly agreed to pay, but only one bottle at a time. On the occasion of the first payment, Petrone, either doubting the fairness of his victory or influenced by the good cheer, absolved Redfield of further payments.[55]

The General Accounting Office was less jovial about the $500 million price tag. A report in 1967 would imply that LC-39 had been a costly mistake, a conclusion that NASA would strenuously oppose (pp. 432–34).

7

The Launch Directorate Becomes an Operating Center

Growing Responsibilities at the Cape

By the time Apollo 11 put Neil Armstrong and Edwin Aldrin on the moon, Apollo field operations were divided among three NASA installations. Marshall Space Flight Center supervised the development of the launch vehicle, the Manned Spacecraft Center in Houston the spacecraft, and Kennedy Space Center assembled, tested, and launched the combination. The actual construction was done by contractors from all over the United States; but generally speaking, management responsibility was divided as described above, with fairly well defined boundaries and a minimum duplication of effort.

This neat packaging was not achieved in a single bound, but was the result of an evolutionary process accompanied by much discussion, some backing and filling, and a few attempts at empire building. A main step in the process was the elevation of the Launch Operations Directorate (LOD), previously part of Marshall, into the Launch Operations Center (LOC) on a par with Marshall. This was a good two years in the doing, during which time Debus had to meet increased responsibilities with limited manpower and authority. Mindful of his difficulties, his superiors at Marshall proposed in the spring of 1961 to expand LOD's organization to include new offices for program control, financial management, purchasing and contracting, construction coordination, and management services. With President Kennedy's message of 25 May 1961, it became obvious that the manned lunar landing program was going to be a very big project and that NASA's launch team at Cape Canaveral would need corresponding status.

General Ostrander requested Debus to develop organizational proposals; he responded on 12 June 1961 with three plans. The first called essentially for the maintenance of the status quo, the second for a launch organization providing administrative support to launch teams from the NASA centers, and the third for an independent Launch Operations Center to serve all of NASA.* All three called for a single point of contact at the Atlantic Missile Range, an in-house capability for monitoring launch operations, and an independent status in master planning, purchasing and contracting, and financial and personnel management. Debus talked over these proposals with von Braun who in turn discussed them with Ostrander.[1]

The three proposals show Debus leaning over backward to avoid any suggestion of officiousness. He was equally convinced, however, that a successful launch program required an experienced team with full powers at the launch site. He set out this thought some six weeks later in a letter to Eberhard Rees, the Marshall Space Flight Center Deputy Director for Research and Development. The letter was occasioned, not by the reorganization proposals, but by a delay in the assembly of the SA-1 booster at Huntsville. Debus agreed to let the work be finished at the Cape, but made it plain that this set no precedent. Writing to Rees, Debus noted that any MSFC division might prefer to send engineers to conduct the related part of the launch operations. Von Braun had tried this at White Sands and found it wanting. With the Redstone, a permanent launch team had been set up as an integral part of the Huntsville organization, and this had worked well the past nine years. Now, given the complexity of the Saturn, it was the only satisfactory approach.

Placing the responsibility for launch checkout with the Huntsville offices that had designed and built the Saturn could only lead to difficulty. If similar arrangements were made with all booster, stage, and payload contractors, the situation would become impossible.[2] Agreeing to the exception for SA-1, Debus insisted that henceforth Huntsville hardware be shipped in as complete form as possible, and after Huntsville's final inspection. At the Cape, "all participants, including contractor personnel, must be supervised and coordinated by one launch agency." Debus stated that LOD would perform any function "that has been or will become a standard requirement at the launch site."[3]

In the meantime, the Deputy Director of Administration at Marshall Space Flight Center, D. M. Morris, recommended to NASA Headquarters that the Launch Operations Directorate have greater authority and stronger support services under its control. Following on this, Harry H. Gorman, Associate Deputy Director for Administration at Huntsville, wrote Seamans at NASA Headquarters on 26 September 1961 recommending greater financial and administrative independence for LOD. Gorman noted that the

---

*While the public has always tended to identify NASA with manned spaceflight, NASA had from its beginning several unmanned projects. These were managed by such centers as Lewis, Langley, and Ames; in some cases, the vehicles were launched from Canaveral. Completely independent of Marshall, such launches complicated matters for LOD.

distance between Huntsville and Cape Canaveral was producing a communications gap, that LOD's dependence on Marshall impaired efficiency, and that the increased work load falling on LOD and other NASA elements at the Atlantic Missile Range dictated a larger role for LOD. Gorman suggested that LOD assume responsibility for services still performed for it by Marshall offices in programming, scheduling, procurement, and contracting; that it increase its personnel in some existing support elements; and that it lease off-base facilities near Cocoa Beach to house such activities as financial management, procurement and contracts, and construction coordination. He urged the immediate hiring of 75 more employees.[4]

The day following Gorman's letter, Debus completed a second position paper on "Launch and Spaceflight Operations." He noted "the current expansion of NASA activities, the magnitude and complexity of future space programs, the requirement for rapid overall growth potential and the resulting need for clear lines of responsibility and authority"; and he called for a "competent organization of NASA elements."[5] Debus evaluated two plans in a third proposal on 10 October 1961. The first would put administration and management, general technical and scientific fields, facility planning and construction, checkout and launch, and operational flight control under a single launch organization reporting to NASA Headquarters. The second would leave operational flight control and some aspects of checkout and launch under the individual launching divisions of their parent centers.[6]

Von Braun supported the first alternative: "This study brings the NASA-wide launch operations problem very well in focus," he wrote. "I consider Plan I the superior plan for the accomplishment of NASA's objectives [manned lunar landing in this decade] but its implementation will require a ringing appeal to all centers for NASA-wide team spirit in lieu of parochial interest."[7] Seamans insisted that personnel at Headquarters give major attention to the matter in the next two weeks. Debus was later of the opinion that Seamans initiated the entire discussion.[8]

Von Braun was correct in assuming that raising LOD to the status of a separate center would meet serious objections from vested interests in NASA. Harry Gorman's arguments from the administrative standpoint were not seconded in the engineering divisions. Eberhard Rees, for one, leaned against separation; if it should prove necessary, he preferred the alternate plan, wherein a Launch Operations Center would control administration and management, general technical and scientific fields, and facility planning and construction, with the launching divisions of the various centers still controlling their flight operations and some aspects of checkout and launching. Most of von Braun's staff opposed the separation of the launch team from

Huntsville. There was some feeling that they would be working in the factory, while the Debus launch team in Florida would enjoy the action and the spotlight. Heated debates continued through a cold winter.[9]

The Argument for Independent Status

NASA meanwhile began construction of the Manned Spacecraft Center at Houston in late 1961. This center had its own launch team, first called the Preflight Operations Division, later the Florida Operations Group, with launch responsibility for the current manned space program, Mercury. The entire relationship of LOD with the Manned Spacecraft activities in Houston and Florida needed definition. Would Houston or LOD control Apollo launches? Debus believed "that there would be serious problems if the Manned Spacecraft Center thought the launch group was always being loyal to another Center [Huntsville]. What was needed was a launch Center that could be loyal to any Center." To summarize the case for an independent launch center: the Florida operation had to be on a par with Huntsville and Houston; it had to have direct access to Washington rather than through channels at Huntsville; and it had to be the one NASA point of contact with the Air Force Missile Test Center—if it was going to provide launch facilities for Apollo in an efficient and timely manner.[10]

NASA announced on 7 March 1962 that it would establish the center as an independent installation. Debus continued in charge, reporting to the Director of Manned Space Flight, D. Brainerd Holmes, at NASA Headquarters. Theoretically the new Launch Operations Center (LOC) would serve all NASA vehicles launched from Cape Canaveral and consolidate in a single official all of NASA's operating relationships with the Air Force Commander at the Atlantic Missile Range. NASA replaced Marshall's Launch Operations Directorate with a new Launch Vehicle Operations Division (LVOD) in Alabama. However, Debus would be director of both LOC and the new LVOD and Dr. Hans Gruene would also wear "two hats" as deputy director.[11] The creation of the Launch Vehicle Operations Division under Marshall, but with Debus as director, may seem to reflect a reluctance to grant the Launch Operations Center independent status, but was more likely intended to ensure that the Debus team stayed in charge of the Saturn flight program regardless of its tenure at LOC.

According to John D. Young, NASA Deputy Director of Administration, LVOD was "an interim arrangement to provide additional time to carefully consider to what extent, if any, the electrical, electronic, mechanical, structural, and propulsion technical staffs of the present Launch Operations Directorate of MSFC should be divided between MSFC and LOC.''[12] Debus saw the matter in a somewhat different light: ''LVOD was strictly a compromise measure to overcome the problem within von Braun's own group. All of his basic contracts were on incentive fees . . .''; the stage contractors ''complained and not unjustifiably, 'We pamper stages through here [Huntsville], then give them to a crew at LOC who may louse it up.' '' Mistakes made at the Cape could therefore reduce a contractor's payment.[13]

Debus and Marshall's Deputy Director Eberhard Rees, acting for von Braun, signed an interim separation agreement between the Launch Operations Center and the Marshall Space Flight Center on 8 June 1962. Of the 666 persons assigned to launch operations for the fiscal year 1962, 375 went to the Launch Operations Office. Independence Day for the Launch Operations Center was 1 July 1962. This arrangement was to hold until the following year when reorganization plans within both NASA centers transferred the Launch Vehicle Operations Division from Marshall to the LOC on 24 April 1963.[14]

New Captains at the Cape

The Gorman recommendations and burgeoning activity on the Cape sparked an increase in the Debus forces in 1961, well before they became the Launch Operations Center. Lewis Melton, reporting for duty in July 1961, initiated a rapid expansion of LOD's Financial Management Office, which entailed a move to ''off-Cape'' office space in the cities of Cape Canaveral and Cocoa Beach.[15]

On the recommendation of Maj. Raymond Clark and Richard P. Dodd, Debus requested the assignment of Capt. A. G. Porcher, of the Army Ordnance Missile Center Test Support Office at AMR, to LOD. Debus appointed Porcher LOD liaison officer with the Corps of Engineers for construction matters. Clark served in a similar liaison capacity between LOD and the Air Force. A 1945 West Point graduate, Clark had been with the Missile Firing Laboratory in the mid-1950s and was reassigned to the NASA Test Support Office in July 1960. He served on the test support team that represented both the Air Force Missile Test Center and NASA. The Debus-Davis study brought his skills to the fore. During the next two years he would represent LOD in a series of complicated negotiations with the Air Force.[16]

In January 1962, the Launch Operations Directorate established its own procurement office—a task previously handled under the supervision of

Marshall. Gerald Michaud, the first procurement officer, handled contracts for $30 000 000 worth of support equipment for launch complex 37. Michaud, like Melton, had to seek off-Cape office space.[17]

The Materials and Equipment Branch of LOD had worked under the supervision of the Technical Materials Branch at Huntsville until the beginning of 1962, when a joint supply operating agreement went into effect. By June 1962 the LOD branch was operating as an independent NASA supply activity.[18]

In this same period, Debus set up the Heavy Space Vehicle Systems Office with Maj. Rocco Petrone as director. Petrone's responsibility for the Saturn C-5 included facilities, operations, and site master planning. In the third area, he co-chaired, with an Atlantic Missile Range representative, the Master Planning Review Board that regulated the development of Merritt Island and ensured that site development met NASA requirements.

The direct supervision of facilities on LC-39 fell to Col. Clarence Bidgood, a West Point graduate with a master of science degree in engineering from Cornell, and a survivor of Bataan and four years in a Japanese prison camp. Described as a "no-foolishness hard worker," Bidgood had packed a variety of experience into his postwar years that included flood control work and construction of U.S. airfields in England. He began working for LOD in November 1961 and took charge of the Facilities Office in February 1962. Bidgood turned his attention in his initial year to three major functions: the acquisition of real estate on Merritt Island and the False Cape; organization of the Facilities Office for the criteria design and construction of LC-39; and the establishment of requirements for LC-39 by the various individuals, firms, panels, and centers involved in Apollo.[19]

The Launch Support Equipment Office under Theodor Poppel and Lester Owens, Deputy Director, retained the design responsibilities for vehicle-associated support equipment. This group remained at Huntsville in order to coordinate the work of designing and launching the vehicles. At von Braun's suggestion, Debus took Poppel's group under his jurisdiction.[20]

In the enlargement of its staff after 1 July 1962, the Launch Operations Center gave priority to individuals who had performed as administrators in similar areas for LOD; and, for other positions of importance, to Marshall personnel with appropriate skills. Associate Director for Administration and Services C. C. Parker, who had served as Management Office Chief at Anniston Ordnance Depot before joining LOD, interviewed the prospective section chiefs and Debus made his final choice from the candidates recommended by Parker.[21]

As a result of internal growth and the acquisition of the LVOD personnel in May, LOC's personnel strength rose almost 400% between July

1962 and July 1963. More offices were forced to seek quarters in the cities of Cape Canaveral and Cocoa Beach. In the case of Procurement and Contracts, the move from military security at the Cape allowed easier access for outside contacts. The location of Public Affairs at Cocoa Beach facilitated relations with Patrick Air Force Base, the contractor offices, and the press.[22]

Most Launch Operations Center personnel remained on the Cape, where LOD had been a tenant. Some NASA elements continued as tenants in Air Force space for several years. In this period many offices had to get by with inadequate facilities, which impaired morale and reduced productivity. George M. Hawkins, chief of Technical Reports and Publications, pointed out that four technical writers worked in an unheated machinery room below the umbilical tower at LC-34. At one time pneumonia had hospitalized one writer and the others had heavy colds. When it came time to install machinery there, they urgently requested assignment to a trailer. Russell Grammer, head of the Quality Assurance Office, established operations in half a trailer at Cape Canaveral with seven employees. When the staff grew to 13 times that size, his force had to expand into other quarters. The Quality Assurance people worked in such widely scattered places as an old restaurant on the North Cape Road, a former Baptist church on the Titusville Road, a residence on Roberts Road, and numerous trailers.[23]

Organizing the Launch Operations Center

Recognizing the magnitude of Apollo, NASA Headquarters in late 1962 and early 1963 relieved the manned spaceflight centers of certain other responsibilities. Management of the Atlas-Centaur and Atlas-Agena was transferred from Marshall to Lewis Research Center. In February NASA released LOC from responsibility for launching these vehicles and gave it to the Goddard Space Flight Center's Field Projects Branch.[24]

As NASA's agent, LOC generally furnished support and services for all launches, manned and unmanned, conducted by the launch divisions from NASA's several centers. But in its chief role as a launch agent for the Office of Manned Space Flight, its principal business during this period was the planning and designing of launch facilities for Apollo. On 10 January 1963 NASA announced that LOC was responsible for overall planning and supervision of the integration, test, checkout, and launch of all Office of Manned Space Flight vehicles at Merritt Island and the Atlantic Missile Range, except the Mercury Project and some elements of the Gemini Project. What the phrase, "all OMSF vehicles," fails to reveal is that the only other authorized manned spaceflight project at the time was Apollo. Almost all of the work at Houston and Marshall in 1963 was devoted to the manned space program. At the Launch Operations Center, most of the planning and the new construction work was also for manned spaceflight, and this was increasingly Apollo.[25]

Indeed, the first task was to organize for the construction effort. The Webb-McNamara Agreement of January 1963 (see p. 105) had helped clear the air by firmly establishing NASA's jurisdiction over Merritt Island. The question of whether LOC was to become a real operating agency or a logistics organization supporting NASA's other launch teams was still unresolved. The Manned Spacecraft Center's Florida Operations, for instance, still received technical direction from Houston. Debus had no place in this chain of command. The transfer of launch responsibility for the Centaur and Agena vehicles from LOC to Goddard Space Flight Center, while a step toward LOC's concentration on Apollo responsibilities, was a step away from centralization of launch operations. The Launch Vehicle Operations Division remained under Huntsville until April. Several areas of overlapping jurisdiction called for resolution. A few section chiefs were certain that they were best qualified to determine their own functions. As Colonel Bidgood said, "Everybody was trying to get a healthy piece of the action."[26]

The publication of basic operating concepts in January 1963 made LOC responsible "for construction of NASA facilities at the Merritt Island or AMR launch site."[27] The LOC Director was empowered to appoint a manager for each project and, in conjunction with other participating agencies, write a project development plan. Debus was also required to prepare a "basic organization structure" for the approval of Headquarters.

Debus submitted the required proposal early in 1963. It called for five principal offices: Plans and Project Management, Instrumentation, Facilities Engineering and Construction, Launch Support Equipment Engineering, and Launch Vehicle Operations.[28] As so often under Debus, the changes in title did not involve changes in personnel. To the five key posts, he assigned men for whom the new responsibilities would be continuations of their earlier tasks—Petrone for Plans and Projects, Sendler for Instrumentation, Bidgood for Construction, Poppel for Launch Support, and Gruene for Launch Vehicle Operations. These staff elements carried out the major functions of management, design, and construction of launch facilities and support equipment for the Apollo program. Other staff elements (public affairs, safety, quality assurance, and test support) dealt largely with institutional matters. NASA Daytona Beach Operations, established on 23 June 1963 to represent NASA at the General Electric plant there, made up another element reporting directly to the Center Director. On 24 April 1963, Deputy Administrator Dryden approved LOC's proposed reorganization—except for the Daytona Beach office, which was approved subsequently.[29]

Under Petrone were two Saturn project offices, one responsible for the early Saturn vehicles, the other for a larger Saturn to come. Both offices were to plan, coordinate, and evaluate launch facilities, equipment, and operations for their respective rockets. Another office was responsible for projects requiring coordination between two or more programs. Other elements of Petrone's staff were responsible for resources management, a reliability program, scheduling, and range support. These responsibilities, especially for resources management and coordination, gave Petrone substantive control over the development of facilities, a control he showed no reluctance to exercise fully. *Spaceport News*, the LOC house organ that began publication in December 1962, described the role that Petrone would play in the new organization in its 1 May 1963 edition. As Assistant Director for Plans and Programs Management, the paper declared, Petrone

> will function as the focal point for the management of all program activities for which LOC has responsibility. In this capacity, he is responsible for the program schedule and for determining that missions and goals are properly established and met. He will formulate and coordinate general policies and procedures for the LOC contractors to follow at the AMR and MILA [Merritt Island Launch Area].[30]

Bidgood organized his division along functional lines, with titles clearly descriptive of responsibilities—a Design and Engineering Branch, Construction Branch, and a Master Planning and Real Estate Office. Most of Bidgood's personnel came from the former Facilities Office, which he had organized several months earlier around a nucleus of R. P. Dodd's Construction Branch, the Cape-based segment of Poppel's former office. He recruited others from such agencies as the Corps of Engineers Ballistic Missile Division in California.[31] Bidgood was shortly to retire from the Army and to relinquish his LOC post to another Corps of Engineers officer, the less outspoken but equally competent Col. Aldo H. Bagnulo.

Poppel organized the four branches of his division along equipment responsibility lines, extending in each case from design through completion of construction. One branch was responsible for launch equipment (primary pneumatic distribution systems, firing equipment, and erection and handling equipment); one for launcher-transporter systems; a third for propellant systems; and a fourth for developing concepts for future launch equipment.

Something more should perhaps be said to differentiate the last two divisions. In terms of specific launch facilities and ground support equipment, Bidgood was responsible for what was commonly, if inadequately, called brick-and-mortar construction: the vehicle assembly building, launch control center, launch pads, and crawlerway. Poppel supervised construction of the launcher–umbilical tower, crawler-transporters, and propellant and high-pressure-gas systems. Later, the arming tower was assigned to Bidgood. With the exception of the arming tower (later modified and redesignated the mobile service structure), Bidgood's area largely involved conventional construction. Poppel's responsibilities were more esoteric; no one could readily formulate plans and specifications in what were new areas of construction. The two divisions also operated differently. Bidgood's division used the Corps of Engineers for all contract work, from design through construction to installation of equipment. Poppel's division depended on commercial procurement and contracting. Although Bidgood and Poppel, like Petrone, reported directly to Debus, and although the organization chart showed no link between divisions, the functional statements in the "LOC Organization Structure" manual assigned responsibility for coordination of launch facilities to Petrone.[32]

The reorganization also clarified the relationship of Launch Vehicle Operations personnel to MSFC and LOC. Although assigned to LOC for operational and administrative matters, they remained under Marshall's technical direction for engineering. The "development operational loop" that had characterized the old MSFC-LOC relationship remained. This loop implemented the propositions that no launch team could be effective unless it participated in the development of a space vehicle from its inception, and that planners had to consider operational factors early in the design of the space vehicle and maintain this awareness throughout the development cycle. In representing Marshall contractors at Merritt Island and the Atlantic Missile Range, the Marshall Director retained authority to modify any responsibilities delegated to LOC, to interpret Marshall contracts for LOC and the contractors, and to direct the contractors with respect to contract implementation, including instances when disagreements might arise between LOC and the Marshall stage contractors.[33]

During these months, LOC spawned a great number of boards, committees, panels, teams, and working groups. In September 1963, C. C. Parker, Assistant Director for Administration, undertook to delineate the spheres and activities of these groups. Six panels dealt with facilities, propellants, electricity, tracking, launching, and firing. Committees handled incentive awards, grievances, suggestions, honors, automatic data processing, and five distinct areas of safety. Boards oversaw property, architect-engineering selection, and project stabilization. The personnel of these groups rarely overlapped, as distinctive disciplines required expertise of a particular nature. A significant team, by way of example, was the LOC MILA Planning Group, appointed by Debus on 6 February 1963 under the chairmanship of Raymond Clark. It looked into unsolved issues in relations with the Air Force Missile

Test Center, recommended divisions of responsibility among various elements of LOC, and established priorities to assure cooperation.[34] The informality of early operations on the Cape was disappearing in the growth of a mighty endeavor.

In the midst of all this organizational activity, one of the most able men to come to the Cape arrived as Deputy Director of the Launch Operations Center in early spring of 1963. Albert F. Siepert had been NASA's Director of Administration since its beginning in 1958. This 47-year-old Midwesterner had played a key role in the basic organization of NASA and in arranging the transfer of the von Braun team from the Army. Previous to his work with NASA, he had won the Health, Education, and Welfare Department's distinguished service award. A fine administrator and a great extemporaneous speaker—he could organize his thoughts in a few moments and speak without hesitation or repetition—he wanted to work in the field and requested a transfer to one of the centers. At LOC, he became responsible for the organization and overall management of center operations and had the further responsibility of maintaining good relations with local communities, the Air Force, the Corps of Engineers, other NASA field centers, and various contractors.[35]

"Grand Fenwick" Overtakes the U.S. and USSR

In spite of the launchings at the Cape, the development of the Launch Operations Center, the agreements between the Air Force and NASA, the preliminaries for the construction of launch complex 39 and the industrial area on Merritt Island, not all was ultraserious. The *Spaceport News* for 20 June 1963 carried this interesting headline: "The Duchy of 'Grand Fenwick' Takes Over the Space Race Lead." The article told of the premiere of a British movie, a space satire called *Mouse on the Moon*, at the Cape Colony Inn on the previous Friday. Distributed by United Artists, the movie was a sequel to the popular *The Mouse That Roared* of several years before.

The Mouse That Roared had centered around the attempt of the Duchy of Grand Fenwick, a mythical principality near the Swiss-French border, to wage an unsuccessful war against the United States in the hope that the United States would pour millions of dollars into the nation for rehabilitation. Surprisingly, the war turned out to be a huge success for the Grand Fenwick Expeditionary Force. It captured a professor at Columbia University, a native of Grand Fenwick, who had invented the "bomb to end all bombs." By threatening to use the bomb on all the major nations of the world, Grand Fenwick brought universal peace.

In the sequel, *Mouse on the Moon*, Grand Fenwick, faced again with a disaster in its main industry, wine-making, requested a half-million-dollar loan from the United States. Instead the United States granted a million dollars to further Grand Fenwick's space program and show America's sincere desire for international cooperation in space. Not to be outdone, Russia gave one of its outmoded Vostoks. The scientists of Grand Fenwick found that the errant wine crop could fuel this rocket. They sent the spacecraft to the moon, beating both the American and Russian teams. The U.S. and USSR spacecraft landed shortly after the Duchy's. In hasty attempts to get back first, both Russians and Americans failed to rise from the lunar surface. As a result, Grand Fenwick's Vostok had to rescue both crews.

The British stars, James Moran Sterling and Margaret Rutherford, came to Cocoa Beach for the world premiere, as did Gordon Cooper and his family, and many of the dignitaries of the Cape area. For a moment the tensions at the spaceport ceased, and the men caught up in the space race enjoyed a good laugh at their own expense.

Mid-1963: A Time of Reappraisal

"The first and the most truly heroic phase of the space age ended in the summer of 1963," wrote Hugo Black, Brian Silcock, and Peter Dunn in *Journey to Tranquility*. "Two years had passed since President Kennedy's commitment to the moon. They were to the public eye, the years of the astronaut; a period when this strange new breed of man was established as something larger than ordinary life, with gallantry and nerve beyond the common experience." This vision stemmed from the novelty of the situation, the ruggedness of some of the characters among the original seven, and partly, too, from the nature of the Mercury program. "Somehow one man in a capsule, alone in the totally unfamiliar void, more easily acquires heroic status than two or three men facing the ordeal together." The last flight of the Mercury series, by Gordon Cooper in May 1963, the authors concluded, "was the last appearance of the astronaut-as-superman."[36]

That summer marked more than the end of Mercury, as people began to realize for the first time what the moon program really meant. Before that, Kennedy's words had mesmerized them. NASA had gone about its work in an atmosphere of public consent and mute congressional approval. It had decided how to go, where to go, and who should go. The general public accepted the basic lines of the gigantic undertaking. Now the very concept of Apollo began to be questioned. When the great debate that Kennedy had asked for two years before finally got under way, scientists began to see that the space program made distorting demands on skilled manpower, economic resources, and human determination. And they began to ask if it was really worth doing. Did we have to beat the Russians? Was this the most important scientific effort we could perform? Was NASA perhaps traveling too fast? The President himself seemed to have his doubts when he began to suggest joint space efforts with the Russians.

The President had not anticipated NASA in this. In March 1963 the Dryden-Blagonravov agreement on space communications and meteorology suggested that cooperation was feasible.[37] In an address to the United Nations General Assembly on 20 September 1963, President Kennedy stated that joint U.S.-USSR efforts in space had merit, including "a joint expedition to the moon." He wondered why the two countries should duplicate research construction and expenditures. He did not propose a cooperative program, but the exploration of the possibility.[38]

On the next day, Congressman Albert Thomas, Chairman of the House Appropriations Subcommittee on Independent Offices, wrote the President to ask if he had changed his position on the need for a strong U.S. space program. The President replied on 23 September that the nation could cooperate in space only from a position of strength and so needed a strong space program.[39]

Scientists began to talk of other priorities, such as the declining water table in the West and the challenge of oceanography. Lloyd Berkner, to be sure, still took a strong stand for Apollo, chiefly concerning himself with the project as a national motivating force. He had been one of the original promoters of the launching of a satellite during the International Geophysical Year. Berkner's grand vision satisfied many on Capitol Hill. But a majority of scientists still seemed to question the entire program. They felt that the President had proposed the lunar landing in a period of panic that had stemmed from the success of Cosmonaut Yuri Gagarin, first man to orbit the earth, and the disaster of the Bay of Pigs just seven days later. In November 1963, *Fortune* magazine summarized the discussion in an article entitled, "Now It's an Agonizing Reappraisal of the Moon Race." The author, Richard Austin Smith, seconded the President's suggestion to the Soviets for international cooperation instead of the "space race," which Smith had originally advocated. Smith discussed three levels of attack on the manned lunar landing program. First, a practical view held that the investment of money and talent in Apollo was out of proportion to foreseeable benefits. Warren Weaver, Vice-President of the Arthur P. Sloan Foundation, had discussed the many alternatives for educational use of the $20 to $40 billion that the moon race was expected to cost. Second, some scientists who were enthusiastic about space exploration feared that Apollo and other man-in-space programs would swallow up the funds that could go to unmanned programs, which they saw as more efficient gatherers of scientific information. Third, a growing number of scientists had reached the conclusion that no appreciable benefits of any sort would come from the Apollo program. Philip Abelson, Director of the Carnegie Institution's Geophysical Laboratory and editor of *Science*, the journal of the American Association for the Advancement of Science, had recently conducted an informal survey and found an overwhelming number of scientists against the manned lunar project. "I think very little in the way of enduring value is going to come out of putting man on the moon—two or three television spectaculars—and that's that," Abelson stated. "If there is no military value—people admit there isn't—and no scientific value—and no economic return, it will mean we would have put in a lot of engineering talent and research and wound up being the laughing stock of the world." After discussing these three objections to the Apollo program, author Smith admitted that the most persistent justification for the moon race was the matter of prestige. He suggested continuing the space program but abandoning the "crash" timetable in favor of one that placed the moon in its perspective as one way-station in the step-by-step development of space. Apollo with a lower priority could provide benefits, while allowing periodic reappraisal.[40]

Kennedy's Last Visit

On 16 November 1963, President Kennedy made a whirlwind visit to Canaveral and Merritt Island, his third visit in 21 months. Administrator Webb, Dr. Debus, and General Davis greeted the President as his Boeing 707 landed. At launch complex 37 he was briefed on the Saturn program. The President then boarded a helicopter with Debus to view Merritt Island, and flew over the coast line to watch a successful Polaris launching from the nuclear submarine *Andrew Jackson*.[41]

The next week the President died by an assassin's bullet in Dallas. The new President, Lyndon B. Johnson, announced he was renaming the Cape Canaveral Auxiliary Air Force Base and NASA Launch Operations Center as the John F. Kennedy Space Center. With the support of Governor Farris Bryant of Florida, the President also changed the name of Cape Canaveral to Cape Kennedy. The next day he followed up his statement with Executive Order No. 11129. In this he did not mention a new name for the Cape, but did join the civilian and military installations under one name, thus causing some confusion. To clarify the matter, Administrator Webb issued a NASA directive changing the name of the Launch Operations Center to the "John

Fig. 35. Dr. George E. Mueller briefing President Kennedy in pad 37 blockhouse, November 1963. Note the periscopes. L to R: George Low, Kurt Debus, Robert C. Seamans, Jr., James E. Webb, Kennedy, Hugh L. Dryden, Wernher von Braun, Maj. Gen. Leighton I. Davis, and Senator George Smathers.

Fig. 36. Seamans, von Braun, Kennedy, November 1963.

F. Kennedy Space Center, NASA," and an Air Force general order changed the name of the air base to the "Cape Kennedy Air Force Station." The United States Board of Geographic Names of the Department of the Interior officially accepted the name Cape Kennedy for Cape Canaveral the following year.[42]

People at the Cape seemed to approve the naming of the spaceport as a memorial to President Kennedy. Up to that time, the Launch Operations Center had only the descriptive name. Debus wrote a little later: "The renaming of our facilities to the John F. Kennedy Space Center, NASA, is the result of an Executive Order, but to me it is also fitting recognition to his personal and intense involvement in the National Space Program."[43] Many in the Brevard area, however, felt that changing the name of Cape Canaveral, one of the oldest place-names in the country, dating back to the earliest days of Spanish exploration, was a mistaken gesture. After a stirring debate in the town council, the city of Cape Canaveral declined to change its name.*

Washington Redraws Management Lines

On 30 October 1963, NASA announced a revision of its Saturn flight program, eliminating manned Saturn I missions and the last 6 of 16 Saturn I vehicles.† NASA discarded the "building block" concept and introduced a new philosophy of launch vehicle development. Henceforth the Saturn vehicles would go "all-up"; that is, developmental flights of Saturn vehicles would fly in their final configuration (without dummy stages).

George E. Mueller, Holmes's replacement as Director of the Office of Manned Space Flight, made the "all-up" decision.‡ Mueller came to his new position from a vice-presidency at Space Technology Laboratories. STL provided engineering and technical assistance to the Air Force on its missile programs, including Minuteman, where the all-up concept was first employed. Despite some mishaps—the first attempt to launch a Minuteman from an underground silo at the Cape (30 August 1961) had resulted in a spectacular explosion—Mueller was confident that all-up testing would save NASA many months and millions of dollars on Apollo.[44] At the OMSF

*Although efforts to have Congress restore the name "Canaveral" to the Cape failed, Governor Reubin Askew signed a bill on 29 May 1973 that returned the name on Florida State maps and documents. On 9 October 1973 the Board of Geographic Names, U.S. Department of the Interior, did likewise for federal usage.

†The Saturn C-1, C-1B, and C-5 were renumbered Saturn I, Saturn IB, and Saturn V in 1963.

‡Pronounced "Miller." Holmes and Webb had clashed over the amount of NASA's funds that Apollo should receive. Holmes wanted to concentrate almost all of NASA's resources on the lunar mission while Webb, supported by Vice President Johnson, preferred a more balanced program that would provide a total space capability including weather, communications, and deep-space satellites. When President Kennedy sided with Webb, Holmes departed in mid-1963.

Management Council Meeting on 29 October 1963, Mueller stressed the need to "minimize 'dead-end' testing [tests involving components or systems that would not fly operationally without major modification] and maximize 'all-up' systems flight tests." Two other aspects of Mueller's all-up concept directly affected the Cape. The OMSF Director wanted *complete* (emphasis is Mueller's) systems delivered at the Cape to minimize KSC's rebuilding of space vehicles. And future schedules would include both delivery dates and launch dates.[45]

Two days after the Saturn announcement, NASA published a major reorganization that combined program and center management by placing the field centers under Headquarters program directors rather than general management. Previously, center directors had received project or mission directives from one or more Headquarters program directors, while direction for general center operations came from Associate Administrator Seamans. Following the 1 November reorganization, NASA gave the responsibility for both overall management of major programs and direction of NASA field installations to three Associate Administrators: Mueller, Raymond Bisplinghoff, and Homer Newell. The three Manned Space Flight Centers—Marshall, Manned Spacecraft, and KSC—would report to Mueller.[46]

KSC realigned its organization on 6 February 1964 to conform with the new NASA structure. At the same time, administrative and technical support functions were separated, in an attempt to strengthen both; and the number of offices reporting directly to Debus was reduced, with more authority and responsibility given to the assistant directors. Henceforth in the Office of Manned Space Flight at NASA Headquarters and in the three Manned Space Flight Centers, the functional breakout in all Apollo Program Management Offices would be: program control—budgeting, scheduling, etc.; systems engineering; testing; operations; and reliability and quality assurance. At KSC Rocco Petrone as Assistant Director for Program Management was also head of the Apollo Program Management Office.[47]

Data Management

On 29 October 1964, the year of the reorganization, in his weekly report to Debus, Petrone stated that his office was preparing a KSC regulation for implementation of the instructions received from Headquarters entitled "Apollo Documentation Instruction NPC [NASA Publication Control] 500-6." This instruction required the following action from each center: identification, review, and approval of all documents required for management of the Apollo program; "an Apollo document index," cataloguing all recurring interorganization documentation used by the Office of Manned Space Flight and the contractors; a "Center Apollo documentation index";

a "documents requirement list," listing all documents required from a contractor—this list would "be negotiated into all major contracts of a half million dollars or over" and would be part of the request for quotation; and a "document requirement description," classifying every item on the "document requirements list," its contents and instructions for preparation.[48]

This instruction, Petrone believed, could provide a strong management tool and eliminate many unnecessary documents. The procedure would classify and catalogue documents and make them readily available to anyone who had immediate need for them. It would force many contractors, especially those who had not previously dealt extensively in government contracts, to clarify in writing the exact nature of their roles in the Apollo program. Throughout the entire program, specific delineation of each phase would bring greater clarity to the respective tasks.

At various times Kennedy Space Center put out Apollo document trees—charts showing the relationship of key documents. On 3 November 1965, for instance, Petrone was to authorize the "KSC Apollo Project Development Plan" under three categories of documents: Apollo Saturn IB Development Operations Plan, Apollo Saturn Program Management and Support Plan, and Apollo Saturn V Development Operations Plan. Within the second category were ten areas of concern to management: program control, configuration management, reliability and quality assurance, vehicle technical support, administrative support, logistics, data management instruction, training, general safety, and instrumentation support. The other two categories had 31 and 44 topics respectively! Typical of those that appeared in both the Saturn I and Saturn V listings were the space-vehicle countdown procedures, the prelaunch checkout plan, and the launch operations plan.[49]

With even its paper work organized, the Debus team had come a long way from the Launch Operations Directorate of 1960 to the John F. Kennedy Space Center of February 1964. Many problems with the Air Force had been resolved, without undue antagonism resulting. Land on Merritt Island had been purchased for the manned lunar landing program, plans laid for launch facilities and an industrial area, and construction had begun. The center had recruited a roster of engineering and administrative personnel and devised a workable organization.

The new organization did not mean an improvement in every respect. It involved the development of a bureaucracy that was incompatible with the informal, personalized approach of the old days on the Cape. Then the engineers had inspected their instruments, worked on them, sometimes built them; they labored with their hands. Now, they monitored contractors.[50] Meetings with department heads and even the Director had been highly informal. Now secretaries scheduled the meetings, each of which required a detailed agenda, and division heads presided with the formality of a college

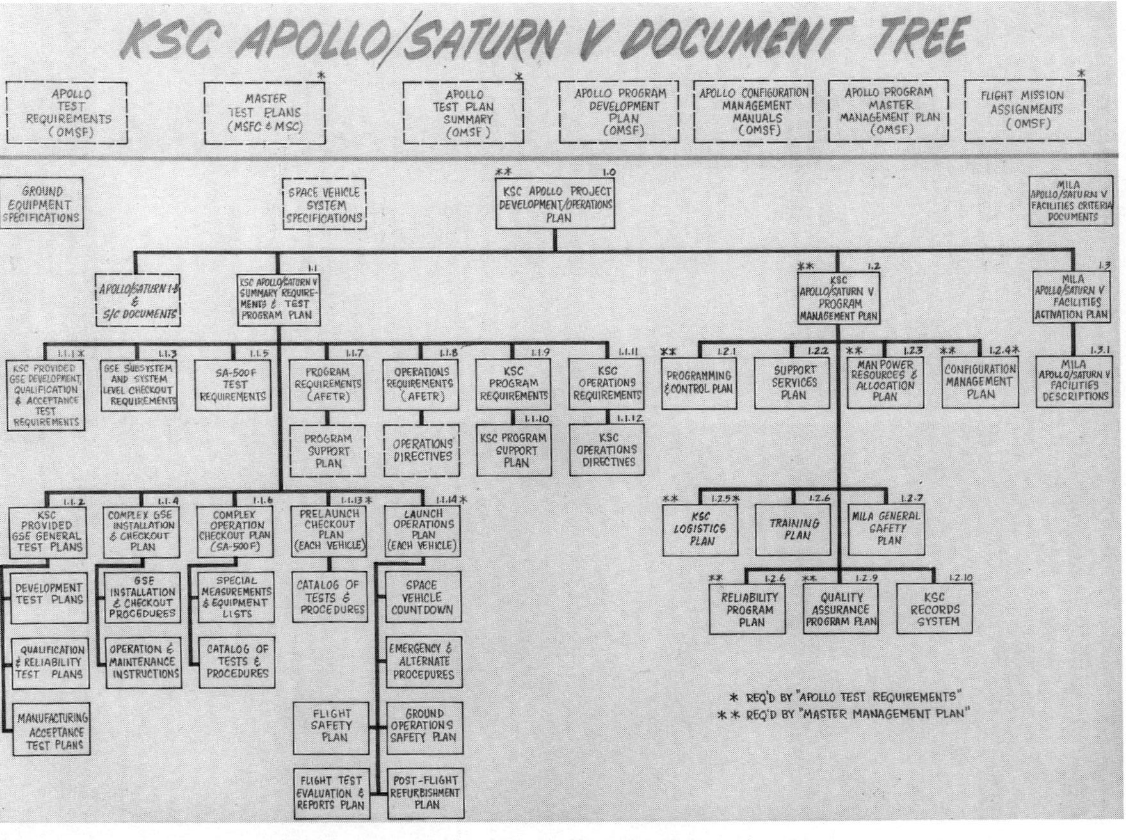

Fig. 37. A document tree for Apollo–Saturn V, December 1964.

dean at a faculty meeting. Because the men who launched rockets were a sentimental crew, there were frequent references to the good old times. But to launch a rocket that would put a man on the moon, they recognized, required an extensive organization.

8

Funding the Project

The Budgetary Process

Always a complicated process for a governmental agency, funding presented new mazes of complexity to the Launch Operations Directorate (LOD) during 1962. The normal budgetary process can be simplified as follows: study of needs for the coming fiscal year (this would ordinarily take place almost 12 months before the start of the fiscal year); presentation to the parent agency, which fits the request into its total proposal; submission to the Bureau of the Budget* for analysis and incorporation into the President's budget request, which is then tendered to Capitol Hill; hearings before congressional committees; discussion and votes within the committees; voting in both houses of Congress; perhaps a joint committee to resolve differences between House and Senate; an authorization act by Congress setting the limit for each item and the total amount; an appropriation act that establishes the actual amount of money the agency will receive; release of funds by the Bureau of the Budget; and, finally, disbursement of funds by the agency to its constituent subdivisions.

This intricate process was further complicated for LOD in 1962. First, the directorate was in process of evolving into an independent center—LOD became the Launch Operations Center (LOC) halfway through 1962 and halfway through this chapter. Second, it had to fend off a flanking attack from the Air Force to retain jurisdiction over the newly acquired land on Merritt Island (see pp. 98–104). Third, it had to plan and budget new facilities and equipment for a still undefined space vehicle to meet the Kennedy deadline of a moon landing "within the decade." And fourth, where it had been dealing in millions of dollars, now it had to request hundreds of millions. While the mobile concept had been accepted, the mode of transport—barge, rail, or crawlerway—had not been determined. In many instances, moreover, LOD had to telescope the work of several years into one, by forecasting the financial implications of a concept from the drawing board to end use.

*Office of Management and Budget since 1970.

One should remember, further, that the lunar landing program had not established itself as an unquestioned part of the American scene. It had to be defended continually. "People frequently refer to our program to reach the moon during the 1960s as a national commitment," Lyndon Johnson wrote. "It was not. There was no commitment on succeeding Congresses to supply funds on a continuing basis. The program had to be justified and money appropriated year after year. This support was not always easy to obtain."[1]

The preparation of project documents for budget submissions to Congress began with a statement of anticipated requirements in three categories: construction of facilities, research and development, and administrative operations. The administrative budget was easier to prepare because it changed less from year to year. The construction budget, for building new facilities or modifying existing ones, was the largest of the three for fiscal 1963 and 1964, when some 90% of the moonport construction was funded. Later, when construction neared completion, the research and development and administrative operations accounts rose sharply. This chapter will deal primarily with the budgets for fiscal years 1963 and 1964, and construction of facilities will therefore be the major topic. The LOD staff did most of the work preparing the documents each year, although NASA Headquarters could be counted on to send broad guidance—and frequent proddings.[2]

The budgetary cycle began, usually in the spring, with statements of requirements. This was no easy task since the lunar rocket changed repeatedly, progressing within a year's time from the Saturn C-2 to the C-3, the C-4, and the C-5. No one could foresee all required facilities and ground support equipment during early planning stages, although the facilities were easier to estimate than the support equipment. Such information as was available went to the Budgeting Office, and thence to the Facilities Office. Using the requirements stated by various elements of NASA and contractor firms, the Facilities Office put together the complete project documents.

A project document could cover a single facility or, as in the case of launch complex 39, a group of facilities. NASA policy demanded that each project document contain a complete statement of requirements necessary to begin operations. The document defined the scope of each requirement, including such specific factors as square footage, and justified the requirement and furnished cost estimates. The prescribed format called for five basic paragraphs covering real estate, site preparation, construction, equipment, and design; this was to be supplemented, when appropriate, with siting plans and sketches.[3] Under this procedure, the purchase and improvement of land, as well as the design, construction, and complete equipment of the facility located on it, could be dealt with in one budgetary action. The user, in theory at least, had only to walk up to the door of the completed facility, turn the key, walk in, and begin operations—a procedure that gained the label "turn-key concept."

Normally, the construction of facilities (CoF) budget included only those projects that would cost a quarter of a million dollars or more. Less expensive projects came under either the administrative operations or the research and development budget. The CoF budget funded projects within a given fiscal year—say fiscal 1963 starting 1 July 1962—rather than over several years, but the Directorate could actually spend the money over a longer period. The NASA Administrator had to approve exceptions to this policy, and did so only when the indeterminate nature of a facility rendered estimates on a fully funded basis impractical.[4]

Numerous launch schedules required different contractors and large numbers of individual structures or items of equipment. Not all of these needed to arrive at the same time, nor in the same fiscal year. Some projects had a long lead time. Air conditioning normally had to precede the installation of delicate computer equipment. The scheduling of events was thus a continuous and detailed task.

Planners had the difficult task of estimating costs of new and unprecedented facilities and ground support equipment. No one had built anything like the vertical assembly building or the launcher-umbilical tower (mobile launcher). The result in many instances had to be simply educated guesswork by LOD personnel, contractor engineers, and members of the Army Corps of Engineers who had worked on earlier, smaller projects.

Originally, the Resources Office, under the direction of C. C. Parker, submitted the budgets. But for the manned lunar projects, the governing influence on substantive matters during almost all phases of the programming and budgetary operations in 1962, as well as later, was Rocco Petrone, then chief of the Heavy Vehicle Systems Office. Since Petrone's office had to make sure that facilities and ground support equipment would be ready in time to meet the deadlines, he had an almost proprietary interest in the identification, cost, and justification of requirements.

When assembled into one package, the project documents constituted LOD's fiscal year construction of facilities program. That program was first submitted as a preliminary budget and later, after adjustments, as a final budget. After NASA Headquarters reviewed LOD's program and incorporated it with those from other installations, the total NASA budget went to the Bureau of the Budget, and then to Congress. During committee hearings, representatives from LOD sometimes testified on the requirements and costs specified in the project documents.

After passage of the authorization act, the Launch Operations Directorate usually submitted an updated series of project documents. These balanced the amounts of money authorized against requirements, taking account of changes that had occurred since the submission of the budget. LOD also revised these documents individually whenever changed requirements made adjustment necessary.

Using the information it received from LOD's submissions and relating it to the amount of money authorized by Congress, NASA Headquarters prepared a program that indicated the approximate amount of money it planned to release to LOD. Knowing that, LOD then prepared a Program Operating Plan that set forth its financial procedures and indicated how it proposed to use money within prescribed ceilings. After Congress passed an appropriation act, the Bureau of the Budget apportioned money incrementally and released it to NASA periodically according to phases of development or a time scheme. NASA Headquarters then released money to LOD at intervals for each project. As one official put it, Headquarters "spoon-fed" LOD. It rarely released all the funds appropriated for a project for a specific fiscal year during that year. The periodic method allowed the agency to spread the money as needed over several fiscal years. The process also involved the occasional transfer of funds from one budget line item to another and from one appropriation source to another. Congress placed a limitation on such transfers (usually 5%). NASA Headquarters tended to restrict itself further.[5]

Fiscal 1963

The history of the FY 1963 budget estimates for the construction of launch facilities, begun in late 1960 and continuing well past the beginning of the fiscal year itself, reflects the evolving organization, mission, and operational concepts of the Launch Operations Directorate. The initial estimates predated President Kennedy's announcement of the manned lunar landing program and had their basis in the Saturn C-1 vehicle program. Although these estimates did not include provision for a third Saturn launch complex, LOD suggested that it would need approximately $65 million should the number of launches increase enough to require a third complex. In such case the complex would be a duplicate of LC-37.[6]

In February 1961, NASA Headquarters called for preliminary FY 1963 estimates based on the ten-year plan approved by the Administrator. The LOD portion was to cover only the support services furnished all NASA activities and projects at the Atlantic and Pacific Missile Ranges.[7] The President's 25 May 1961 announcement, however, altered the tempo and direction of planning, as did NASA's subsequent selection of Merritt Island as the site for the manned lunar landing program and the change in plans from the C-2 vehicle to the C-3.[8]

While the Debus-Davis Report of July 1961 (pp. 80, 89) had concerned itself chiefly with the selection of a launch site for Apollo, it proved to be a key document also in fiscal planning. In a series of meetings during the hectic month of July 1961, LOD personnel submitted detailed budgetary figures on their areas of responsibility to Petrone's office. This they were able to do, based on their experience with previous programs. Bertram Greenglass consolidated and qualified the final report; in doing so, he foreshadowed the role he would later play as Petrone's alter ego on program control matters. A 1955 graduate of New York University, Greenglass had begun his association with rocketry at Redstone Arsenal in 1956. His rise from Army Private First Class to a high NASA position by the age of thirty was meteoric.[9] When Petrone moved up to Apollo program management for launch facilities, Greenglass would serve as his comptroller, handling contract management, manpower, and funding.

The decision, announced in August 1961, to acquire new land for the lunar program mandated a revision of the FY 1963 program for construction of facilities. Intensive planning marked the remaining months of 1961. NASA Headquarters applied pressure on LOD, particularly in the form of frequent telephone calls, to produce FY 1963 project documents for budgetary purposes. The Facilities Office, responsible for engineering and construction, prepared the CoF project documents.

While its own planning continued apace during September and October, LOD held frequent meetings with Air Force representatives of the Atlantic Missile Range. Using the Debus-Davis Report as a guide, this joint group developed a range development plan for the lunar program. The plan contained rough cost estimates for support facilities, but did not include requirements for a new Saturn launch complex.

Following these meetings, LOD staff sections held a series of lengthy meetings of their own during November and December. Using the range development plan as a basis, LOD refined the estimates for support facilities and also developed requirements for the advanced Saturn launch complex. Based on the technical data that emerged from both series of meetings, J. F. Burke and C. J. Hall of the Facilities Office developed and evaluated a series of project documents for the FY 1963 CoF program. Bidgood, Parker, and Petrone approved these documents before passing them on to Debus and Huntsville for approval en route to Washington.[10]

On 13 December, LOD gave NASA Headquarters some of the details of its FY 1963 requirements for the advanced Saturn launch complex, based on "the presently known C-3 vehicle," but capable of handling larger vehicles at increased cost. The estimate for the launch complex reached $167 million, exclusive of land acquisition. The proposed complex consisted of three major operating areas: a vertical assembly and checkout area, an intermediate area, and a launch area. Major requirements included a vertical assembly building, a launch control center to be located within the VAB, a transport system, a stationary ordnance arming tower, and two launch pads.[11]

Reorienting itself to the Saturn C-5 program, and considering that NASA had not yet chosen between the three mission modes, LOD in early 1962 redefined its CoF program for the next fiscal year and prepared 14 detailed project documents, with cost estimates and justification for each. All of the facilities and ground support equipment described in the project documents were still in the study or design phase; and much of the technical data furnished in the budget, though based on the best information available at the time, later proved unsatisfactory.

These 14 documents, constituting LOD's total construction of facilities budget for FY 1963, asked for $359 963 000. Eight of these 14 requests, representing 98% of the total, pertained directly to the manned lunar landing program.[12] The largest single item sought $176 550 000 for launch complex 39. LOD stated that it would "provide the necessary capability for launching the Advanced Saturn vehicle."[13] Yet the huge outlay for LC-39 represented only about 40% of the total complex cost, and covered only long-lead-time items that had to be started promptly to meet operation dates.

Subcommittee Hearings at the Cape

NASA's budget went to Congress in February 1962. A month later, the House began hearings. Before the end of March, Congressman Olin E. Teague (D.-Tex.) decided to take his Subcommittee for Manned Space Flight of the House Committee on Science and Astronautics to Florida for on-the-spot hearings on 23 March. Teague wanted to "educate the subcommittee's members" and to "attempt to justify the money that's spent here before Congress."[14] NASA Headquarters and Houston representatives also attended the hearings, along with officers of the Air Force Missile Test Center.

After Debus outlined the Launch Operations Directorate's organization, mission, and operational concepts, Petrone described the FY 1963 funding requirements, together with total facility requirements for the manned lunar landing program. The initial reaction of the subcommittee was that LOD should spread its budget requests over several fiscal years. Some subcommittee members questioned the basis for LOD's budget figures. "You've got a great big ball of money, and it is very easy for someone to come along and cut it, really cut it," one unidentified congressman observed.[15]

The subcommittee members deliberately asked pointed and critical questions to fortify themselves so that they could justify the budget before the appropriations committee later on. In the day-long conference, the subcommittee stressed saving money, and the Directorate emphasized precise scheduling. "I'm sure the Doctor [Debus] feels that we are friendly," Chairman Teague justifiably remarked, for he was one of the most influential friends the space program had in Congress. "We don't want to delay this program one minute. . . . If you can give us your program timing . . . I think we can pave a smooth road to the appropriations committee." But he added, "If we don't take any action, I think the appropriations committee will."[16]

The estimated total cost for LC-39, including FY 1963 and later increments, Petrone told the subcommittee, was $432 million. LOD was trying "to evaluate, not sell," the program. The budget figures were honestly arrived at. "We've got to live with them for years to come," Petrone declared.[17] Program timing was based on schedules that had to be met for the manned lunar landing program and had to be responsive to NASA Headquarters schedules. To the men of LOD, time was critical. To the congressmen, however, the amount of money spent in fiscal 1963 was the critical issue.

Debus explained that the LOD budget was made out against scheduled facility completion dates and launch schedules. These provided a little leeway for some slippage, but on requirements that were pressing, slippage "would hurt very, very much." They wanted launch complex 39 ready by January 1965, Petrone said, since they hoped to launch the first Saturn C-5 in March or April of that year. As an example of facility scheduling, Petrone said that LOD expected the erection of steel for the VAB to begin in March 1963.[18]

The programming of funds tied in so closely to the scheduling of facilities that a slippage in one resulted in a slippage of the other, and the hard fact was that Congress rarely appropriated funds in time for use at the beginning of the fiscal year. In response to a subcommittee question as to when LOD began receiving funds after the fiscal year started, Petrone answered that in the preceding year it had been October. Drawing upon his

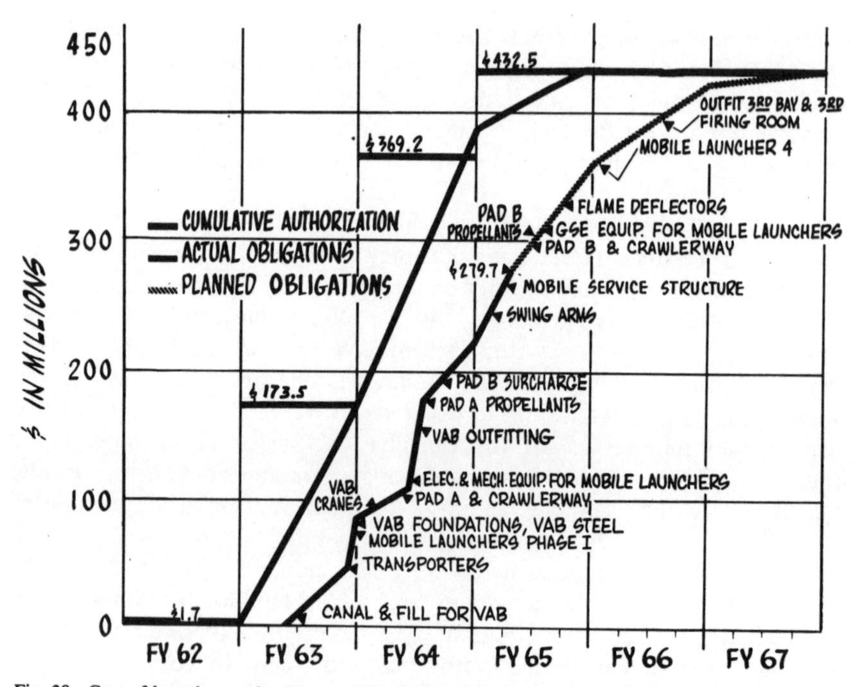

Fig. 38. Cost of launch complex 39, as of November 1964. Authorizations (by Act of Congress) are on the left line. NASA obligations (contracts, purchase orders, etc.) are on the right, with the latter half being predicted.

long experience in construction, Colonel Bidgood added that he had never seen money "hit the market before the first of October." Contingency authority from the Appropriations Committee helped little in new or increased programs, since such authority permitted expenditures only for normal operating costs at constant rates. LOD had to have funds on hand before awarding construction contracts.

The subcommittee asked for a comparison of relative costs between mobile and fixed facilities. As against the $432 million for a mobile complex with four pads and a launch rate of 36 per year, Petrone said, fixed facilities would require nine pads costing $900 million. Additionally, mobile facilities made possible significant savings in manpower costs both in LC-39 and in the industrial area, even with a launch rate of only 24 per year. The biggest savings came in the reduction in the number of supervisors and other personnel at the higher grade levels. Dollar savings in manpower, the subcommittee observed, were the strongest argument for mobile facilities.

Subcommittee members then mentioned the possible impact of new developments, such as the atomic rocket motor, on the design of LC-39 facilities. Rather than sink a lot of money into this complex, might it not be better to wait and see what the future held in the next five years, and thereby save money in FY 1963? Debus explained that one of the basic decisions made early in the planning stages was to base the design on "the state-of-the-art and its most likely development." One of the basic presumptions was the use of liquid propellants. LOD had to be ready by 1965 and could not wait on the possibility of new developments. With the existing state-of-the-art, LOD could be ready in 1965. The subcommittee's view was that no decision should be so binding as to deny LOD flexibility to take advantage of new technology.[19]

In response to a question about the Department of Defense's role in funding the manned lunar landing program support facilities, Debus explained that LOD was limiting its funding to the new Merritt Island area and to LC-34 and LC-37 on the Cape. NASA Headquarters and the Department of Defense would coordinate downrange stations, including ships. LOD was coordinating other support requirements for the Merritt Island area with the officials of the Air Force Missile Test Center. Two of these support items—utility installation in the new area (causeways, roads, water, and power) and launch-phase range instrumentation—although in effect Air Force requirements, were included in LOD's budget. The range instrumentation that LOD would fund, in agreement with AFMTC, extended to a radius of 105 kilometers. NASA Headquarters would fund data acquisition and tracking requirements beyond that distance. So far as cost-sharing for maintenance and operations of the two areas was concerned, LOD was to handle funding on Merritt Island and the Air Force on the Cape.

When Dugald Black of the Manned Spacecraft Center presented facility requirements for checking out and testing the Apollo spacecraft, members of the subcommittee interrupted his presentation several times with questions regarding quality control and the overlap of functions and facilities. The MSC representatives explained that MSC would develop the spacecraft at Houston, but would check and test it at Merritt Island. Just as the Mercury and Gemini programs had overlapped in some instances, so the Gemini program would overlap Apollo. Facilities to support the Apollo and Gemini spacecraft had to be available simultaneously. The scheduling of the operations and checkout building, for example, was extremely tight. It had to be finished early enough to install and inspect equipment before the spacecraft arrived in October or November 1963.[20]

The scope and expense of the checking and testing requirements for the spacecraft led one member of the subcommittee to question the program, at least momentarily. Debus reminded him that "this is a research and development facility and is only slightly operational." This prompted Chairman Teague to observe:

> That doesn't take away the argument that this entire operation is an expensive one. It gets more expensive with the buildings and personnel, and everything swelling in size. Perhaps we are trying to do too much in too much of a hurry. If we are subject to so many changes in so many places so that we are watching every nut and bolt right up to the time we are ready to shoot. . . .[21]

Teague's cautious remark ended the long, productive day.

Progress in Washington

In response to a March 1962 request from NASA Headquarters, LOD reviewed its entire fiscal 1963 program. In May six projects were identified that required an early release of funds to avoid slippage in construction schedules: modifications to launch complex 34, support facilities in the Cape area, Apollo mission support facilities, launch complex 39, advanced Saturn or Nova support facilities, and utility installations in the Merritt Island area. Two months later NASA Headquarters was to issue allotments from fiscal 1962 construction of facilities funds for the final planning and design of these projects.[22]

On 23 May the House voted 343–0 to authorize a NASA budget of $3 671 million for FY 1963 and an additional $71 million for FY 1962. In the Senate, William Proxmire of Wisconsin offered several amendments to the authorization bill. One called for competitive bidding practices to the "maximum practicable extent."* The Senate rejected it by a vote of 23 to 72. He then called for the establishment of a Space Manpower Commission to assess the impact of the lunar landing program on the nation's supply of scientific personnel. The Senate defeated this, 12 to 83. In the end, by voice vote, the Senate authorized a NASA budget of $3 750 million. After the differences between the two houses were resolved, Congress passed the NASA Authorization Act; it totaled $3 744 million. The Launch Operations Center was allowed $328 333 000 for construction of facilities in fiscal year 1963, including $173 550 000 for launch complex 39. The total was $3 000 000 less than the center had requested.[23]

Because the congressional authorization was less than the amount sought by NASA, and because of newly generated requirements, the Office

---

*This would later prove ironic when competitive bidding cost a Wisconsin firm a contract that had seemed to be securely in hand (p. 272).

of Manned Space Flight proposed that NASA request a supplemental FY 1963 grant of $70 million for construction of facilities. The supplemental, D. Brainerd Holmes, head of the Office of Manned Space Flight, felt, was the only possible way to hold to existing schedules. To support this proposal, the Launch Operations Center—now an independent field installation with its own budgeting agency—prepared another series of project documents and forwarded them on 8 September 1962 as its FY 1963 CoF Resubmission and Supplemental.[24] These documents reflected new or revised requirements that had come to light since the February budget submittal. In October Associate Administrator Seamans decided not to submit the request.

Updating LC-39 Requirements

As a separate action, LOC submitted to NASA Headquarters the same project document for LC-39 that had been prepared for the "resubmission and supplemental," since it reflected updated requirements and priorities that demanded prompt accommodation. Based on a reevaluation of the scheduled launch rate, the project document provided for a redistribution of funds among line items, but no additional funds, and asked for an early allotment.[25] The updated LC-39 document reduced the number of high bays in the VAB from six to four (but with provisions for subsequent expansion) and changed the arming tower from a stationary to a movable structure. It substituted two crawler-transporters for the two rail-mounted launcher-transporters and introduced requirements for the special roadways needed by the crawlers. Because of the larger size of the Saturn, LOC increased the elevation of the launch pads to 12 meters and the pad diameter from 610 to 915 meters.

Trimming the VAB, after reevaluating the scheduled launch rate, lowered the FY 1963 incremental estimate from $92 882 000 to $75 590 000. The adoption of the crawlerway system reduced the transfer estimate by $3 million; however, making the arming tower movable increased the cost of its prime mover, the crawler-transporter. The figure for each of the three proposed pads almost tripled, going from $5 588 000 to $15 930 000, largely because of higher elevation and increased diameter. The largest addition, $22.8 million, would go for two launcher–umbilical towers, plus steel for a third. The changeover from the rail-mounted integral units to the dual unit (mobile launcher and crawler) accounted for most of the addition.

In submitting this updated project document for LC-39, LOC pointed out, as it had in its budgetary submission, that the FY 1963 funding covered only those facilities required to meet the initial phase of the launch schedule, and that it would request funds to complete LC-39 in subsequent fiscal years. Only the VAB remained from the earlier list. The new list included the crawler roadway, the crawler-transporters, and the launcher-umbilical towers. LC-39 was to provide the necessary capability to launch the Saturn C-5 vehicle and other advanced configurations such as the C-1B.

In summation, then, the updated project document for LC-39 requested FY 1963 funding for portions of the VAB, one launch pad, the crawler roadway from the VAB to one launch pad, two crawler-transporters, two umbilical towers and steel for a third, and a number of minor items. It also included study, design, and initial procurement of checkout and control equipment, flame deflectors, firing accessories, instrumentation and connecting cabling, general support equipment, and design of the arming tower. It did not list a separate figure for the launch control center since designers still planned to place it within the VAB.[26] LOC communicated its plan for the funding of LC-39 facilities with FY 1963 construction of facilities money to NASA Headquarters on 18 September 1962.[27]

A week later, both Houses of Congress approved the NASA appropriation. Public Law 87-741 appropriated $3.67 billion for FY-63. NASA's actual expenditures during fiscal 1963 were to total $2.55 billion, less than half what the government spent on agriculture, and $200 million less than spent on public assistance programs. The NASA figure represented 2.75% of the total national budget expenditures of $92.6 billion.[28]

Even though the appropriations bill became law on 3 October 1962, the Launch Operations Center could not begin to award contracts until NASA Headquarters allotted the funds. In late October, Debus addressed a rather sharply worded letter to D. Brainerd Holmes, the Director of OMSF, about the delay in the receipt of fiscal 1963 construction of facilities funds at the Launch Operations Center. Debus pointed out that nearly $10 million in construction had proceeded through design, advertising, bidding, and the selection of contractors, but that contracts could not be awarded until Headquarters released the funds. There appeared to be "insufficient effort at the Washington level," Debus felt, "for the prompt forwarding of funds to LOC after Headquarters received them from the Bureau of the Budget."[29]

The letter had the desired effect. On 6 November, Associate Administrator Seamans officially approved initial funding for construction of facilities at launch complex 39 to the amount of $22 080 000. On the same day, Frederick L. Dunlap, Chief of Budget Execution at Headquarters, formally transferred funds to the Launch Operations Center. All was not yet well, however. Seamans had approved $4 780 000 for site development and utility installations; $11 000 000 for equipment, instrumentation, and support systems (specifically for two transporters); and $6 300 000 (plus $500 000 previously allocated from fiscal 1962) for design and engineering services. The document approved no funds for facility construction and modifications.[30] Finally, on 27 December, a teletype message from Headquarters notified Lewis Melton, Chief of the Launch Operations Center's Financial Management Office, that fiscal 1963 funds allotted for launch complex 39 now totaled $167 850 000.[31]

Thus, as calendar 1962 came to an end, the Launch Operations Center had the money to start construction on launch complex 39. It had already put some of the money to use. A NASA-DoD intragovernmental purchase order on 13 November had provided funds to the Corps of Engineers for site preparation and design and engineering services, including the design of the vertical assembly building.[32]

Changing requirements and priorities for LC-39 made further adjustments necessary in the distribution of money for particular items. Since LOC could carry forward construction of facilities funds to subsequent fiscal years, it continued to update its FY 1963 CoF program, often reprogramming some construction for later fiscal years. Actually, the FY 1963 CoF account was to remain active through calendar 1968. Both the redistribution and reprogramming actions required congressional notification, a much simpler procedure than the lengthy budgetary process. LOC's financial planners worked simultaneously on several fiscal year CoF programs. Between September 1962 and January 1963, LOC transmitted to NASA Headquarters the aborted FY 1963 supplemental, the FY 1964 CoF budget, and preliminary estimates for FY 1965. It was in 1962, in fact, that LOC did most of its budgetary homework to obtain the appropriations needed for the later construction of lunar launch facilities. The shift in emphasis from design to construction would not be apparent until mid-1963.

During this period of intensive budgetary preparation, the Office of Manned Space Flight took a major step toward determining the launch lineup. On 15 October 1962, after two years of considering various proposals, NASA announced its first "official flight schedule" for the Saturn vehicles. The first Saturn C-1B would go up in August 1965, and the first Saturn C-5 in March 1966; the first manned Saturn C-1B in May 1966, and the first manned Saturn C-5 in June 1967. Compared to earlier assumptions, this set back both initial manned launches by about six months. It also set the 1966 launch rate at five Saturn C-1B vehicles (three of them manned) and four developmental launches of the Saturn C-5. For 1967 the launch rate was four Saturn C-1Bs (all manned) and six Saturn C-5s (four manned).[33] "Until further notice," Holmes announced, "these schedules were to be used by

OMSF and the Centers for scheduling and financial planning." Meanwhile, LOC was getting ready to submit its project documents for the FY 1964 budget.

The Fiscal 1964 Program

Preparation of the FY 1964 budget, like that for FY 1963, had begun before the Launch Operations Directorate became an independent installation. During March 1962, NASA Headquarters called for the FY 1964 preliminary budget and in August began issuing guidelines for detailed estimates. In November 1962, two months after forwarding its FY 1963 supplemental request, LOC submitted its CoF project documents for the FY 1964 budget, salvaging some of the requirements stated in the FY 1963 supplemental. The Office of Manned Space Flight programs included 20 individual project documents, some for new requirements. The FY 1964 submittal was based on the 15 October flight schedules.[34]

As in FY 1963, most of the money LOC requested for its FY 1964 CoF programs ($333 130 600) was for the manned lunar landing program. LC-39 alone accounted for $225 967 000, or about two-thirds of the total request, with other Apollo requirements, such as support facilities, LC-34, and LC-37, making up a substantial portion of the remainder. Combined with the previous fiscal year CoF request, the FY 1964 program brought the total request for LC-39 to $339 517 000.

The months of intensive study of the lunar program requirements enabled LOC to state its requirements in considerable detail and, for the first time in an LC-39 project document, to include a description of the mobile concept. Of the eight major facilities for LC-39, FY 1963 construction funds had provided for the basic VAB structure (including an area set aside for the launch control center). In FY 1964 funding, LOC requested funds for outfitting the VAB (including the launch control center), two additional launch pads with associated facilities on complex 39, extension of the crawlerway to the additional pads, three additional launch umbilical towers (less steel for one funded in FY 1963), propellant services, and the mobile arming tower, as well as minor modifications and additions at launch complexes 34 and 37.[35]

Many projected buildings in the Merritt Island industrial area, such as the headquarters building, budgeted for $9 309 000, were simpler to design than the facilities of LC-39. There were exceptions: the central instrumentation facility, which consisted of two buildings—a large structure in the industrial area and an auxiliary structure located about a mile north to avoid radio interference from equipment operating in the primary structure. LOC asked $31 508 000 for these facilities. Almost three-fourths of this would go for equipment, principally for telemetry and tracking.[36]

Another complex of major importance, the technical and support facilities needed by the Manned Space Center for preflight operations with the Apollo spacecraft, presented two big questions: funding and siting. The Manned Spacecraft Center of Houston wanted to include the complex in its budget, but LOC demurred. Originally planned for the Cape area, the two centers agreed on 28 August 1962 to site the facilities in the Merritt Island industrial area. This complex, initially called "Apollo Mission Support Facilities," consisted primarily of six structures: one for operations and checkout, the others for various support systems.[37]

LOC had requested $22 510 000 for them in FY 1963. Plans at that time restricted them to the Apollo spacecraft. During August and September 1962, changes in the Apollo and Gemini schedules and in the choice of spacecraft fuels led to a reevaluation of the need for separate Gemini spacecraft facilities. This resulted in the combining of some Apollo and Gemini requirements.[38] On 15 October 1962 the Manned Spacecraft Center submitted to LOC an estimate of $23 273 983 for these buildings. LOC forwarded the proposal to Headquarters on the same day.[39]

With the submission of its FY 1964 budget on 1 November 1962, the Launch Operations Center had accomplished the basic budgetary tasks for the construction that would be required by the manned lunar landing program.[40]

"What Is It Going to Cost?"

The Manned Space Flight Subcommittee of the House Committee on Science and Astronautics held hearings on the FY 1964 program for the construction of launch facilities in May 1963. Only representatives of NASA Headquarters were present to defend LOC's program. In his formal questioning for the subcommittee, Lt. Col. Harold A. Gould, a technical consultant to the House committee, focused on costs.

"A total of $444 million had already been made available for LC-39 and support facilities," Gould observed. With 40% of LC-39 programmed for fiscal 1963 and 50% of it being programmed for fiscal 1964, Gould asked, "What is the total cost of this complex going to be?"

William E. Lilly, Director of Program Review and Resources Management of NASA's Office of Manned Space Flight, estimated that the total complex would run "very close to a half billion dollars." He then furnished an "exact figure" of $481 576 000. Congressman Emilio Q. Daddario, the subcommittee's acting chairman and long-time watchdog of LOC's budget, disregarded Lilly's "exact figure" and converted the "close to a half billion dollars" statement to "over $500 million," a phrase he used repeatedly in pressing his questions. In 1962, Daddario reminded Lilly, his subcommittee had received an initial estimate closer to $400 million. In January 1963 while at Cape Canaveral, the subcommittee had heard an estimate of $432 million. "Now you give us an estimate of over $500 million," Daddario stated. He wanted to know why the current figure was "over $100 million beyond that originally estimated."[41]

Aided by Capt. John K. Holcomb, NASA's Assistant Director for Launch Operations, Lilly marshalled several answers, including the adoption of the crawler transfer system in lieu of the rail system, an increase in the size of the launch pad and the number of pads required, and NASA's turnkey procurement policy. Daddario saw most of these explanations as being more valid for FY 1963 than for FY 1964 estimates, and brushed them all aside. Lilly offered the further explanation that, in order to be able to present a firm estimate as soon as possible, NASA had stressed the need for advance design funds. Daddario showed little regard for what he called Lilly's "cloudy logic," pointing out the estimate had gone from $432 million to "over $500 million" within a matter of months.

"I can't really give you a definitive answer," Lilly confessed, "of why the difference between $400 million, $432 million, and $500 million." It helped matters little for Lilly to add: "Our estimate, of course, is always based on the best information that is available. I could not say that the $500 million will be the final figure." Holcomb added that as a result of having actual designs and firm design criteria, "now we know pretty much what we are planning to do." But Daddario would not be assuaged and asked whether every starting estimate given to the subcommittee was going to be 25% out of line.

Daddario said that he and other members of the subcommittee expected some changes from original estimates but were less concerned about the amount than the percentage of increase and the embarrassment of having to report this to the full House. Congressman Edward J. Patten suggested that a 12% increase would not be too far out of line, but when costs increased 20–30% "this committee finds itself then in an embarrassing position of explaining this increase to the other members of Congress. I doubt that they will take the explanation you have given us as being a proper one."

As a final thrust, Lilly said that he had some doubts that the $432 million figure given to the subcommittee at Cape Canaveral in January 1963 was the "officially approved estimate" of the Office of Manned Space Flight. Daddario parried by asking why, if a higher figure had been available in January, it "was not given to us at that time."[42]

Only slightly less tenacious was Florida Congressman Edward J. Gurney's questioning regarding the pace of committing and obligating FY 1963 funds.* NASA representatives told the subcommittee that as of 31 March 1963, NASA had committed only $38 million and obligated only $18.9 million for LC-39, out of the FY 1963 budget of $163.5 million. Gurney wanted to know why $217 million was needed for FY 1964 when "you haven't even been able to scratch the surface on last year yet," even though the fiscal year was nearly over.

The basic delay in obligations, Lilly explained, was the time required for design. Once the design was completed, the "big money" would go out for construction. "I think you will find that the money will move much faster from this point on," Lilly assured Gurney. NASA would obligate the remainder of FY 1963 funds for LC-39 by August 1963. NASA had laid out its plans for the obligation of funds month by month, and by the end of FY 1964 only $10 million of the combined FY 1963 and 1964 funds would remain unobligated.[43]

Testimony regarding LC-39 next centered on the number of pads and their cost. Colonel Gould asked Lilly to explain why the FY 1963 figures for LC-39 varied from those shown in the FY 1964 budget. Lilly answered that the revised figures were the result of a more comprehensive analysis of operational requirements and that NASA had adjusted figures for equipment, instrumentation, and support systems after completing engineering studies.[44]

Other information given to the subcommittee on the FY 1964 program indicated that NASA was still thinking of on-pad time in terms of "possibly one week"; that each mobile launcher (which Holcomb aptly described as "partly launch pad and partly umbilical") would cost about $12 million, compared with $1 million for the less complicated umbilical towers used on complexes 34 and 37; that five launchers were required in order to service four bays in the VAB and to provide time for refurbishing after each launch; that the cost for the design and engineering of LC-39 would be roughly $37.6 million; that about 35% of the items in the FY 1964 increment of facilities were under design as of May 1963; that the operational target date for bay 1 and pad 1 in LC-39 was 1 December 1965; that facility construction lead times for FY 1964 were 25 months; and that the estimated cost of the crawler roadway was $982000 per kilometer, with almost 13 kilometers of roadway required from the VAB to three pads.

At the subcommittee's request, Holcomb explained the implications of an operational capability date of 1 December 1965. It meant, Holcomb

---

*Funds were *committed* when financial management certified that funds were available and would be reserved for a particular purpose. Funds were *obligated* when a contract was signed for specific work to be done. The former was internal to NASA, the latter was legally binding on the agency.

said, that the construction of the first bay and its initial outfitting had to be completed by the end of May. Between May and December, an extensive checkout of the complete facility was to be made. "When we say that we have an operational capability beginning in December, we mean at that point we are able to bring in the first flight article, put it on the [mobile launcher] in the building, and check it out for our first launch in early 1966."[45]

The House hearings made clear that many problems regarding launch facilities for the Apollo program still confronted NASA. Most of the projects for which LOC requested FY 1964 funding, as well as the projects for which LOC had obtained FY 1963 funds, had undergone such drastic revision, when individually updated beginning in late 1962, that a discussion of them in terms of fiscal year budgets became academic. Through reprogramming actions, NASA postponed some of the construction requirements originally proposed for FY 1963; others, proposed for subsequent years, were paid for with FY 1963 funds. As a result, the year in which construction was budgeted often bore little relationship to the year of actual construction. With the passage of time, the budget documents diminished in importance as a barometer of actual construction. Instead, such documents as program operating plans and the periodic reports of the Corps of Engineers became the real indicators of construction.

Extensive criticism of NASA marked the congressional discussion of the FY 1964 budget for the first time since the agency's creation in 1958. Most barbs flew at the moon program, as congressmen argued that the Soviets seemed to have lost interest in a moon race, or that certain contractors were moving too slowly. Many Republicans thought the moon program detracted from more important military objectives in space. A Senate GOP policy committee stated on 10 May: "To allow the Soviet Union to dominate the atmosphere 100 miles above the earth's surface, while we seek to put a man on the moon could be . . . a fatal error." General Eisenhower had, as President, denied the existence of a "space race." Now he stated on 12 June 1963 that "anybody who would spend $40 billion in a race to the moon for national prestige is nuts." At a hearing of the Senate Aeronautical and Space Sciences Committee on 10 June, Dr. Phillip Abelson of the Carnegie Institution repeated the contention of many scientists that manned space exploration had limited scientific value. He thought its alleged importance utterly unrealistic. The rush to get to the moon, Abelson insisted, took scientific resources that the nation might use more wisely on other important objectives, and thus lessened our national security.[46]

In spite of this attack on the lunar program and several attempts to reduce the budget by amendment, the Senate by a voice vote and the House by a vote of 248 to 125 authorized $5.35 billion for NASA on 28 August

1963. During fiscal 1964, NASA was actually to spend $4.17 billion—a billion and a third less than either Agriculture or Health, Education, and Welfare. The NASA expenditure represented only slightly over 4% of the total national budget expenditures.[47]

Between authorization and appropriation, President Kennedy spoke before the United Nations General Assembly and suggested a joint U.S.-Soviet voyage to the moon. In spite of his assurances to Representative Albert Thomas of Texas, the chairman of the subcommittee considering NASA's budget, that to be able to deal from a position of strength the U.S. should continue the space program, not all members of Congress agreed. Senator Fulbright proposed a 10% cut for NASA in view of the needs of education and welfare—but lost. Senator Proxmire sought to strike out a $90 million addition made in committee, and this time won by the margin of 40–39.

As finally approved by both chambers on 10 December 1963, less than three weeks after President Kennedy's assassination, the bill appropriated $5.1 billion to NASA for fiscal 1964 and barred use of funds for joint lunar expeditions with any other country without congressional approval. President Johnson signed the bill on 19 December with reservations about the joint venture proviso. He thought it unnecessary and asserted it would impair our flexibility.

The manned lunar landing program had gotten through its most difficult Washington summer.

9

Apollo Integration

An Integration Role for General Electric?

To Congress, the moon program meant money. To the American people, it was a contest of American skills pitted against the Russians or the mysteries of space. But for NASA, perhaps the biggest challenge was organization.

In retrospect, one of the major reasons for the program's success was the ability of a lieutenant colonel and an aerospace engineer to sit down and work out a solution to a problem that, coupled with a few thousand more solved problems, could put a man on the moon. But someone had to get the lieutenant colonel and the engineer into the same room. This carried over on a far larger scale to the thousands of items of equipment that came together on the launch pad for the moment of truth at countdown. A fitting, designed in Huntsville and manufactured in California, had to connect precisely with a fitting designed in Texas and fabricated in New York. At KSC the ground support and electrical support equipment alone totaled more than 34 000 items. Each connection was an interface—eventually the most overworked word on Merritt Island—and keeping track of every interface, bringing together all the parts into a unified whole, was called integration.

Compared to earlier programs, Apollo-Saturn required drastically more coordination. During the 1950s, the Missile Firing Laboratory's contacts were limited to the Eastern Test Range, a few support contractors, and Huntsville. The Apollo program added scores of contractors, labor unions, and government organizations. The new relationships brought conflicts. There were differences of opinion with contractors and struggles for power among the NASA centers—divisive tendencies that were balanced by the unifying urge of the lunar goal.

NASA Headquarters, unable to handle the many integration requirements of Apollo by itself, sought help from an outside source—the General Electric Company. NASA asked GE to do three things: develop checkout equipment for launch operations; assess reliability, which was largely the reduction and analysis of data from various tests; and perform the

Organization	Function	Item	Program	Center	Headquarters
CHRYSLER		S-I/S-IB STAGES	SATURN I & IB	MSFC	NASA HDQRS MSF
NORTH AMERICAN AVIATION (ROCKETDYNE)		H-1 ENGINE	SATURN I & IB	MSFC	NASA HDQRS MSF
DOUGLAS		S-IV STAGE	SATURN I & IB	MSFC	NASA HDQRS MSF
PRATT AND WHITNEY		RL-10 ENGINE	SATURN I & IB	MSFC	NASA HDQRS MSF
DOUGLAS		S-IVB STAGE	SATURN I & IB	MSFC	NASA HDQRS MSF
NORTH AMERICAN AVIATION (ROCKETDYNE)		J-2 ENGINE	SATURN I & IB	MSFC	NASA HDQRS MSF
INTERNATIONAL BUSINESS MACHINES		INSTRUMENT UNIT	SATURN I & IB	MSFC	NASA HDQRS MSF
BENDIX		INSTRUMENT UNIT	SATURN I & IB	MSFC	NASA HDQRS MSF
RADIO CORPORATION OF AMERICA	COMPUTER	GROUND SUPPORT EQUIPMENT	SATURN I & IB	MSFC	NASA HDQRS MSF
GENERAL ELECTRIC	ELECTRICAL CHECKOUT EQUIPMENT	GROUND SUPPORT EQUIPMENT	SATURN I & IB	MSFC	NASA HDQRS MSF
BOEING	SATURN V LAUNCH VEHICLE INTEGRATION			MSFC	NASA HDQRS MSF
BOEING		S-IC STAGE	SATURN V	MSFC	NASA HDQRS MSF
NORTH AMERICAN AVIATION (ROCKETDYNE)		F-1 ENGINE	SATURN V	MSFC	NASA HDQRS MSF
NORTH AMERICAN AVIATION (SPACE AND INFORMATION DIV.)		S-II STAGE	SATURN V	MSFC	NASA HDQRS MSF
NORTH AMERICAN AVIATION (ROCKETDYNE)		J-2 ENGINE	SATURN V	MSFC	NASA HDQRS MSF
DOUGLAS		S-IVB STAGE	SATURN V	MSFC	NASA HDQRS MSF
INTERNATIONAL BUSINESS MACHINES		INSTRUMENT UNIT	SATURN V	MSFC	NASA HDQRS MSF
BENDIX		INSTRUMENT UNIT	SATURN V	MSFC	NASA HDQRS MSF
GENERAL ELECTRIC	ELECTRICAL CHECKOUT EQUIPMENT	GROUND SUPPORT EQUIPMENT	SATURN V	MSFC	NASA HDQRS MSF
RADIO CORPORATION OF AMERICA	COMPUTER	GROUND SUPPORT EQUIPMENT	SATURN V	MSFC	NASA HDQRS MSF
TELECOMPUTING CORPORATION	STRUCTURE SYSTEMS STUDY		SUPPORTING RESEARCH & TECHNOLOGY	MSFC	NASA HDQRS MSF
MIDWEST RESEARCH INSTITUTE	ENVIRONMENTAL CONTROL STUDY		SUPPORTING RESEARCH & TECHNOLOGY	MSFC	NASA HDQRS MSF
GENERAL ELECTRIC	STABILIZATION CONTROL SYSTEMS STUDY		SUPPORTING RESEARCH & TECHNOLOGY	MSFC	NASA HDQRS MSF
REPUBLIC	STRUCTURE SYSTEMS STUDY		SUPPORTING RESEARCH & TECHNOLOGY	MSFC	NASA HDQRS MSF
LOCKHEED	STRUCTURE SYSTEMS STUDY		SUPPORTING RESEARCH & TECHNOLOGY	MSFC	NASA HDQRS MSF
FAIRCHILD STRATOS CORPORATION	MICROMETEOROID EXPERIMENT			MSFC	NASA HDQRS MSF
GENERAL ELECTRIC		MISSISSIPPI TEST OPERATIONS	FACILITES	MSFC	NASA HDQRS MSF
BOEING		MISSISSIPPI TEST OPERATIONS	FACILITES	MSFC	NASA HDQRS MSF
AEROJET GENERAL/AETRON			FACILITES	MSFC	NASA HDQRS MSF
SVERDRUP AND PARCEL			FACILITES	MSFC	NASA HDQRS MSF
PAUL HARDEMAN INC.			FACILITES	MSFC	NASA HDQRS MSF
CORPS OF ENGINEERS			FACILITES	MSFC	NASA HDQRS MSF
LEAR SIEGLER			FACILITES	MSFC	NASA HDQRS MSF
TELECOMPUTING		SLIDELL	FACILITES	MSFC	NASA HDQRS MSF
MASON – RUST		MICHOUD	FACILITES	MSFC	NASA HDQRS MSF
AERO SPACE LINES		AIR TRANSPORTATION	LOGISTICS	MSFC	NASA HDQRS MSF
MECHLIG BARGE LINE		BARGE	LOGISTICS	MSFC	NASA HDQRS MSF
MILITARY TRANSPORT SERVICE		SHIPS	LOGISTICS	MSFC	NASA HDQRS MSF
UNITED STATES AIR FORCE				PARTICIPATING AGENCIES	NASA HDQRS MSF
UNITED STATES ARMY				PARTICIPATING AGENCIES	NASA HDQRS MSF
CORPS OF ENGINEERS MAPPING				PARTICIPATING AGENCIES	NASA HDQRS MSF
ATOMIC ENERGY COMMISSION				PARTICIPATING AGENCIES	NASA HDQRS MSF
UNITED STATES NAVY				PARTICIPATING AGENCIES	NASA HDQRS MSF
DEPARTMENT OF STATE				PARTICIPATING AGENCIES	NASA HDQRS MSF
DEPARTMENT OF DEFENSE				PARTICIPATING AGENCIES	NASA HDQRS MSF
FEDERAL AVIATION AGENCY				PARTICIPATING AGENCIES	NASA HDQRS MSF
UNITED STATES COAST GUARD				PARTICIPATING AGENCIES	NASA HDQRS MSF
FEDERAL COMMUNICATION COMMISSION				PARTICIPATING AGENCIES	NASA HDQRS MSF
UNITED STATES COAST AND GEODETIC SURVEY				PARTICIPATING AGENCIES	NASA HDQRS MSF
BELLCOMM MANNED SPACEFLIGHT SYSTEM – ENGINEERING					NASA HDQRS MSF

APOLLO PROGRAM GOVERNMENT – INDUSTRY FUNCTIONAL MATRIX

Fig. 39. The government-industry team behind Apollo.

NASA HDQRS MSF				
Center	**Element**		**Function**	**Contractor**
KSC/MSFC	GOSS		LAUNCH INFORMATION EXCHANGE FACILITY	AMERICAN TELEPHONE & TELEGRAPH
KSC	LAUNCH COMPLEX 39		LUT	PAUL SMITH CONSTRUCTION-INGALS IRON WORKS
KSC	LAUNCH COMPLEX 39		CRAWLER TRANSPORTER	MARION POWER SHOVEL COMPANY
KSC	LAUNCH COMPLEX 39		VAB	MORRISON-KNUDSON-PERINI-HARDEMAN
KSC	LAUNCH COMPLEX 39		VAB	U.S. STEEL
KSC	LAUNCH COMPLEX 39		VAB	BLOUNT BROTHERS
KSC	LAUNCH COMPLEX 39		OPERATIONS & CHECKOUT BUILDING	P. HARDEMAN & MORRISON-KNUDSON
KSC			LAUNCH OPERATIONS SUPPORT	CHRYSLER
KSC			LAUNCH OPERATIONS SUPPORT	
KSC			LAUNCH OPERATIONS SUPPORT	CHRYSLER
KSC			LAUNCH OPERATIONS SUPPORT	RADIO CORPORATION OF AMERICA
KSC			LAUNCH OPERATIONS SUPPORT	TRANS-WORLD AIRLINES
KSC			LAUNCH OPERATIONS SUPPORT	LING-TEMCO VOUGHT
KSC			LAUNCH OPERATIONS SUPPORT	BENDIX
KSC			LAUNCH OPERATIONS SUPPORT	DOW CHEMICAL
KSC			LAUNCH INSTRUMENTATION	INTERNATIONAL TELEPHONE & TELEGRAPH (FEDERAL ELECTRIC DIVISION)
KSC			LAUNCH CONTROL CENTER	VARIOUS
KSC			COMMUNICATION	MOLECULAR RESEARCH
KSC			COMMUNICATION	NORTHROP
KSC			LIQUID HYDROGEN TRANSPORTATION SYSTEM	UNION CARBIDE LINDE COMPANY
KSC			LIQUID HYDROGEN FACILITY	AIR PRODUCTS AND CHEMICAL
KSC			LAUNCH EQUIPMENT	GENERAL ELECTRIC
MSC	COMMAND MODULE		COMMAND MODULE	NORTH AMERICAN, SPACE DIVISION
MSC	COMMAND MODULE		GUIDANCE AND NAVIGATION	AC SPARK PLUG
MSC	COMMAND MODULE		GUIDANCE AND NAVIGATION	MASSACHUSETTS INSTITUTE OF TECHNOLOGY
MSC	SERVICE MODULE		NORTH AMERICAN AVIATION (SPACE AND INFORMATION DIVISION)	
MSC			SPACECRAFT INTEGRATION	NORTH AMERICAN, SPACE DIVISION
MSC	LEM		LEM	GRUMMAN
MSC	LEM		GUIDANCE AND NAVIGATION	MASSACHUSETTS INSTITUTE OF TECHNOLOGY
MSC	LEM		GUIDANCE AND NAVIGATION	AC SPARK PLUG
MSC	ACE		GENERAL ELECTRIC	
MSC	SPACECRAFT SUPPORT		SPACE SUIT	UAC HAMILTON STANDARD
MSC	SPACECRAFT SUPPORT		PARA GLIDER	NORTH AMERICAN, SPACE DIVISION
MSC	SPACECRAFT SUPPORT		REAL TIME COMPUTER COMPLEX	INTERNATIONAL BUSINESS MACHINES
MSC	SPACECRAFT SUPPORT		LUNAR LAND TRAINER	BELL AIRCRAFT
MSC	SPACECRAFT SUPPORT		APOLLO CM TRAINER	NORTH AMERICAN, SPACE DIVISION
MSC	SPACECRAFT SUPPORT		LEM TRAINER	GRUMMAN
MSC	SPACECRAFT SUPPORT		SPACE MEDICINE	UNITED STATES AIR FORCE
MSC	SPACECRAFT SUPPORT		RECOVERY FORCE	DEPARTMENT OF DEFENSE
MSC	GROUND OPS SUPPORT SYSTEM		INTEGRATED MISSION CONTROL CENTER	PHILCO
MSC	GROUND OPS SUPPORT SYSTEM		IMCC COMPUTER	INTERNATIONAL BUSINESS MACHINES
MSC	GROUND OPS SUPPORT SYSTEM		TERMINAL LANDING	PHILCO
MSC	LITTLE JOE		LAUNCH VEHICLE	GENERAL DYNAMICS
MSC	LITTLE JOE		PROPULSION	AEROJET
OTHER PARTICIPATING CENTERS	GSFC	GOSS	NETWORK OPERATIONS AND MAINTENANCE	INTERNATIONAL BUSINESS MACHINES
OTHER PARTICIPATING CENTERS	GSFC	GOSS	EQUIPMENT	COLLINS RADIO
OTHER PARTICIPATING CENTERS	GSFC	GOSS	MANNED SPACE FLIGHT NETWORK	NASA TRACKING SITES
OTHER PARTICIPATING CENTERS	GSFC	GOSS	NEAR SPACE INSTRUMENTATION FACILITY	DEPARTMENT OF DEFENSE TRACKING SITES
OTHER PARTICIPATING CENTERS	JPL	GOSS	DEEP SPACE INSTRUMENTATION FACILITY	TRACKING SITES
OTHER PARTICIPATING CENTERS	ARC		PHYSICAL AND LIFE SCIENCES	
OTHER PARTICIPATING CENTERS	FRC		FREE FLIGHT TRAINER	
OTHER PARTICIPATING CENTERS	La RC		MATERIALS AND STRUCTURES	
OTHER PARTICIPATING CENTERS	Le RC		ENGINE DEVELOPMENT	
OTHER PARTICIPATING CENTERS	WS		EXPERIMENTAL FLIGHTS	

APOLLO PROGRAM GOVERNMENT – INDUSTRY FUNCTIONAL MATRIX (CONT'D)

Fig. 39—Continued

integration role. D. Brainerd Holmes, head of the Office of Manned Space Flight (OMSF), defined that term in congressional testimony:

> General Electric Co.'s job is to . . . study and make sure that there is proper integration. By that I mean that the signals flowing across the various interfaces between pieces of equipment being built at various places in the country, are compatible. This is necessary whether it be electrical signals . . . or whether it be hydraulic flow that goes from a small quarter-inch tubing into a 2-inch pipe, or just a straight mechanical integration.[1]

GE teams at the centers and at stage contractor plants would provide OMSF with the information to coordinate the various pieces of Apollo. Within OMSF, James Sloan, the Director for Integration and Checkout, monitored the contract with GE.*

Opposition to the GE contract appeared almost immediately. Directors of the Marshall Space Flight Center and the Launch Operations Directorate (LOC) believed that GE's proposed mission would infringe on center responsibilities. At Huntsville 10 April 1962, the two set up a common front to restrict GE's role. Stage contractors shared the feeling; North American, Boeing, and Douglas officials were loath to have a competitor supervise their operation. Petrone expressed opposition to the GE management role at a 15 May meeting with GE representatives. The group discussed appropriate and—as Petrone emphasized—*inappropriate* areas of GE activity. The following week OMSF sent Petrone's office a revised work statement more in line with LOD's position. Holmes clarified two important points at the 29 May OMSF Management Council Meeting. GE would work for the centers with Sloan's Integration Office coordinating the effort. GE would not give work directions to stage contractors.[2]

Controversy continued during the summer. Lengthy portions of the July and August Management Council meetings were given over to discussions of GE's proper role vis-à-vis the field contractors and stage contractors. At the Cape the Launch Operations Center (as of 1 July) prepared a list of seven tasks considered suitable for GE. The GE contract was the sole topic of discussion at a two-day meeting in late August. Officials from LOC, Marshall, and the Manned Spacecraft Center at Houston met on the 29th "to

*Signed on 26 February 1962, the contract eventually totaled more than $615 million, a large portion of which went into checkout equipment at KSC.

ensure that the tasks for GE written by each Center were properly and adequately integrated so as to minimize GE's overall integration role and minimize interference from [NASA] Headquarters." The three centers concentrated on checkout problems and agreed that they would not require "any overall integration guidance from either GE or Headquarters."[3] Sloan and GE's top Apollo program managers joined the session on the 30th. The latter were dismayed to learn that the centers had rejected GE's checkout concept and had relegated GE to a support role.

NASA officials were in agreement about what GE should not do, but could not formalize a positive statement of the company's role. The issue generated three lengthy discussions at the 21 September Management Council meeting, and Holmes was disappointed at the lack of understanding. After a Cape visit in early November, Walter Lingle, NASA's Deputy Associate Administrator for Industry Affairs, told Holmes that the centers could not work with GE. Debus expressed surprise when Holmes called him about this report. The LOC Director admitted that, while reliability and checkout roles were set, there were still loose ends, and "there seemed to be an absence of a clear description of what GE is supposed to do." At the November Management Council meeting, there were further complaints about GE statements that suggested a management role for the company.[4]

Due to the broad nature of the contract and because it appeared to place the General Electric Company in the position of supervising or directing other NASA contractors, the House Committee on Science and Astronautics gave the GE contract considerable attention during the authorization hearings on the fiscal 1964 budget. In March 1963, the Manned Space Flight Subcommittee conducted hearings at GE's Daytona Beach, Florida, office.[5] NASA was still undecided about GE's role three months later, and the issue, added to Congress's first attack on the Apollo program (p. 170) and the Webb-Holmes dispute (note, p. 148), caused considerable unhappiness. After a visit to the Daytona office in early July, von Braun and Debus thought they had reached a satisfactory arrangement for GE work at Marshall and Merritt Island. However, von Braun notified Debus on the 9th that the plan had apparently fallen through. Joseph Shea, OMSF's Deputy Director for Systems, still wanted GE's assistance in integrating Apollo activities. NASA finally resolved the dispute in August. The centers and stage contractors prevailed; GE would not manage space vehicle development. OMSF would rely on a review board to help control and integrate the Apollo program, using GE as a management consultant and data processor. GE retained the reliability assessment and checkout roles.[6]

Intercenter Panels

The centers had begun coordinating their work on Apollo months before the GE integration role was proposed. In November 1960, the Space Task Group initiated Apollo technical liaison groups. Rapid program advances, following President Kennedy's 25 May 1961 address, prompted closer relations. In October 1961, von Braun and Robert Gilruth established the MSFC-STG (later MSC) Space Vehicle Board to resolve all space vehicle problems such as design, systems, research and development tests, planning, schedules, and operations.[7] Four panels were initially set up to integrate the efforts of Apollo and Saturn working groups. These panels served as "idea-exchange platforms," where centers could discuss their plans before pursuing them in depth. The panels also established a formal level of agreement, a means of obligating each center to a course of action. Over the next two years, the panels provided working-level communication between the centers. Von Braun indicated their importance in a December 1963 letter to George Mueller, Holmes's replacement as chief of Manned Space Flight: "The intercenter panels have proved to be the only effective medium of working out technical problems in detail which cut across Center lines."[8]

A Launch Operations Panel was among the four panels initially established by the Space Vehicle Board. The charter stated that the panel would:

- Ensure the compatibility of the launch vehicle and spacecraft ground support equipment.
- Ensure that adequate space and facilities are available at the launch site for checkout and mating of the launch vehicle and spacecraft.
- Integrate the overall space vehicle countdown and operational plan.
- Define and resolve tracking and data requirements during launch.
- Define and establish the overall ground safety plan for pad operations.
- Review all areas of the space vehicle for compatibility and possible interface problems with launch operations.[9]

Saturn C-5 and Apollo design decisions and the selection of stage contractors crowded the MSFC and MSC calendars during the remainder of 1961, delaying the inauguration of the panels for five months. In early February 1962, the Preflight Operations Division at Houston asked LOD to join in an Apollo coordinating committee patterned after a Mercury group. Petrone's Heavy Space Vehicle Office rejected the suggestion, citing the October agreement between von Braun and Gilruth. Petrone proposed, instead, a Launch Operations Panel meeting to discuss Apollo requirements.

When Petrone repeated his proposal the following month, Preflight Operations acceded to such a meeting on 15 March. The 27 members who attended the first session agreed to set up sub-panels that would exchange technical information at the working level. The panel would consider problems raised by the sub-panels; concur, where appropriate, with sub-panel conclusions or agreements; evaluate unresolved problems; and assign new tasks and deadlines.[10]

Attendance at the second meeting on 20 June 1962 nearly doubled, as representatives from NASA Headquarters, General Electric, and the stage contractors joined the discussion. The group organized seven sub-panels: electrical; facilities and complexes; launch preparations; propellants and gases; firing accessories and mechanical support equipment; trajectories and flight safety; and instrumentation, tracking, and data acquisition. As Petrone reported back to Debus, "It is now possible for *all* operating level personnel in respective areas of responsibility to directly resolve technical problems on an expedited basis in groups of reasonable size."[11] During the following year, the Facilities and Complexes sub-panel met monthly, the others less frequently. Seventy-five NASA and contractor representatives attended the fourth meeting of the full panel 1 August 1963. Although spacecraft requirements were the major topic, the participants also discussed the role of a proposed Panel Review Board.[12]

An OMSF-directed Panel Review Board had emerged from conversations between Wernher von Braun and Joseph Shea in May 1963. Previously, when panel matters required adjudication, the three center directors had met as a review board. Von Braun considered the arrangement unsatisfactory because, in striving for compromise, the directors had sometimes passed up the best solution; OMSF's participation on the board might help correct this. Shea welcomed von Braun's offer. OMSF had found itself exercising little influence over the panels; further, the board could control the proliferation of integration groups. The number of intercenter panels had increased to seven, and there were ten other groups handling OMSF-center interface matters.* Many agreed with Robert Gilruth's complaint, "there are too many meetings."[13] During its first session, held at Cape Canaveral in August, the Panel Review Board abolished two groups, placed several more under existing panels, and created a Documentation Panel to control the growing stacks of paperwork.[14]

---

*The seven intercenter panels were Launch Operations, Mechanical Design Integration, Electrical Systems Integration, Instrumentation and Communications, Flight Mechanics, Crew Safety, and Mission Control Operations. The ten other groups that had sprung up were the Integration Review Board, System Checkout Design Review Board, Reliability Assessment Review Board, Apollo Engineering Documentation Board, Policy Review Board for GE Project Effort, Systems Review Meeting, Communications and Tracking Steering Panel, Communications and Tracking Working Group, Systems Description Steering Committee, and the Apollo Reference Trajectory Working Group. Except for the Launch Operations Panel, the activities of these groups and panels go beyond the limit of this work.

New Contractors with New Roles

The Army Ballistic Missile Agency of the 1950s had represented the arsenal concept of weapons development—a largely self-sufficient government research and development program.* Although Pan American had provided limited support at the Cape, the Missile Firing Laboratory had been a government show. The Launch Operations Directorate, short of manpower at the start of the Saturn I program, resorted to "level of effort" contracts, under which companies such as Hayes International and Chrysler's Space Division supplied skilled technicians for a specified number of man-years. LOD assigned the technicians to particular tasks, directly supervised them, and approved their performance. Such contracts were not universally popular, and the terms *body shop, flesh peddling,* and *meat market* were sometimes used. LOD retained technical responsibility, and civil servants continued to work directly with hardware.[15]

A major change came in mid-1960 when MSFC awarded Douglas Aircraft Corporation a "mission" contract to build the Saturn I's S-IV stage and check it out at the Cape. LOD exercised responsibility for the launch vehicle and supervised the contractor, but Douglas was responsible for accomplishing a clearly defined task. In doing so, the company supervised its own employees. The following year NASA awarded Chrysler a mission contract to build, check out, and test 20 S-I stages for the Saturn I. Chrysler's role was subsequently expanded to include technical support for Saturn I and IB launch operations. The latter involved such things as the environmental control systems, umbilical arms, propellant operations, postlaunch refurbishment of support equipment, logistics, ground electrical networks, and telemetry checkout. On the early launches of the Saturn I block II series, Douglas technicians checked out the upper stage while a Chrysler crew worked alongside KSC engineers on the S-I stage. SA-8 in early 1965 marked the first flight of a Chrysler-built booster with the contractor assuming responsibility for stage checkout. It also marked the end of an era for veterans of the Missile Firing Laboratory. Henceforth, KSC civil servants would no longer operate launch equipment, but would act more like traditional managers.

The transition to mission contracts was not always easy. LOD officials, accustomed to level-of-effort contracts, considered Douglas Aircraft uncooperative. In turn, the California firm, used to the Air Force's

---

*The Air Force in the 1950s represented the opposite position: contractors performing R&D for a government agency. For more detail on this subject, see H. L. Nieburg, *In the Name of Science* (Chicago, 1960); and *Government Operations in Space,* the Thirteenth Report by the Committee on Government Operations, House of Representatives, 89th Cong., 1st sess., House report 445, June 1965.

broad guidelines, resented NASA interference. An early difference of opinion involved the loading of Saturn I propellants. Looking ahead to Saturn V operations, LOD planned remote, automated controls for the Saturn I. Douglas officials accepted the LOD position regarding checkout and main loading operations, but wanted manual control of the S-IV stage's final slow fill. After meetings in March and May of 1961, LOD thought the matter was resolved. However, when Orvil Sparkman visited Douglas's Santa Monica, California, plant in September, he was surprised:

> The Douglas S-IV GSE to be utilized at Sacramento [the contractor's test area] is designed and built with a complete disregard for instructions contained in the three referenced memorandums [minutes of March and May meetings mailed to Douglas as official working documents]. Not only are these panels designed for manual propellant servicing, but no attempt was made by Douglas to incorporate standard nomenclature developed by Douglas and LOD. . . . It is the intention of the contractor to furnish equipment of the same design at AMR.[16]

Douglas officials and Sparkman agreed that the control networks for SA-5 (the first two-stage Saturn I launch) could not be completed until the loading issue was resolved. The dispute was settled in LOD's favor at an October meeting of the Propellant and Gases Panel, but only after Marshall's intervention.

LOC's peculiar relationship with the stage contractors caused difficulties during the next two years. The stage contractors, still working under contracts with Marshall, looked to Huntsville for direction and contract management. The launch team's efforts to monitor contractor operations, suggest equipment modifications, or obtain information on contractor requirements were relayed by the contractor to his home office and from there to Marshall. Douglas officials pointed up the awkwardness of the arrangement during the SA-5 launch preparations when they questioned the launch team's right to reject company work. Douglas officials refused to yield until Col. Lee B. James, Saturn I–IB Project Manager in Huntsville, notified company management that LOC was responsible for the quality of S-IV stage equipment at the Cape.[17]

Relations with Marshall Space Flight Center

The launch team's separation from Marshall in July 1962 did not significantly alter the close ties between the two centers. Debus, believing that interfaces were best managed by locating responsible design elements in close physical proximity, was pleased that the Launch Vehicle Operations Division (LVOD) was both an operating element of LOC and an engineering element of Marshall. He wrote:

> Through this arrangement launch operations requirements are fed back into the design organization and become incorporated in design criteria. For example, the Astrionics Division Electrical Systems Integration Branch of MSFC which is responsible for design of vehicle associated (active) GSE and checkout equipment incorporates into the design the operational requirements obtained from LVO; thus the interfaces are a responsibility of the group.[18]

Theodor Poppel's design group, responsible for much of the launch equipment, remained in Huntsville where it could readily exchange information with launch vehicle engineers. One area of potential strife—the center's relations with contractors—was eliminated in August 1964 when the two centers reaffirmed Marshall's primary responsibility for Saturn vehicle development, but delegated to KSC the responsibility for preparation of support equipment and vehicle checkout. As a result, Hans Gruene's Launch Vehicle Operations team dropped its formal ties with the Huntsville organization. The agreement also gave KSC contract authority to supervise stage and support equipment activities at the Cape.[19] Seven months later Debus and von Braun signed a series of clarifying and implementing instructions, which included the provision that:

> Design of components and equipment to be installed in the complexes at the Cape are responsibilities of each of the three MSF centers [Marshall, Houston, and KSC] resulting from decisions that have already been made and which are continuously coordinated through the workings of Intercenter Panels and the system of Interface Control Documents. The design and construction of facilities in which this equipment will be placed is the responsibility of KSC.[20]

Marshall subsequently stopped contracting for launch checkout, and KSC negotiated its own contracts.

Coordination between KSC and Marshall got a boost in 1964 when their communications lines were organized into the launch information exchange facility (LIEF). Communications had been primitive by modern standards, with LOC personnel commuting between Huntsville and the Cape, and commercial wires carrying the daily message load. With the

Saturn program, the need for a better system became apparent. A huge increase in information flow was expected with the launching of larger vehicles; engineers cited 88 telemetry measurements on the Redstone versus an anticipated 2150 on the Saturn V.[21]

NASA Headquarters approved LIEF in August 1963, and the system met KSC expectations. The new communications network provided the backup support of designers to operations personnel in the analysis of unexpected problems, expedited transmission of additional information on demand, and made available the resources of the development agency throughout the checkout period. LIEF employed the voice, teletype, and facsimile circuits already linking the two centers, and a tape-to-tape transceiving system that carried digital engineering data and launch vehicle computer programs via a NASA automatic fascimile switchboard in New Orleans. More sophisticated equipment was added in time, eventually putting Huntsville displays on the scene for KSC launches.[22]

Relations with the Manned Spacecraft Center

While KSC's relations with Huntsville were relatively good, its early coordination with Houston was another matter. During 1962–1964, KSC officials frequently complained that the Houston center was tardy with its spacecraft-related requirements for the launch facilities. Some KSC officials believed their counterparts were less than frank in their dealings. This feeling gave way slowly as KSC gained an appreciation for Houston problems.

Information from Houston came slowly for two reasons. First, spacecraft design was dragging, and the July 1962 decision to rendezvous in lunar orbit imposed new assignments, including development of the lunar excursion module. The lunar module contract, won by Grumman Aircraft in November 1962, initiated one of Apollo's most difficult projects, which by 1967 threatened to delay the entire program. The addition of a rendezvous and docking capability to the command-service module required two years of extensive study. Configuration work on the two vehicles culminated with the mockup review of North American's block II spacecraft on 30 September 1964. Secondly, the Manned Spacecraft Center did not have enough experienced spokesmen on the intercenter panels. Many of the center's engineers were occupied with the Mercury and Gemini programs. Houston's Apollo team, understaffed for the large tasks it faced, allotted priority to its North American and Grumman relations. A reluctance to share information that might lessen a center's authority also contributed to Apollo's coordination difficulties. All three centers, however, shared in this sin of omission.[23]

In August 1962, LOC had a detailed concept for Saturn V operations but only a general understanding of Apollo spacecraft needs. Early that month Debus, Petrone, and Poppel journeyed to Houston for a discussion of requirements. The two centers agreed that a spacecraft checkout center would be constructed in the Merritt Island industrial area, checkout of the spacecraft at the assembly building and later on the pad would be controlled from the launch control center, and Houston would not need a computer or display console on board the launch umbilical tower.[24]

The disagreement about servicing the spacecraft (first expressed at the Management Council meeting in May 1962—see page 128) continued for several more months. At a Launch Operations sub-panel meeting in October 1962, MSC insisted that pad facilities provide access to the Apollo spacecraft from all sides. Design of the command and service modules was too far along to modify this requirement. The Houston engineers did not care whether LOC built the 360° service capability into the launch umbilical tower's swing arms or made the arming tower mobile. Neither alternative appealed to LOC, but Petrone informed Houston in early November that a mobile arming tower would provide the necessary pad access.[25]

While conceding that matter, LOC won a dispute over the responsibilities for establishing criteria in the industrial area. LOC's concept paper on launch operations stated, "LOC will provide design, contracting, and construction monitoring services for facility construction . . . based on MSC functional and technical requirements." The Florida Operations launch team of the Houston center interpreted this to mean that LOC would provide the services based on "design and specification requirements or criteria developed by MSC." Debus objected to Houston's providing fully developed criteria for the spacecraft facilities and won Holmes's support at a meeting in October 1962. Subsequently, the LOC director and G. Merritt Preston, chief of Florida Operations, agreed that Houston would provide rough criteria while LOC selected the architect-engineering firm and approved the final design.[26]

A bigger problem—one that dragged on for several years—concerned submission of spacecraft data. In October 1962, Petrone wrote Houston's Apollo Project Office that spacecraft requirements were "urgently needed" so that LOC could proceed with the criteria studies for the assembly building, launch pad, and mobile launcher. He restated LOC's needs the following month and frequently thereafter.[27] Unfortunately, the Houston engineers could not ascertain all their spacecraft requirements. In October 1962, they projected a need for one 6-meter console in the firing room of the launch control center. By early 1963, this had grown to thirty-five 48-centimeter racks and two 6-meter consoles. A year later Houston was still uncertain about the checkout equipment for the mission operations room; in February 1964 a Houston representative asked if the Manned Spacecraft Center could simply indicate what spacecraft functions had to be performed and the approximate locations for the test consoles.[28]

Problems in achieving a final design for the command and service modules delayed LOC's design of the mobile service structure well into 1964. By September 1963, the design of the tower was nearly a year behind schedule, and the growing number of spacecraft requirements increased the likelihood of a top-heavy, overweight tower. The contractor, Rust Engineering, undertook a weight reduction program, redesigning the service platforms and modifying the lower structure. Petrone reported in December that Rust had the tower's weight and wind-load factors back within the limits of the initial criteria. Seven months later, the design work completed and construction bids on hand, there were two more changes: a KSC decision to relocate ground servicing equipment at the base of the arming tower, and a late list of cabling requirements from Houston. KSC made the necessary modifications within a month.[29]

Since the lunar module had started late, a delay in its requirements was expected. After the data became available in January 1965, launch engineers modified their facilities to accommodate the third spacecraft module. The changes affected the electrical and fluid systems of the mobile launcher, office space in the assembly building as well as the second level of platform B in the high bays, and platform 3 of the mobile service structure. KSC altered the pad area to provide space for the lunar module's ground support equipment and additional power receptacles.[30]

Range Safety

The question of safety was always paramount at KSC and usually involved much intercenter negotiation, as well as long study sessions with the Air Force. The possibility of the space vehicle colliding with the umbilical tower during launch touched off a study in mid-1962. The LOC group concluded that the Saturn I's proposed emergency detection system would not catch all possible failures in time to signal an abort. If engine number 1 of the first stage failed, attitude and rate mechanisms in the detection system would not sense a rocket drift that could result in a collision with the tower. An initial experiment with backup television coverage (the SA-3 flight of 16 November 1962) was disappointing; flame and dust kept astronaut D. K. Slayton and Marshall's John Williams from seeing the rocket as it climbed by the face of the tower. Petrone concluded from film of the liftoff that ground level visibility would always be sharply limited by blast and flame. He recommended placing a television camera at the top of the umbilical tower to look down between the tower and the vehicle.[31]

The Crew Safety Panel (one of the intercenter panels) took charge of this study in early 1963. LOC's chief representative on the panel, Emil Bertram, examined several proposals for ground support instrumentation including color television, an electronic "beat-beat" system based on the Doppler principle, and the placement of sensing wire on the umbilical tower. The panel finally settled on television and field observers. The launch team had to overcome further problems with the television during the latter Saturn I flights; for example, the intensity of light at liftoff burned holes in the camera's vidicon tube. The panel, satisfied with the coverage by 1965, approved an abort advisory system for LC-34. (With no manned flights scheduled, LC-37 did not require a similar system.) Since the light intensity bleached out colors, the system employed four black-and-white cameras. Two cameras, pointing downward from the 72-meter level of the umbilical tower, covered the space between the tower and the rocket. Three hundred meters away on opposite sides of the launch vehicle, zoom-lensed cameras mounted on 5-meter towers provided a profile of early flight. The four cameras formed part of the complex's operational television network. Telescope sites, located around the perimeter of the complex, supplemented the TV. Gordon Cooper, an influential voice on the Crew Safety Panel, and other astronauts helped man the observation posts. LC-34's operational intercom system gave the posts instant communication with the blockhouse. The coverage proved satisfactory, and a similar arrangement was prepared for LC-39.[32]

While the establishment of the abort advisory system went smoothly, the matter of who held abort authority during the first ten seconds of flight (until tower clearance) proved more troublesome. KSC officials believed the launch operations director was in the best position to command an abort. The astronauts objected, arguing that the launch director might abort the mission at an undesirable moment for them or the spacecraft. Eventually the astronauts won the argument. As information came to the launch director during the first seconds of flight, he would assess the situation. If an abort appeared necessary, the director could trigger the "Abort Light" on the flight panel in the spacecraft. If the "Thrust O.K." light indicated a malfunction or if the astronauts sensed a problem, the crew could manually activate the launch escape system.[33]

Range safety matters caused considerable disagreement between NASA and the Air Force before the issues were ultimately resolved. The Air Force had exercised responsibility for range safety at the Cape since launching the first rocket back in 1950. The basic concern was to prevent an errant rocket from landing in a populated area. Accordingly, when NASA scheduled a mission, the Air Force wanted details on the flight plan: launch azimuth, trajectory, and impact point. Range safety policies required that the launch vehicle have at least one tracking aid and two digital range safety command receivers on each active stage. The receivers had to be compatible with range instrumentation. If a destruct signal was received from the ground, the receivers would cut off the flow of fuel to the engines and then detonate small explosive charges to rupture the propellant tanks. The propellants would then mix and their explosive force be consumed before vehicle impact.[34]

The command receivers were activated prior to liftoff. The range safety officer sat at a group of consoles located in the range control center of the Cape Kennedy Air Force Station. The display had been developed in the 1950s and it remained relatively unchanged during the succeeding 15 years. The consoles received tracking data on the vehicle from the Eastern Test Range tracking system. This information was processed by a digital computer, and the display showed both the present location of the vehicle and its impact point if thrust were terminated.[35]

The plot included a set of lines that followed the planned path of the vehicle. These so-called "destruct" lines indicated the maximum deviation of the impact point from the trajectory that could be allowed without endangering life or property. As long as the impact point remained within the destruct lines, no action was required. Should a failure occur or the destruct lines be crossed, the safety officer first sent an arming signal to the receivers aboard the vehicle. This performed the dual function of initiating thrust termination and preparing the destruct system for activation. After an appropriate built-in delay, a second signal was transmitted. It caused the detonation of the explosives in the propellant dispersion system. Within seconds the vehicle would be transformed into tumbling, burning chunks of scrap.[36]

The Air Force's authority in matters of range safety was reaffirmed in the Webb-McNamara Agreement of 17 January 1963. Essentially, the agreement confirmed the authority of the Air Force to require flight termination and propellant dispersion systems on NASA vehicles as well as those of the military, and this authority extended from liftoff through orbital insertion. The agreement was supplemented by the Air Force Missile Test Center–Launch Operations Center agreement of 5 June 1963, which gave NASA the responsibility for ground safety within the confines of KSC but left flight safety with the Air Force.

LOC acknowledged the Air Force's responsibility for range safety, but in a letter of 10 May 1962, General Davis noted that "there are occasional differences of opinion on what constitutes reasonable safety practices" and asked for Debus's comments on Air Force policy. In his response,

Debus hesitated to cite specific disagreements since many rules were undergoing review and change. However, he did list a few areas where NASA and its contractors felt uninformed as to how the Air Force reached its decisions. One area concerned the computation of destruct areas; a second was the amount of trajectory data required on a new program. Debus also questioned the rationale for a dual destruct capability in all powered stages.[37]

This last matter involved KSC in a lengthy debate which found the Manned Spacecraft Center and the Air Force at odds over the latter's insistence on including a destruct system in the Apollo spacecraft. The dispute began in March 1962, when Houston requested a waiver—spacecraft engineers did not want the astronauts carrying a destruct package with them to the moon. The Range Safety Office proposed to restrict Apollo flights severely if the spacecraft did not carry a destruct system. Neither side altered its position in the next twelve months. When the NASA centers and the Eastern Test Range discussed Apollo–Saturn V safety requirements in May 1963, Houston again asked to fly the Apollo spacecraft (including the S-IVB stage) without a destruct capability. Engineers cited the possibilities of an errant signal triggering the systems or of an explosion during docking.[38] The Air Force stood firmly by the requirements of the range safety manual: "Both engine shutdown and destruct capability are required for each stage of the vehicle."[39]

The sparring over the destruct systems soon took on the trappings of international diplomacy. On 9 May Dr. Adolf Knothe, LOC's range safety chief, warned Debus that a crisis could develop. Although no agreement had been worked out by June, Knothe and his assistant, Arthur Moore, began damage probability studies to justify omission of a destruct system. Their calculations indicated that an explosion of the three launch-vehicle stages, triggered by the range safety officer, would also destroy the lunar and service modules with their propellants. (In the meantime the launch escape system would have pulled the astronauts' command module away from the explosion.) Their plan employed a shaped charge on the front end of the S-II stage to explode the S-IVB stage. The results were inconclusive, however, and the Air Force stressed the possibility of a spacecraft falling back onto the Cape. Range officials contended that a spacecraft destruct system would not endanger the mission; NASA could design the system with a jettison capability.* Knothe recommended a detailed destruction probability study by

*An abort during the latter phase of the launch sequence (between approximately T + 3 minutes into the flight when the launch escape tower jettisoned and T + 10 minutes when the spacecraft entered orbit) would depend upon the service module propulsion system to separate the command and service modules from the Saturn. As B. Porter Brown, Houston's representative at the Cape, indicated, "the Manned Spacecraft Center will be most reluctant to carry a destruct system that can in any way jeopardize the capability of this module to perform its abort function" ("Apollo Program Information Submission," 23 August 1963). Since the space vehicle would have cleared the Cape before the launch escape tower jettisoned, the Air Force was willing to discard the service module's destruct system at that time.

the Lear-Siegler Corporation but saw "no absolutely objective answer to this dilemma."[40]

The Air Force countered LOC's calculations with a July presentation on a liquid explosive, Aerex. Impressed with Aerojet-General Corporation's product, NASA engineers gathered in Houston two weeks later for a North American briefing on a destruct system using the liquid explosive. Afterward, Moore sounded out spacecraft officials. There was still misunderstanding between the two centers in August when Christopher C. Kraft, Jr., chief of Houston's Flight Operations Division, moved to break the impasse. His call for an Apollo Range Safety Committee, modeled on a Gemini group, included AFMTC participation. LOC and MSFC vetoed Air Force representation until NASA had achieved a common front.[41]

At the first meeting of the Range Safety Committee, Knothe reviewed safety problems including the Range requirement for dispersion trajectories on all propelled stages.* The destruct systems on the S-I and S-II stages caused no concern, and Knothe believed that Aerex might prove acceptable for the S-IVB and spacecraft. Houston, however, was sharply divided over the destruct requirements, with the astronauts leading the opposition. The committee put the matter aside until the Manned Spacecraft Center could reach an understanding within its own ranks.[42]

In October, Kraft managed to add Air Force representatives to the Range Safety Committee. In the minutes of the 22 October meeting, he noted: "It was apparent at the meeting that the Range Safety Office is just as concerned that their regulations do not hamper the program as we are that we are not hampered by range safety."[43] Kraft's note foreshadowed the agreement reached with the Eastern Test Range the following month. North American would prepare a destruct system for the service module. The spacecraft could fly early tests without the destruct capability since the service module tanks would contain little fuel. The decision, however, did not bring the matter to a close. Marshall and KSC officials were visibly upset in March 1964 when North American Aviation presented five spacecraft destruct systems, none of which incorporated the designs of the Saturn stage destruct system.† When von Braun and Debus raised the issue at an Apollo Review Board, Mueller, head of Manned Space Flight, asked the KSC chief

*The dispersion trajectories marked the right of way for space vehicle flight. The boundaries on the flight corridor were formed by permissible lateral and vertical deviations. The deviations were necessary because of inevitable variations from standard—two rockets of the same model would have different thrust because of slight differences in alignment of the engines and in propellant weight. The wind effect was another factor that could never be fully accounted for. By taking into consideration the normal deviations from standard in relation to probability curves, LOD gave the Range Safety Office 99.73% assurance that the launch vehicle, in normal flight, would stay within the corridor. Any deviation outside the boundaries indicated a malfunction and the safety officer destroyed the vehicle.

†MSFC and KSC personnel thought the destruct systems should be standard throughout the space vehicle. They viewed MSC's research for a different destruct arrangement as lack of confidence in the Saturn system.

to seek elimination of the destruct requirements. Over the summer of 1964 KSC officials met with Air Force officers, including Lt. Gen. Leighton I. Davis, who had moved from the Missile Test Center to the command of the National Range Division. KSC stressed among other things the weight penalty. A 120-pound service module destruct system would require nearly 7500 more newtons (1700 pounds) of thrust or a reduction in the weight of the S-IC stage. When Mueller submitted a formal request for waiver in September, General Davis directed the Range to go along.[44]

Summary

Integration matters at KSC required a great deal of attention during the early years of Apollo. KSC officials worked closely with Marshall, Houston, and the stage contractors in shaping the launch facility to Apollo-Saturn dimensions. While an integrating role for General Electric was rejected, intercenter panels provided an effective means of coordination. The increased workload altered KSC's relations with its contractors. The launch center took on the direction of contract work previously performed for Marshall or Houston. In turn contractors assumed more responsibility under mission contracts. The Apollo coordination brought its share of disagreements—witness the dispute over a destruct charge on the command and service modules. By 1965, however, most of the conflicts were resolved. KSC had achieved a good working order between its government team and contractors, and relations with other organizations were reasonably well defined.

10

Saturn I Launches (1962–1965)

Testing the Booster (SA-2–SA-4)

After the launch of the first Saturn rocket on 27 October 1961, the rest of the research and development schedule went like clockwork. The nine remaining launches of the Saturn I program (April 1962–July 1965) set a record for consistent performance while receiving a minimum of recognition. The launches coincided with America's first successes in manned spaceflight and all eyes were on the astronauts. When one of them was cradled out into space in a Mercury shot, the nation paused to participate by television in the liftoff, flight, and recovery.

While no human passengers lent drama to the Saturn I flights, Saturn team members had much to be proud of. The ten launches proved the clustered booster concept, the hydrogen-propelled upper stage, and the Cape's ground facilities. In 1964, in what was to become a historic collaboration, the Saturn rocket and Apollo vehicle were mated for the first time, with both SA-6 and SA-7 flying an Apollo "boilerplate" model.* The last three Saturn vehicles carried Pegasus, a satellite flown in low earth-orbit to detect meteroids. Although Marshall Space Flight Center engineers introduced new features in every Saturn I launch, the tests came off without a major failure. The confidence gained from these successes was Saturn I's great contribution to the Apollo program.

SA-2 (25 April 1962)

The second Saturn I, vehicle SA-2, arrived at Cape Canaveral on 27 February 1962. Launch preparations took 58 days. Although there were no serious delays, daily status reports revealed many minor problems:

19 March. A leak has been detected between the injector and the LOX [liquid oxygen] dome on Engine Position No. 4 . . . . Discussions concerning this matter are being

**Boilerplate* means a full-scale model of a flight vehicle flown on research and development missions, without some or all of the internal systems.

held with Rocketdyne and Propulsion and Vehicle Engineering Laboratory personnel.

20 March. Attempts to correct the LOX dome leak, reported yesterday, have failed to remedy the problem. Further discussions are now in progress, to determine whether to buy the "as is" condition or change the engine. A change in the overall schedule will result if the engine has to be changed.

21 March. Discussion between Propulsion and Vehicle Engineering Laboratory, Rocketdyne, and LOD has resulted in a decision to launch without replacement on engine, Position 4.

26 March. Minor difficulties exist in the guidance sub-system; these are under investigation. No interference was noted during the RF [radio frequency] test.

27 March. The service structure was removed from around the vehicle; alignment and RF checks were made and the structure replaced around the vehicle. Minor difficulties were encountered with structure operations.

28 March. Two strain gauges have been found to be damaged (LOX stud and truss member). Attempts will be made to repair the truss member gauge.

30 March. The manhole cover on the top of the S-V-D was found damaged yesterday. A replacement cover has been received from MSFC, which will be installed this afternoon.

6 April. A modification to the fuel density and fuel level sensing lines has been completed.

9 April. Fuel loading test in the manual mode is in progress . . . . During preparations for the fueling test, a leak was detected in the fuel level computer. The computer was removed and sent to the lab for repair . . . . An effort was made to get a spare computer from MSFC. A second computer was sent down by plane Saturday evening [7 April] . . . . It developed that the second computer was not in a sufficient state to be properly calibrated prior to today's operation. Therefore, the primary effort Sunday night was directed toward readying the original computer for the test today.

11 April. LOX tanking test was postponed one day after difficulties developed in the electrical tanking computer

circuit. Attempts are being made to isolate and correct the problem area. The one day delay . . . will not affect the overall schedule. If the test can be satisfactorily performed tomorrow, we will be back on the original schedule by [16 April].

17 April. The fuel loading computer has been repaired and functionally checked satisfactorily.

19 April. A potential problem area exists with respect to three hydraulic systems. If it should be declared by Propulsion and Vehicle Engineering, Astrionics and Quality Laboratories that the three systems must be checked, the launch date [25 April] cannot be met.[1]

Marshall engineers had made one significant change in the SA-2 booster design, placing additional baffles in the propellant tanks to prevent a recurrence of the sloshing experienced in the latter part of the SA-1 flight. The countdown on 25 April went smoothly; the only hold came when a ship strayed into the flight safety zone, 96 kilometers downrange. The successful flight was terminated with a dramatic experiment. When SA-2 reached an altitude of 105 kilometers, launch officials triggered the command destruct button. Project "High Water" released 86000 kilograms of water from the dummy upper stages, giving scientists a view of a large disturbance in the upper regions of the atmosphere. A massive ice cloud rose 56 kilometers higher in a spectacular climax.[2]

SA-3 (16 November 1962)

A tropical storm greeted the SA-3 vehicle's arrival at the Launch Operations Center on 19 September 1962. Three days of rain and high winds delayed erection of the booster, and conditions were still unfavorable when the launch team resumed work on the 21st. Aeronautical Radio Incorporated engineers, hired by NASA to review Saturn operations, reported: "The erection operation was safely performed but is rather hazardous, with technical personnel climbing around on top of the horizontal booster to install hoisting equipment. This operation was performed on the slick plastic covering of the S-I stage in a wind of up to [37 kilometers per hour]." The Aeronautical Radio team considered the preparation prior to stage erection (removing the end ring segments) "a relatively slow, inefficient, and dangerous operation, with a considerable amount of trial and error," and recommended more familiarity with the instruction handbooks. During the eight-week checkout, the Washington, D.C., firm found other shortcomings such as "the use of metallic hammers to urge recalcitrant components into place." The observers noted that proper tools were not always handy, "and expediency sometimes prevailed." They concluded, however, that the "efficiency and dedication" of Hans Gruene's Launch Vehicle Operations Division* was instrumental in the success of the Saturn test.[3]

SA-3 lifted from Cape Canaveral on 16 November 1962. Debus asked von Braun not to invite outside visitors, as the United States armed services were still on alert for the Cuban missile crisis. The rocket incorporated a number of important new features. The first two Saturns had used 281 000 kilograms of propellant, about 83% of the booster's capacity. Marshall, wanting information for the new Saturn IB program, flew SA-3 with a full propellant load to test the effects of a lower acceleration and a longer first-stage flight. The flight also tested the retrorockets that would separate the two live stages on SA-5, the first launch of the upcoming block II series. SA-3 flew three other important prototypes: the ST-124 stabilized platform, a pulse code modulated data link, and an ultrahigh-frequency link. The stabilized platform was a vital part of the Saturn guidance and control system, containing gyroscopes and accelerometers that fed error information to the control computers, which provided steering signals to the gimballed engines. The data link's importance lay in its ability to transmit digital data, a vital ingredient in plans for automation of checkout and launch procedures. The ultrahigh-frequency link would be used to transmit measurements, such as vibration data, that could not be handled effectively on lower frequencies.[4]

SA-4 (28 March 1963)

SA-4 set records for the shortest launch checkout (54 days) and the longest countdown holds (120 minutes) of the block I series. At T – 100 minutes on launch day, test conductor Robert Moser called a 20-minute hold while the launch team adjusted the yaw alignment of the ST-90 gyro guidance platform. Readings from a ground theodolite showed that the platform was not properly aligned on the launch azimuth. An operator oriented the Watts theodolite on a geodetic survey line and then turned the head of the instrument to the launch vehicle. The alignment prism in the ST-90 platform reflected a light directed from the theodolite. If the platform was aligned properly, the reflection from the prism appeared in the center of the theodolite's scope. In this case, the problem was with the theodolite and not the gyro platform.

---

*See chap. 7. From 1 July 1962 to 24 April 1963, LVOD was a division of MSFC. Since Debus and Gruene served as Director and Deputy Director of both the Launch Operations Center and LVOD, this was an administrative distinction with little or no bearing on launch activities.

The final hold came at T – 19 minutes as a result of a LOX bubbling test. Andrew Pickett's propulsion group performed the test late in the countdown to verify the flow of helium to the LOX suction ducts of the eight engines. The decreasing temperature of the LOX indicated a proper flow of helium, but the propulsion panel did not register a signal that the LOX bubbling valve was open. Without the signal the terminal sequencer would shut down. Pickett's team, along with Isom Rigell's electrical engineers, improvised a bypass for the valve signal on the sequencer. The propulsion team assured a proper LOX temperature for the Saturn and then initiated the bypass manually as the sequencer brought the vehicle to liftoff.[5]

In SA-4's most important test, officials deliberately shut down the number 5 engine 100 seconds after liftoff. Booster systems rerouted propellants to the seven other engines. Contrary to some predictions, the shutdown engine remained intact and the imbalance of hot gases on the engine compartment heat shield had no ill effect. The SA-4 vehicle simulated all block II protuberances on the dummy second stage, e.g., fairings and vent

Fig. 40. SA-4 ready for launch from LC-34, March 1963.

ducts, to determine the aerodynamic effects of a live second stage. Block II antenna designs were also flown. The SA-4 vehicle employed a new radar altimeter and two experimental accelerometers for pitch and yaw measurements. After the successful flight, the von Braun team in Huntsville looked confidently toward two-stage missions.[6]

Pad damage from the first four launches did not surpass expectations. Restoration cost an average $200000 and took one month. LVOD officials were particularly interested in assaying pad damage after the launch of SA-3. One of the mission's goals was to determine the effect on the pad of an increased propellant load with the consequent slow acceleration and longer exposure to rocket exhaust. The damage was comparable to the first two launches. The only effect readily attributable to the slower acceleration was increased damage to the pedestal water deluge system (the torus ring) and a warping of the flame deflector.[7]

The LOX fill mast at the base of the rocket had to be replaced after each launch. The 21-meter cable mast assembly extending up alongside the rocket also crumpled during each of the first two launches. After watching the long aluminum fixture collapse the second time, officials replaced it with an umbilical swing arm. The Huntsville engineers converted a swing arm intended for the SA-5 launch and shipped it to the Cape in early August. At LC-34, Consolidated Steel and Ets-Hokin-Galvin began work on the new umbilical tower two weeks after the SA-2 shot.* The swing arm, mounted in August, suffered very little damage in the SA-3 launch.[8]

A Second Saturn Launch Complex—LC-37

Block I—the first four Saturn launches—had gone up from launch complex 34. With the block II launches (SA-5 through SA-10), the program would move to new facilities at launch complex 37. The second complex had originated with the Hall Committee study of 1959, which found that an explosion would render LC-34 useless for a year (page 30). On 29 January 1960 Debus asked Dr. Eberhard Rees to approve a second Saturn complex. Since LC-37 would serve primarily as insurance for LC-34, no major design

*Saturn construction became rather complicated at times. LOD personnel observed that the column splices connecting the new construction to the existing 8-meter base were not consistent with Maurice Connell & Associates design drawings. In a letter to the Corps of Engineers, Debus stated, "Upon investigation, it appears as though the Jacksonville District Office had instituted changes in the original design without the concurrence of LOD, who has the design responsibility." The fabricator of the first phase steel had apparently erred in the column's angle of slope. The Corps solution, using one-inch diameter interference body bolts, was satisfactory; but the construction teams were using one-inch high-tension bolts, which had only two-thirds the necessary strength. Debus requested that the Corps get LOD's approval in future modifications.

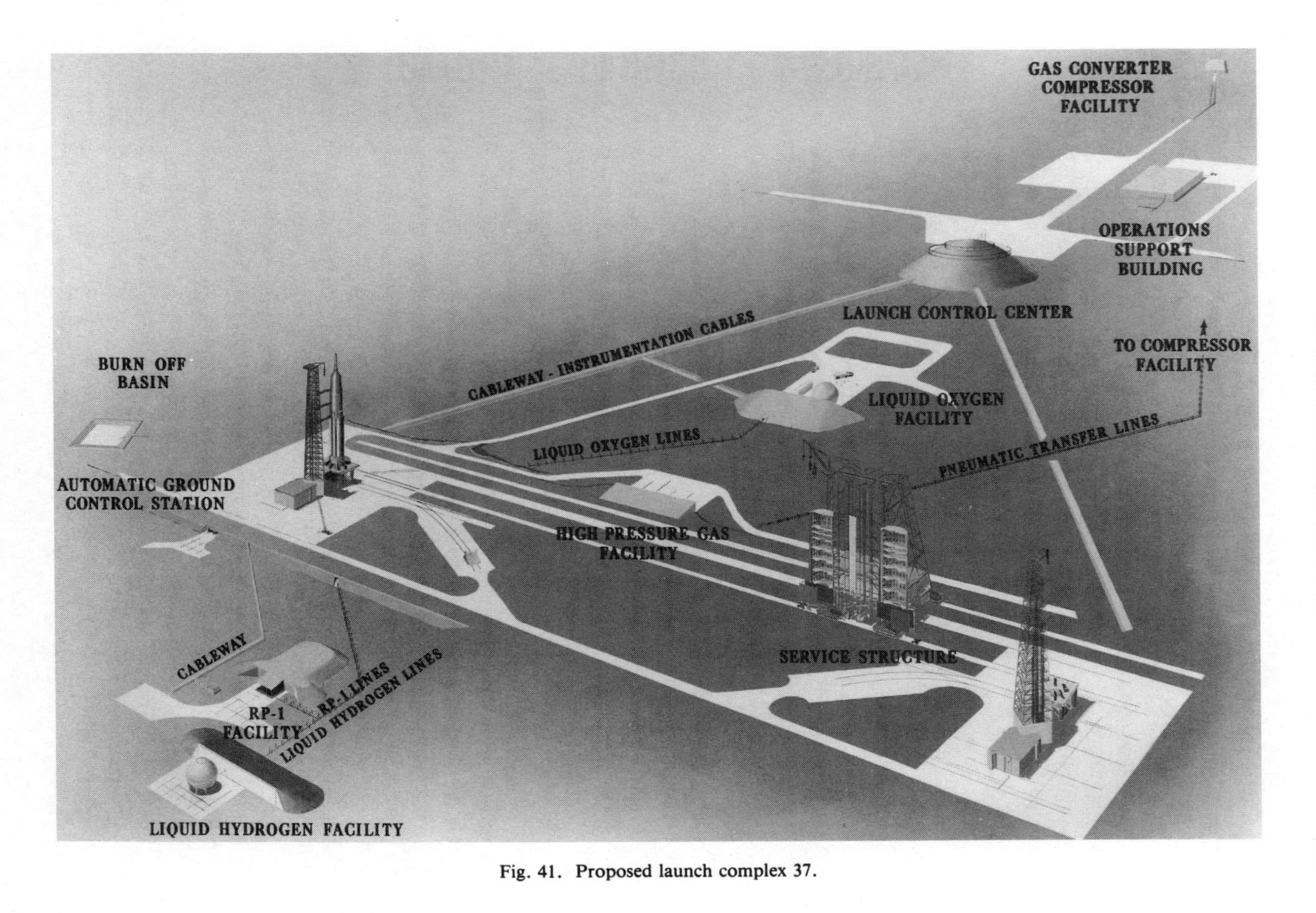

Fig. 41. Proposed launch complex 37.

Fig. 42. LC-37 under construction, January 1963.

changes were anticipated. Taking into account the rising costs on complex 34, Debus estimated the price of LC-37 at $20 million (roughly one-third more than LC-34's costs as of January 1960). In his report, Debus warned Rees that LC-37 would likely be sited at the undeveloped north end of Cape Canaveral, 1220 meters north of LC-34 and 425 meters from the Atlantic Ocean. A complex at that location "would require utility capacities of unusually large magnitudes and the cost to Saturn, as the initial [user] could be excessive."[9] In February 1960, representatives from the Missile Firing Laboratory, Army Ballistic Missile Agency, and the Air Force Missile Test Center estimated demands for water, power, roads, communications, and instrumentation at LC-37 and discussed the cost of extending these to the proposed site. Eventually, development for LC-37 included a new electrical power substation and transmission lines, a 3785000-liter water reservoir, and a pumping network, at a price of $2.5 million.[10]

Hoping to have LC-37 ready for backup duty by January 1962, MFL originally set a mid-1960 deadline for criteria on the launcher, umbilical tower, and propellant systems.[11] Debus's decision to put a new service structure on LC-37 dashed these plans. Harvey Pierce, a Connell engineer, had prompted the change. Pierce had played an important role in designing LC-34 and more recently on the Hall Committee. On 26 February Pierce had written Debus about some inherent shortcomings in the inverted U service structure and recommended the formation of a study group.[12]

By mid-April 1960 Albert Zeiler was directing a two-pronged investigation into problems encountered with LC-34's service structure and concepts for a larger one. The latter reflected NASA's decision to build LC-37 for both C-1 and C-2 versions of the Saturn.[13] The service structure committee met periodically over three months to review 21 concepts proposed by NASA officials and private industry. No proposal proved fully satisfactory; attractive features from several were combined in the final recommendation. The committee concentrated on a half-dozen aspects of the service structure design, posing these alternatives:

- Mobile or fixed structure?
- Bridge crane or stiff-leg derrick for hoisting?
- Protection for the launch vehicle from wind loading or absorption of the rocket's wind loads into the service structure?
- Open or closed service platforms?
- Launch stand above or below ground?
- Collapsible or fixed umbilical tower?

The fixed service-structure designs were attractive since they offered economy and good utilization. The committee, however, feared the effects of a pad explosion on a fixed structure. The fixed design also posed a difficult engineering problem. Long cantilevered platforms with elaborate retracting mechanisms were needed to keep the main structural frame outside the rocket's drift cone (the safety allowance for effect of surface wind at launch). At an 11 July meeting in von Braun's office, Marshall officials discussed the effects of wind drift, thrust malalignment, and loss of one engine on the clearance requirement for a service structure or umbilical tower. The participants agreed to a 12-meter clearance between the vehicle's center line and the nearest obstruction at the 91-meter level. About the same time the Zeiler committee opted for a mobile service structure.[14]

The hoisting matter was settled in favor of a stiff-leg derrick mounted on top of the service structure. Although a bridge crane offered more flexibility, its use in the upper reaches of the service structure would obstruct the vertical escape trajectory of a manned payload. In the final design a 40-ton mobile crane positioned at a lower level assisted the 60-ton main hook on the stiff-leg derrick.

The question of wind loads arose because the Saturn was not self-supporting in high winds. One alternative was to design "hard point" connections between the vehicle and service structure platforms. This would require additional structural members on the rocket, increasing its empty weight. It would also add considerable stress to the service platform. The committee chose a design enclosing the launch vehicle in a 76-meter silo of

Fig. 43. The LC-37 service structure at pad B.

Fig. 44. The LC-37 service structure in the open position, February 1963.

five sections that eliminated wind loads and protected the rocket from flying debris. The silo design also solved the service platform problem. The committee recommended a minimum of ten adjustable work platforms in the structural steel frame silo. Air conditioning would provide the necessary ventilation during propellant loading.

The committee rejected a plan to put the launch stand below ground with the flame diverted into side trenches. Doing so would reduce the height of the service structure, but the higher costs of subsurface facilities, due to Cape Canaveral's high water table, were unacceptable.[15]

In designing the umbilical tower, the major concern was separation of the umbilical connections from the launch vehicle at liftoff. The committee studied jointed, collapsing towers; towers supported by cable catenaries (a curved cable suspended from two poles); and pivotal reclining structures. The size and weight of the Saturn umbilical connections—propellant piping, pneumatic lines, instrumentation circuitry, and electrical power lines—rendered all those concepts impractical. The committee recommended a free-standing umbilical tower, with ties to the service structure for support against high winds. Swing arms, entering the silo enclosure through cutouts in the platform mating edge, would connect the umbilicals to the launch vehicle.[16]

In August 1960 the launch team approached von Braun about adding a second pad to the LC-37 complex. The additional pad would provide a backup for Saturn C-2 launches and reduce launch time by one-third, since it would eliminate the month needed for pad repairs. Von Braun directed Debus to add a second pad on LC-37 if funds could be secured. Before the meeting adjourned, General Ostrander, Office of Launch Vehicle Program Director, arrived. After reviewing the proposal, Ostrander agreed to provide $700 000 for the initial modification work.[17]

Further revision of LC-37 plans occurred in early 1961. In January Debus heard of an Air Force–sponsored study on blast potentials of the Atlas-Centaur rocket. The Arthur D. Little Company findings, Debus informed von Braun, "indicated a problem of considerable magnitude with Saturn complex siting."[18] Since there was little data on liquid hydrogen's explosive characteristics, the calculations were tentative. The Little report, however, reinforced the Hall Committee's conclusions. On 12 January Debus asked Petrone, as Saturn project coordinator, to investigate the explosive potential of liquid hydrogen and determine the cost of extending pad distances beyond 183 meters. The distance between pads was subsequently increased to 365 meters.[19]

Two Florida firms won the LC-37 design contract: Connell and Associates prepared the service structure and umbilical tower designs, while Reynolds, Smith, and Hills handled the subsurface facilities. The architects' design work extended from February to July 1961. During the same period, Gahagan Company dredged thousands of cubic meters of sand from the Banana River onto the LC-37 site. Vibroflot machines began their work at the complex in mid-July. Blount Brothers Construction Company of Montgomery, Alabama, won the pad B construction bid in August 1961 and started work the following month. The project was 45% complete on 30 March 1962, when the Corps of Engineers awarded Blount Brothers a contract to build pad A.[20]

The new construction soon overshadowed the older Saturn facility. LC-37 was nearly three times larger than LC-34. The two umbilical towers rose 82 meters from a 10-meter-square base. Stability of the towers in high winds presented a challenge to the designers. The large number of electrical, propellant, and pneumatic lines running up through the lofty structures gave the tower surface a wind resistance nearly equivalent to a solid wall. At the base of each tower stood a four-story building (one floor was underground) containing a generator room, high-pressure-gas distribution equipment, and a cable distribution center. The building would later house digital computers for the automated checkout.[21] Hydrogen burn ponds were an added feature on LC-37. The gaseous hydrogen boiled off from the LH_2 storage tank and the S-IV stage and flowed several hundred meters through pipes to the burn pond. The LC-37 launch control center, or blockhouse, was similar to LC-34's, but half again as large. By far the most imposing of LC-37's facilities was a 4700-ton, 92-meter-high service structure, containing four elevators, nine fixed platforms, and ten adjustable platforms that allowed access to all sides of the vehicle. The six semicircular enclosures could withstand 200-kilometer-per-hour winds. When completed in 1963, the self-propelled, rail-mounted structure was the largest wheeled vehicle in the world.[22]

Erection of a special assembly building was a third construction project for Saturn I in 1962. Some novel building designs were rejected before deciding on a conventional hangar configuration. The new hangar AF was in the Cape industrial area, a short distance from the Saturn dock. A bridge crane in the hangar's main bay provided a lift capability for the initial upper stage checkout; lean-tos on both sides provided extra office space. The Launch Vehicle Operations Division performed some preliminary checkout work in hangar AF, but half of its big bay was soon given over to Gemini and Apollo spacecraft operations.[23]

The Troubled Launching of SA-5, January 1964

The block II version of Saturn I (SA-5 through SA-10) represented a sizable increase in launch requirements over block I. Additional RF links,

Fig. 45. The industrial area on the Cape. Hangar AF is in the upper left. The causeway (under construction) leads to Merritt Island in the distance.

Fig. 46. Mating spacecraft modules inside Hangar AF, March 1964.

calibrations, and systems tests in the two-stage rocket nearly doubled launch checkout time (see table 1).

The greatest change in the block II rocket was the addition of a hydrogen-fueled second stage. Douglas Aircraft Corporation had won the contract for the S-IV stage in April 1960, five months after NASA adopted the Silverstein Committee's recommendation to use liquid hydrogen in the Saturn's upper stages. The 13-meter stage had six Pratt & Whitney RL-10 engines, the same power plant that NASA intended to use in the Centaur rocket. Confidence in the S-IV stage originally stemmed from expectations that Centaur tests would prove the effectiveness of the engine long before SA-5. As things worked out, the first successful Centaur launch came in November 1963, more than two years behind schedule and only two months ahead of the SA-5 launch.

SA-5 differed in other ways from its Saturn predecessors. Engineers had increased the 340500-kilogram capacity of the S-I first stage by more than 31%. Each H-1 engine had been uprated to its intended 836600 newtons, giving that stage its full thrust. Marshall had also attached eight fins to the base of the S-I stage, four stubby fins and four longer ones that extended 2.7 meters from the rocket. These provided additional aerodynamic stability (a decision prompted by possible use in the Dyna Soar program). The guidance and control instruments for both stages flew in an experimental instrumentation unit above the S-IV stage. The payload for SA-5 was a Jupiter nosecone.[24]

The postlaunch celebration for SA-4 was barely over when Hans Gruene's Launch Vehicle Operations team turned its attention to the block II series. The first order of business was fitting LC-37B with a dummy SA-5 vehicle. The dummy stages were erected and mechanical support equipment tests completed by the end of April 1963. In the first two propellant flow tests, the transfer system kept the hydrogen below 20 kelvins (−253° C). Chemical analysis revealed contaminants, but the liquid hydrogen cleared up on the third test, saving the launch team a detailed investigation. There were a number of routine problems such as leaking LOX lines, freezing LOX vent valves, and inoperative gauges. Only one major change was required, a modification of the baffles in the S-I stage LOX tank. There was time for this since the SA-5 launch date had been moved from August to December.[25] After the wet tests were completed in late June, NASA flew the S-IV dummy stage back to California aboard a modified B-377 aircraft.*

Gruene's launch team erected the Saturn booster on 23 August and during the next 30 days performed mechanical system tests, calibrations for

*Because of its enlarged fuselage, the plane was popularly known as the "Pregnant Guppy."

Fig. 47. The Pregnant Guppy, a modified B-377 aircraft used to airlift Saturn stages, July 1963.

the instruments, and telemetry and RF tests. The only serious difficulty was one that apartment dwellers can appreciate—four service structure elevators that were frequently out of order and usually crowded. In the upper levels of the 90-meter-high, open structure, elevator cables were exposed to rain and wind. Maintenance problems were inevitable. In September 1963 the combined load of facilities contractors (outfitting the service structure) and SA-5 launch technicians strained the elevators' capacity. Gruene informed Debus on the 12th that "elevator usage is now critical and may become intolerable when checkout activities require more personnel."[26] Gruene hoped to finish outfitting the service structure after the normal workday to alleviate the problem.

Fig. 48. Transporting the SA-5 first stage to pad 37B.

Fig. 49. Erecting SA-5. The live S-IV stage is being lowered into position, replacing the dummy spacer, which is on the ground (left).

In Sacramento, Douglas engineers completed four weeks of post-static checkout of the S-IV on 10 September. The second stage was removed from its test stand, loaded aboard the B-377, and flown to Cape Canaveral. Douglas personnel gave the stage a thorough inspection, including the use of a sound probe to detect debonding of tank insulation. The probe was moved back and forth over the outer surface of the stage, its signal reflecting back from the inner side of the tank skin into the probe's sensing device. An oscilloscope showed both the output signal and the echo. Welds or any other irregularities stood out clearly. Heavy winds and rain that struck the Cape the following week did not halt S-IV activities in hangar AF. However, out on pad 37B the telephones failed, the service structure elevators were temporarily shut down, and the launch team lost three days of work.[27]

Operations reached a hectic pace in mid-October. After the S-IV stage was erected on the 11th, Robert Moser's office revised the launch schedule to give Douglas a week longer for S-IV checkout and modifications and a combined LOX-LH_2 tanking test. Moser maintained the 6 December launch date by compressing the time allowed for launch vehicle tests in November. Even with the extra week, Douglas found the test requirements more than it could handle in a 16-hour day. On 17 October the California firm asked for around-the-clock operations until the propulsion tests were completed. Gruene, hard pressed to support the S-IV stage operations, reluctantly agreed.[28]

The Cracked Sleeves

Although the S-IV erection was the major activity on 11 October, that day's status report also mentioned the discovery of a cracked sleeve on the "S-I engine position #3 hydraulic package, yaw actuator, low pressure return line."[29] The sleeve, a centimeter-long metal cylinder, was an integral part of end-fitting assemblies on hundreds of pneumatic and hydraulic line joints in the first stage. Technicians replaced the sleeve on the 15th and continued the check of the hydraulic actuator. The incident, however, caused concern in Huntsville where Chrysler personnel had reported similar sleeve failures after pressurization tests. A special investigation of S-I engines on the 22d found 12 more cracked sleeves. These sleeves and the affected tubing were replaced during the next two weeks. The cracked sleeves apparently had little to do with the decision in late October to delay the launch another five days. Gruene blamed the delay on contamination in the S-IV engine and time lost for a hurricane alert.[30]

The assassination of President Kennedy slowed operations for three days, but the revised schedule was still being maintained in late November. A cryogenic tanking test on the 26th started well enough. There were only minor problems as the team went through the various phases of S-I LOX loading: the 15% slow fill, the fast fill, topping off, and replenishment. It was evening when liquid hydrogen began to flow to the S-IV stage. Albert Zeiler, arriving at LC-37 to watch the last portion of the test, heard an explosion but could not immediately contact Andrew Pickett, the Chief of the Mechanical and Propulsion Division. Inside the blockhouse, a technician at the periscope saw fire on the pad. Television monitors picked up the flames, but gave only a vague idea of the fire's extent and location. Pickett terminated the hydrogen loading. A visit to the pad revealed the cause of the explosion. Gaseous hydrogen had leaked from a ruptured bellows in the hydrogen vent line that ran from the rocket to the burn pond. The rupture was probably caused by water seeping back into the pipe from the burn pond and then freezing when the cryogenic hydrogen entered the line. The escaping hydrogen had collected beneath the metal plates covering the vent line trench. Purging the vent line with helium quickly extinguished the fire. The launch team then detanked the propellants, leaving damage assessment for the following day.[31] The fire caused Robert Moser to reschedule the cryogenics test for 6 December, put operations on a seven-day week, and predict a one-week delay for the launch.

Although there were problems on the next cryogenic test, launch was still expected before Christmas. On the 10th, however, the launch team detected its fourth cracked sleeve in two days. The discovery of seven more cracked sleeves the following day caused Marshall to postpone the launch for a month despite a successful simulated flight test on the 13th. In the interim the launch team replaced all of the sleeves* in critical pneumatic and hydraulic circuits.[32]

The cracked sleeves were not the last of SA-5's problems. During the simulated flight test, D. C. McMath's RF and telemetry section had experienced radio interference in the 400- to 450-megacycle band. Results of an RF check on 23 December provided no holiday cheer as three of SA-5's four command destruct receivers responded to an Air Force Range signal, 42 megacycles above that used for the Saturn destruct command. Although

*The sleeve failure was attributed to a change in specifications and the longer length of SA-5 checkout. SA-5's sleeves had been cast at a different temperature from previous sleeves and one result was the appearance of carbon pockets in the stainless steel cylinders. These carbon pockets reduced the "long-life" factor (measured in seconds of operational life for some rocket hardware). MSFC eventually scrapped 22 000 defective sleeves.

Fig. 50. The service structure moving back from SA-5, November 1963.

McMath was anxious to unscramble the signal-mixing, further testing had to wait two weeks for complete external RF silence. January tests appeared to place the source of signal mixing within the service structure, but when the structure was removed on the 14th, the interference continued. Suspicion next turned to the umbilical tower, and the possibility "that RF signals transmitted from the vehicle are being mixed [there] to produce the interference." The launch was nine days away when the RF section finally ran a satisfactory test on the 18th. Even so, the source of trouble was not definitely identified. Since some team members still considered the UDOP tracking station a possible source of interference, McMath recommended removal of the UDOP power amplifier.[33]

All's Well That Ends Well

The last weekend in January, America's television networks prepared live coverage of the SA-5 shot scheduled for Monday the 27th. An incident on Friday had threatened to postpone the launch: during a static firing test at Sacramento an S-IV stage had exploded, damaging the test stand and support equipment. After evaluating the accident, NASA officials decided the likelihood of an S-IV engine failure was sufficiently low to proceed with the SA-5 launch.[34]

Col. Lee James, Marshall's Saturn I-IB project manager, and Ted Smith, Douglas director of S-IV stage development, were among the 200 who gathered at the LC-37 blockhouse on Sunday evening for the start of the SA-5 countdown. Robert Moser was test supervisor for the operation; KSC's John Twigg and Douglas's John Churchwell served as test conductors for the S-I and S-IV stages. There were three holds during the night: 3 minutes for network checks, a 17-minute hold for battery verification, and a 27-minute hold to change an accelerometer. Shortly after sunrise the launch team discovered a leak in the S-IV main LOX line that took 48 minutes to correct.[35]

The countdown proceeded satisfactorily despite these minor problems. S-I LOX loading began about 8:30 a.m. and went smoothly through the fast fill. When LOX reached the 93% level in the first stage tanks, the propellants team switched to the LOX replenish system (used to ensure a controlled slow flow). Instead of continuing its rise, the S-I stage "mass readout" (the percentage of LOX in the tanks) began to fall. Launch officials quickly realized that the replenish system was not supplying LOX to the S-I stage. Leroy Sherrer, Oxidizer Section chief, first thought a frozen valve might be the cause of the failure. Finding the replenish facility in order, Sherrer's group moved up the LOX line toward the pad. W. C. Rainwater's Ground Support Equipment Section started from the other end of the line, the base of the rocket. In less than an hour, the teams found the blockage—a "blind" flange (plate without an opening) left in the replenish line from a previous pressurization test. Safety precautions and venting problems precluded the immediate removal of the aluminum plate, and Debus reluctantly scrubbed the mission.[36]

A tired Rocco Petrone informed 150 newsmen of the launch postponement. He admitted that failure to remove the flange was a human error, but refused to single out anyone. "It was a routine procedure that we've done many times before. This time we didn't do it. We make mistakes."[37] Debus had an even less pleasant task—explaining the mishap to five members of the House Subcommittee on Manned Space Flight, down from Washington for an inspection. The KSC director assured the visiting Congressmen that in future operations the launch team would tag flanges with red flags, as they presently did with all electrical work. In this way any deviation from the operational flight configuration would be flagged and a record kept by test supervisors. Debus rescheduled the launch for Wednesday morning, the 29th.[38]

There was one unplanned interruption in the second countdown, a 73-minute hold due to RF interference on the C-band radar and command destruct frequencies. At 11:25 a.m. SA-5 lifted off into a 37-kilometer-per-hour wind and a heavy sprinkling of clouds. Painted designs on the rocket's skin aided nine unmanned and four manned cameras to track pitch, yaw, and roll movements for the first 1000 meters. Six camera-equipped tracking telescopes, located along the Florida coast and on adjacent Grand Bahama Island, provided higher-altitude photographic coverage. Radars fed information to three computer-operated flight position plotting boards located in blockhouse 37. Another KSC computer, linked for the first time to an Eastern Test Range vehicle impact prediction computer, transmitted real time (very nearly instantaneous) vehicle position data to Marshall, as well as to Goddard Space Flight Center, NASA's communications center in Maryland. Telemetry aboard the SA-5 transmitted 1183 separate measurements back to seven receiving stations in the Cape area; the ground stations relayed this information by radio and hardwire* to data processing machines in hangar D.[39]

**Hardwire* meant any system of electrical wiring over which signals passed, as distinguished from radio transmission.

Fig. 51. The launch of SA-5, 29 January 1964.

Fig. 52. The launch of SA-5, moments later.

Fig. 53. Damage to pad 37B from the launch of SA-5. The short cable mast (top) carried electrical and pneumatic lines to the first stage. Access plates have been opened in the support arms (lower R and L) to inspect the pneumatic system.

The launch vehicle carried eight movie cameras and a television system to record stage separation and ignition of the S-IV engines. The separation of the two stages began at T + 147.2 seconds, 6 seconds after the first stage inboard engine had shut down and 0.2 second after the outboard engines had cut off. The first action was the firing of small S-IV ullage rockets which forced propellants toward the engines. As booster retrorockets fired to slow the S-I stage, explosive bolts disconnected the two stages. The S-I and eight camera capsules fell into the Atlantic 800 kilometers downrange from Cape Canaveral. The S-IV engines then burned for 8 minutes, placing 16965 kilograms in orbit, the heaviest payload in history.[40]

A nationwide audience viewed the SA-5 launch on television and received a remarkably clear picture of booster engine shutdown at 60000 meters altitude. Immediately following the launch, President Johnson telephoned his congratulations to the launch team in blockhouse 37. He told Wernher von Braun that he hoped his recent gift of a Texas hat would still fit the MSFC director. Von Braun contrasted the day's success with the Explorer I launch six years earlier and praised the Douglas Company for its role in developing the S-IV stage. Although the achievement of earth orbit was not even a secondary goal, Robert Seamans said the mission left "no question" that the United States had surpassed the Soviet Union in "ability to take large payloads into orbit." George Mueller, NASA's Associate Director for Manned Space Flight, described the launch as "the first step to the moon."[41]

The Remaining Block II Launches, SA-6–SA-10

SA-6 (28 May 1964)

Later Saturn I missions brought new requirements and major launch problems, but none of the subsequent operations dragged on like SA-5. Launch preparations for the remaining five Saturns averaged 91 days, 70 days less than the SA-5 operation. An Apollo boilerplate, duplicating the weight and external configuration of the fully equipped spacecraft, flew on the May 1964 launch of SA-6. Boilerplate 13, the payload for SA-6, was one of 30 spacecraft built by North American for preliminary Apollo tests. The Manned Spacecraft Center had already launched several boilerplates at White Sands Proving Grounds to test the spacecraft for land and water impact, parachute recovery, pad aborts, and water egress and flotation. SA-6 demonstrated the spacecraft's structural compatibility with a Saturn launch vehicle.[42]

The checkout of boilerplate 13 had begun in December 1963 when G. Merritt Preston, Director of Houston's Florida Operations, sent George T. Sasseen to North American's Downey, California, plant with a 40-man team. Sasseen's counterpart on the North American staff was project engineer Robert Gore. For two months the NASA–North American team subjected boilerplate 13 to a series of rigorous tests, from assembly line inspections to simulated flights. After the spacecraft was transferred to Florida, there were more tests in hangar AF. By early April the spacecraft team was ready to stack the boilerplate atop the Saturn I vehicle. During the next six weeks, the team resolved problems in the spacecraft cooling systems and in the mechanism for jettisoning the launch escape tower. Much time was spent checking telemetry and the 116 instrumented measurements that recorded structural and thermal responses.[43]

The 20 May launch date was postponed after liquid oxygen damaged a wire mesh screen during a test, causing fuel contamination. Six days later, a countdown proceeded satisfactorily until T – 115 minutes, when a compressor in the environmental control system failed. The air conditioning gone, the temperature in the rocket's guidance system soon exceeded tolerance and the launch was scrubbed.[44]

On 28 May it seemed that Launch Vehicle Operations might postpone the third attempt. Liquid oxygen vapors, vented from the S-IV stage, obscured the line of sight from a ground theodolite to an optical window in the SA-6's instrument unit. Winds blew the vapor away after a 38-minute hold, but adjusting a LOX replenish valve forced another hour's delay. Then in the last minutes of countdown, the sighting problem recurred. This time LOX vapors from an umbilical tower "skid vent" blanketed the optical window. Since stabilized platform alignment control was essential to the launch, the automatic sequencer included this function among its checks. If the theodolite did not have a clear sighting, the sequencer would shut down at T−3 seconds. Quick action by two launch team members saved the day. With the count stopped at T−41 seconds, Terry Greenfield, Electrical Systems Branch chief, removed the stabilized platform reference from the sequencer's functions by "jumpering out" several electrical wires. Meanwhile, Milton Chambers, Gyro and Stabilizer Systems chief, improvised a way to maintain the platform in its proper flight azimuth through manual control. The count resumed 75 minutes later. Ironically, the vapors blew away from the optical window during the final 40 seconds of countdown.[45]

SA-7 (18 September 1964)

Since 1954 Redstone, Jupiter, Pershing, and Saturn rockets had employed a 33-pound multichanneled tape recorder, commonly called a "black box," for inflight commands such as inboard engine cutoff, ullage rocket ignition, and fuel pressure valve openings. It was replaced on SA-7 by a computer that could be corrected during flight. SA-7 also marked the close of Saturn I research and development tests. Following the seventh successful launch, NASA officials declared the Saturn I launch vehicle "operational."[46]

SA-7 set two precedents in Kennedy Space Center launch operations. In early July technicians found a cracked LOX dome on engine 6 of the S-I stage. It was the first time the launch team had to replace a Saturn engine. The experience was not novel for long. NASA officials, attributing the cracks to the same "stress corrosion" that had plagued SA-5 sleeves, returned all eight engines to the Rocketdyne plant in Neosho, Missouri. The removal of each 725-kilogram engine took KSC and Chrysler mechanics about ten hours. As the supervisor described it: "We had to disconnect all electrical cables, unhook the hydraulic systems from the outboard engines, and disconnect LOX and fuel suction lines, the turbine exhaust, purge lines, networks and measuring cabling. It was quite a job."[47]

Replacing the engines in the S-I stage set the launch back from late August to mid-September. Hurricanes Cleo and Dora cost another half-week's work. Although Cleo struck the Cape on 28 August with 110-kilometer-per-hour winds, SA-7 was unharmed inside the service structure.* A surprise visit by President Johnson on 15 September coincided with the first countdown demonstration test, an exercise added to the launch

*NASA officials estimated that the two hurricanes cost about $250 000 in terms of property damage and manhours for storm preparation and cleanup. Water damage was extensive at the LC-39 construction sites. Hangar AF on the Cape was another casualty; a leaky roof resulted in a lot of soggy artwork and photo-processing gear for Technical Information's Graphics Section.

Fig. 54. Rocco Petrone briefing Maj. Gen. Leighton I. Davis, Administrator James Webb, and President Lyndon Johnson in the LC-37 control center, September 1964.

checkout after the blind flange incident on SA-5. Robert Moser's Technical Planning and Scheduling Office had decided to run, as the last test, a full countdown of the fully fueled Saturn (with a mission abort just prior to scheduled umbilical ejection). The test would become a focal point of launch operations in later Saturn missions. Its first performance went smoothly, as did the launch on the 18th.[48]

SA-8, 9, 10 (16 February through 10 July 1965)

Each of the last three Saturn I's carried a Pegasus satellite enclosed within a boilerplate service module. The satellite's function was to determine the incidence and severity of meteoroids in the region where Apollo astronauts would orbit the earth. As Pegasus was not an integral part of the Apollo program, its use raised an administrative question—who would be responsible for launch and inflight control? NASA Headquarters placed Huntsville in charge of configuration changes during launch operations. Debus was assigned mission responsibility through earth-orbital insertion. He then turned over Pegasus direction to a representative from the Headquarters Office of Advanced Research and Technology.[49]

Fig. 55. Countdown demonstration test of SA-8 on pad 37B, May 1965. The service structure is moving away. The launch escape system (the top-most part of the space vehicle) was flown, but not activated, on this mission.

As the manufacture of the SA-9 booster progressed more rapidly than the SA-8, the next two Saturn shots were fired out of sequence; the SA-9 launch preceded SA-8 by three months. Problems with the Pegasus satellite delayed the erection of SA-9 until late October 1964. Once operations were under way, the launch team experienced little difficulty. SA-9 roared off its launch pedestal on 16 February after two technical holds: one involved the recharge of a battery in the Pegasus; the other came when the Eastern Test Range's flight safety computer suffered a power failure. Pad damage from the rocket exhaust was described as "the lightest of any to date."[50] There was some water damage, however, from a broken torus ring. The ensuing cascade of water flooded the launcher and adjacent electrical support equipment.[51]

Contractor teams dominated LC-37 during launch preparations for SA-8. The operation marked Chrysler Corporation's assumption of responsibility, under broad guidance, for first-stage operations. The company's launch team also participated in overall space vehicle testing. Douglas officials directed S-IV stage checkout, IBM conducted tests on its instrumentation unit, and Bendix Corporation provided ground support. After 86 days of space vehicle checkout, SA-8 launched Apollo boilerplate 26 (with Pegasus 2 inside) on a successful 25 May flight.[52]

The SA-10 operation was conducted in haste. NASA officials had decided to begin LC-37 modifications for the Saturn IB rocket in August. If Kennedy Space Center could not launch the rocket by 31 July, its flight would have to come after the IB series. Under the pressure of this deadline, Chrysler and Douglas undertook 24-hour operations.

The SA-10 countdown proceeded without a technical hold, a near perfect finish to a highly successful series. The NASA–Saturn-contractor team had demonstrated the soundness of the Saturn I rocket and its launch facilities. A confident launch team looked forward to the next challenge: Saturn IB.[53]

11

Ground Plans for Outer-Space Ventures

The Task

By late 1962 NASA had made most of the basic decisions affecting the development of launch facilities and was ready to build the moonport. Contractors would start construction on the main buildings at launch complex 39 and in the industrial area, eight kilometers to the south, as soon as sufficient design information was available, and install equipment as construction proceeded far enough to allow safe access. At the same time, engineers were expanding and modifying the existing facilities at launch complexes 34 and 37 for earth-orbital tests of Apollo spacecraft launched by Saturn IBs.

Designers, meanwhile, were working on the final stages of the Apollo spacecraft. This complicated the design and equipment of facilities at the Launch Operations Center. The basic dimensions, weights, and operating principles of the rocket and spacecraft were known, but questions remained about specific sizes, types, quantities, flow rates, pressures, or even methods of use. Answers to many such questions awaited completion of designs at Huntsville and Houston. Policy makers had to make commitments on the basis of the best information available, knowing that costly and time-consuming changes might well become necessary.

Any large construction project passes through several common stages: selecting and preparing the site, choosing or developing the equipment for use in the operation, planning the external structure, and constructing and equipping the facilities. The pressure of time was such that, during the erection of the Apollo launch facilities, what would ordinarily be consecutive steps were often simultaneous.[1]

By working backwards from the earliest launch date (March 1966) and estimating the time required for vehicle assembly and checkout, the date when the basic launch facilities had to be in operation could be found. Working backwards further and estimating the time required for construction and outfitting yielded the date for the start of construction. Such computations showed, in 1962, that little time remained for development of criteria and detailed design.

The requirements of the manned lunar landing program found the Launch Operations Center facing some new problems, while some old problems were becoming more acute. The new were the size and complexity of the Saturn V vehicle; the need for unprecedented reliability, flexible launch rates, and a short recovery time between launches from the same pad; and the use of the mobile concept. These, in turn, raised old questions about the marshy composition of Merritt Island and the possibility of hurricanes.

Central to LC-39 would be an assembly building, where the Saturn V vehicle would be put together. The Saturn's size was such that the vehicle could not be transported as a unit from its place of construction, but had to be assembled and checked in a vertical attitude near the launch site. The major components were three stages, an instrument unit, and the Apollo spacecraft.

The design of the assembly building had to allow for stacking the 110-meter Apollo-Saturn space vehicle on top of its 14-meter-high movable launch platform. The structure would be taller than any building in Florida. To handle the stages of the vehicle, bridge cranes had to span 45 meters and lift 121 metric tons to a height of 60 meters. The architect-engineers faced complex problems, particularly since the structure had to be capable of withstanding hurricane winds.

To make room for the assembly and checkout of the various stages of three or four vehicles of this size simultaneously required an enormous building. The planners decided to have four high bays or checkout areas, each big enough to handle all stages of the Saturn V and the spacecraft in a stacked position—that is, completely assembled in an upright position ready for launch. The planners could foresee no situation that would require working on more than four rockets at one time; but if requirements changed, they could add more high bays at a later date. Additional low bays would accommodate preliminary work on single stages.

URSAM Makes Its Debut

In August 1962, a Launch Operations Center committee asked the Corps of Engineers to select an architect-engineering firm to complete the criteria for the vertical assembly building, or the VAB as it came to be called. The Corps formed a selection board representing its South Atlantic, Southeastern, North Atlantic, and North Central Divisions, as well as the Jacksonville District Office. The selection board submitted a list of five firms. From these the Chief of Engineers selected a New York combine made up of a quartet of companies—Max Urbahn (architectural); Roberts and Schaefer (structural); Seelye, Stevenson, Value and Knecht (civil, mechanical, and electrical); and Moran, Proctor, Mueser and Rutledge (foundations).[2] From the first name in each of the company names—Urbahn, Roberts, Seelye, and Moran—came a new acronym, URSAM.

The idea for the joint venture emerged in early 1962 when Max Urbahn and Anton Tedesko, of Roberts and Schaefer, discussed the possibility of designing the lunar launch center in Florida. Tedesko had directed the design of launch complex 36, the basic plans for the Minuteman facilities, and facilities at Chanute and Vandenberg Air Force Bases. Urbahn's firm, working in joint ventures with Seelye, Stevenson, Value and Knecht, had designed the intercontinental missile launching station at Presque Isle. Urbahn and Tedesko invited A. Wilson Knecht to join them; and Philip C. Rutledge of Moran, Proctor, Mueser and Rutledge, a firm that had designed foundations for more than forty projects in Florida, became the fourth partner.[3]

By March 1962 the combine had organized as URSAM. Although essential aspects of the Apollo launch facilities were yet to be determined, Urbahn and his associates set out to prove they could do a superior job in designing any concept ultimately selected. During the next five months, URSAM furthered its cause in a series of exploratory discussions at the Cape, Atlanta, Jacksonville, and Huntsville. On 10 August the Corps of Engineers asked the firm for a proposal on VAB design work. If URSAM's presentation appeared satisfactory, the Corps was prepared to offer the combine a criteria contract. Beyond that lay the possibility of the design contract. Shortly after the presentation in Jacksonville, URSAM received word that it had won a $99000 criteria contract.[4]

In a day-long orientation session held at the Launch Operations Center in late August, 21 persons, representing the Launch Operations Center, URSAM, the Corps of Engineers, and such contractors as Douglas Aircraft, were introduced to the projected building program.[5] Col. Clarence Bidgood of NASA Launch Operations Facilities opened the session with a discussion of the requirements of the VAB. He stressed practicality, insisted that a large portion of the criteria was available, and requested an early decision on the arrangement of the high bays: Should they be back-to-back or in-line? R. P. Dodd, of the LOC Facilities Branch, explained the basic premises of the VAB design. He said that initially the building would have four high bays. No hazardous operations, such as propellant loading or simulated altitude testing, would take place in the building. N. Gerstenzang, also of LOC's Facilities Branch, outlined the format of the criteria book; R. H.

Summarl of Douglas Aircraft discussed upper-stage checkout; and James H. Deese of Facilities Engineering gave a technical report that included wind loads on the VAB and the launch umbilical tower.

Gerstenzang set up a proposed work schedule from 3 September through 20 October and established 1 January 1963 as the date for foundation bids. He insisted that NASA wanted a free interchange of ideas directly with the architect-engineers during the criteria stage, with the Corps of Engineers as observer and monitor to assist in removing bottlenecks. He requested that the first man in Florida be a soils man from Moran, Proctor, Mueser and Rutledge, the foundations company of the URSAM combine.[6]

After winning the criteria contract, URSAM directors hired retired Col. William D. Alexander as project manager to coordinate the work and ensure firm adherence to schedules. Alexander had served as Chief of Facilities Design for the Air Force's Ballistic Missile Program in his last assignment. He took charge of the VAB design project when a 16-man team from URSAM began work at Cape Canaveral on 10 September 1962. The first major decision called for a back-to-back placement of the four bays (see p. 125).

On 17 September representatives of the Manned Spacecraft Center met with URSAM personnel to establish their guidelines for the VAB. They discussed the number of platforms, the size of the crew to work on each level, and the need for a dust-free room, called a white room. Three work levels would probably be needed, with 40 persons at each working level. The power requirements for the command module and the service module would be the same as in LC-34 and LC-37, but the requirements for the lunar excursion module would be double that of the service module. Houston wanted to bring representatives of North American into the discussion so that they would understand the anticipated checkout procedures.[7]

URSAM prepared preliminary draft criteria for the vertical assembly building based upon rough notes, sketches, and abstracts, which included a description of primary and supporting functions of the project, the estimated total number of occupants, functional flow lists of equipment, and the description of utility requirements within and adjacent to the VAB. When URSAM released this draft on 24 September, Bidgood's office immediately solicited comments from all agencies of the Launch Operations Center, as well as related offices at Huntsville.[8]

On 22 October 1962, URSAM submitted a 96-page report of descriptive material and 54 drawings, along with two scale models. Included were estimates of what each component would cost and when the bills would come due. URSAM sent copies to NASA Headquarters, LOC, the Corps of Engineers, Marshall Space Flight Center, and the Manned Spacecraft

Center. During the following month, many individuals of the Launch Operations Center offered criticisms, pointed out problem areas, and recommended changes.[9]

URSAM and the Design Contract

On 4 December the contract to design the vertical assembly building, launch control center, and adjacent permanent facilities was awarded to URSAM for $5 494 000. The New York firm had already begun work on the project and proposed to complete it by 23 September 1963. URSAM put a hundred men of its own staff to work on the VAB design and hired an equal number to supplement their efforts. The team of designers produced 2700 general drawings and a grand total of 18 000 shop detail drawings (one-third of them for the structural steel).[10]

URSAM divided the project into elements that could be designed individually and placed under contract at early dates. This step resulted in seven different contracts for procurement of equipment and for construction. In chronological order they were: first, preparation of the mobile launcher and crawler erection sites and the barge canal terminus so that the first launcher could be ready as soon as the first high bay in the VAB could receive it; second, foundation work for the VAB, including the piling and floor; third, setting up the structural steel frame for the high and low bays of the vertical assembly building; fourth, procurement of transformers and switching gear for the 69 KV substation; fifth, building of two 250-ton and one 175-ton bridge cranes; sixth, construction of the 69 KV substation; seventh, construction of the VAB, LOC, and utilities. The contracts were scheduled to bring the foundation and the structural steel frame well enough along at the time of awarding the last contracts so that the general contractor could proceed in an orderly manner with the work of completing the facility.[11]

Design Problems—VAB

In designing a building that was to have an enclosed volume of 3.6 million cubic meters (almost as much as the Pentagon and the Chicago Merchandise Mart combined) and an area of 32 000 square meters, URSAM faced a challenge. By using a simple box shape, the designers could obtain a strong building at minimum cost. Further, they could eliminate the need for separate cranes for each bay by putting a transfer aisle between the high

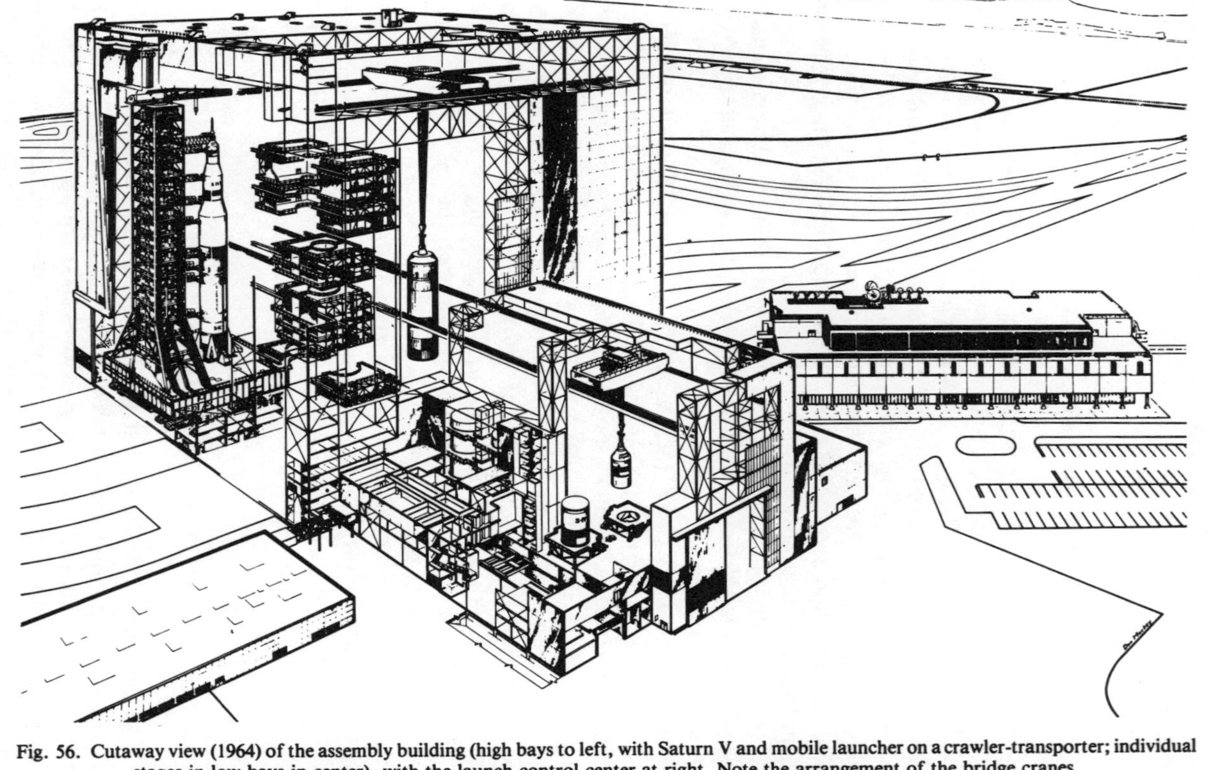

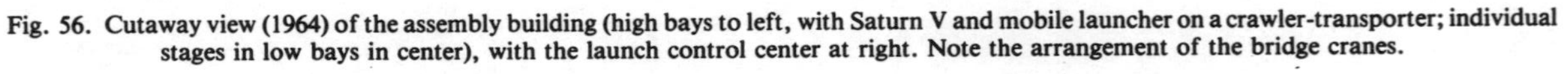

Fig. 56. Cutaway view (1964) of the assembly building (high bays to left, with Saturn V and mobile launcher on a crawler-transporter; individual stages in low bays in center), with the launch control center at right. Note the arrangement of the bridge cranes.

bays. The boxlike layout of the building, 160 meters high, 218 meters long, and 158 meters wide, allowed for an individual door and passageway from each high bay to the crawlerway.[12] According to Anton Tedesko, the following factors influenced the layout and structure: "stiffness against windloads, adaptability to changes, ease of connection with future extensions (planning included provisions for a 50% increase in assembly capacity to accommodate six space vehicles), and above all, adequate working space for those who would assemble and check out the vehicles and an efficient arrangement of that working space."[13]

One of the biggest design problems involved the high windloads that the building would have to withstand. After consulting authorities on wind velocity, URSAM designed the VAB for winds of 200 kilometers per hour. The design had to minimize the building's sidesway, because the work platforms in the high bays were tied into the structure. If the building swayed in high winds, the resulting movement of the platforms might damage a space vehicle. Although the box shape was not the most effective in shedding wind, it kept the sidesway low. The final design held the building's sway to less than 15 centimeters in winds up to 100 kilometers per hour. In higher winds the platforms would be withdrawn from the vehicle.

The designers had to accept certain operational penalties to achieve the required stability at reasonable cost. One of the most apparent was the 58-meter-high framework along the transfer aisle. This framework took 65% of the load from winds blowing parallel to the aisle (north-south), but restricted the passage from the transfer aisle to the high bays. Crane operators would have to lift the first stage up and over the framework to place the booster on a mobile launcher in the high bay.[14]

Since launches from the pad, 4.8 kilometers away, would subject the building to heavy shock waves and acoustical pressures, more than 100 000 square meters of insulated aluminum panels fastened to steel girders would be used to protect the structure on the outside. To create a "sense of airiness in the transfer aisle without admitting a glare or random sunbeams," URSAM recommended a total window surface of 6440 square meters provided by 1.2 by 3.7-meter impact-resistant, translucent plastic panels.[15]

The size of the building and height of the high bay areas presented other unique problems. Three major ones were the design and development of the atmospheric control system, the high-bay doors, and the lifting devices within the building. To provide proper distribution of air inside the building and to prevent condensation, the designers proposed a forced-air ventilation system with blowers at the top of the low bay and exhaust openings at the bottom. Large gravity ventilators in the roof of the high bays would pass sufficient air to replace the entire high bay volume at least once an hour. In order to maintain a comfortable temperature in the office, laboratory, and workshop complex situated within the low bay area, the designers planned a 9000-ton-capacity air conditioning system sufficient to cool 3000 homes.[16] In addition, by using its standby capacity this system would cool the space vehicle and base section of the mobile launcher. Self-contained units would cool individual platform levels in the high-bay section.

The selection of the proper doors to protect the inside of the VAB required much thought. The mobile launcher would enter and leave a high bay through an opening 139 meters high. The opening, shaped like an inverted T, would measure 45 meters wide at the base and 22 meters at the top. The designers settled upon a plan with seven leaves covering the top part of the opening. These leaves, 22 meters wide and 15 meters high, would lift vertically and sequentially to be stacked at the top of the opening. Four motor-driven leaves, among the largest doors ever placed on a building, would slide horizontally to cover the bottom 35 meters of the high bay opening. The eleven leaves weighed from 29 to 66 metric tons; opening them took nearly an hour.[17]

Fig. 57. Sketch of the assembly building, September 1963, showing the doors to the high bays.

Initially there was concern about some of the VAB's lifting devices. Operational requirements called for a 250-ton bridge crane with a hook height of 141 meters that could span a distance of 45 meters. The large crane and its support in the upper reaches of the VAB posed a weight problem for the foundation. As the design progressed, however, this problem disappeared—the foundation and structural strength required for the anticipated windloads provided ample support for the cranes.[18]

Design studies of the building's foundation included wind-tunnel tests of a model and a pile-test program. The latter was initiated under an URSAM subcontract by the C. L. Guild Construction Company in January 1963. Using a sonic hammer, the Guild Company tested the one-meter limestone shelf that lay 36 meters below the surface of Merritt Island and about 12 meters above bedrock. From the results of the wind-tunnel tests and bore samplings, URSAM engineers decided to rest the massive building on a bed of steel pipe piles, each 41 centimeters in diameter. The 4225 pilings, when driven down to bedrock, would total 205 kilometers of steel pipe. In addition the design called for 38 200 cubic meters of concrete as pile caps and floor slab.[19]

During the development of the design, URSAM representatives met regularly with individuals from the Corps of Engineers and the Launch Operations Center. Changes in equipment were frequent, and some of them meant changes in the design of the huge VAB. A change in the dimensions of the mobile launcher, for instance, represented a large—and welcome—weight reduction for the launcher, but also required a major change in the VAB doors.[20] Colonel Alexander urged the URSAM personnel to keep their counterparts in the Corps and the Launch Operations Center acquainted with daily progress and insisted that careful notes be kept on all intergroup discussions. The Facilities Office promised to deliver the final design instruction on 7–8 March 1963. Bidgood wrote: "The design must be frozen at this time to meet the design schedule and the subsequent construction schedule."[21]

Notwithstanding Bidgood's vigorous efforts, modifications of the space vehicle continued to cause problems for the VAB designers. In March, he noted that a recent change had undone 48 sheets of drawings. Shortly thereafter the Manned Spacecraft Center decided to transport the spacecraft in vertical attitude from the operations and checkout building to the VAB, a change that required more height in the low-bay doors. Bidgood refused to adjust the design schedule, stating that "delays in completion of final design as a result of this additional requirement are not acceptable." As late as 27 June, it was discovered that a required platform for S-IC intertank access was omitted from the design. Finally on 3 July 1963, the design agency notified R. P. Dodd, chief of the Design and Engineering Branch, LOC, that no changes or additional requirements could be permitted except as an amendment during the bidding period.[22]

URSAM forged ahead despite all the changes, drawing upon the technical capacity of the four constituent companies as the need arose. When key men had to leave, as three did during the course of the year, replacements were easily recruited. Initially the design chiefs relied on manual calculations for the basic designs, using computers to solve some equations. As their confidence increased, URSAM engineers came to rely more extensively on electronic computations. The task was completed and approved on schedule—23 September 1963.[23]

The original selection of URSAM had not won unanimous approval. When the combine had almost completed its work, a June 1963 article in the *New York Post* criticized the choice on the grounds that the Moran firm had designed the Air Force's $21-million Texas Tower that collapsed off the New Jersey coast in January 1961. In a report to Administrator James Webb on 13 June 1963, R. P. Young, executive officer at NASA Headquarters, discussed the matter at length. He admitted that many deficiencies had shown up in the tower, not all of them related to Moran's design; but in the URSAM combine Moran was working in foundation design, and the firm was "outstanding in that field." Young went on to explore the entire matter of the URSAM contract and the design, which he had discussed at length with Gen. Thomas J. Hayes III, assistant to the Chief of Engineers for NASA Support, who insisted that the Corps had made "a careful and straightforward selection of what they considered the best group of firms to do the job, and they know no reason at this time to believe differently." Hayes also pointed out that the Corps had selected the firm of Strobel and Rengved to make an independent review of structural design; the firm had often worked with the Corps in this capacity before. Hayes admitted the concern engendered by the newspaper article, but noted that a number of competent individuals had reviewed the work and gone away satisfied.[24]

Launch Control Center Design

URSAM also designed the launch control center, which presented far fewer problems than the VAB. The Manned Space Flight Management Council established ground rules for the design of the building in a meeting on 22 June 1962. Originally, the launch control center was to be placed at ground level in the western section of the low bay of the VAB. In October 1962, a suggestion to place it on the roof of the high bay held up the planning. An URSAM estimate that locating the center on the roof of the high

Fig. 58. Sketch of the launch control center, February 1963. The large shutters along the front of the upper floor could be closed quickly, in the event of an emergency.

bay would mean an additional expense of $1 200 000 ended the discussion. The structure was built on the southeastern side of the VAB with a long hallway connecting the two. The original plans called for a steel structure, but the structural engineers recommended concrete as better for acoustical purposes, and the final choice was a 114 × 55-meter, four-story, monolithic, reinforced-concrete building that made extensive use of precast and prestressed elements.[25]

The architects wanted the launch control center to be symbolic. The VAB was to be the factory, and the control center was to be man's window for observing events projecting into the future. The four multilevel firing rooms were rectangular in shape, 28 meters in width and 46 meters in length. Since many checkout requirements were still unknown, the planners emphasized flexibility, eliminating all columns and providing removable floors.[26]

The design of the windows shut off the sounds and pressures of the outside world. Two-centimeter-thick glass windows with adjustable sun visors in special aluminum frames faced the launch area. Infrared lamps outside the windows prevented fogging. The tinted, laminated windows, which covered an area 24 meters long and 7 meters high, filtered out heat and glare, permitting only 28% of the light to enter the room. Transparent glass, separating a viewing section from the rest of the firing room, gave guests a feeling that they were part of the operation. For its efforts on the launch control center, URSAM won the 1965 Architectural Award for the industrial design of the year.[27]

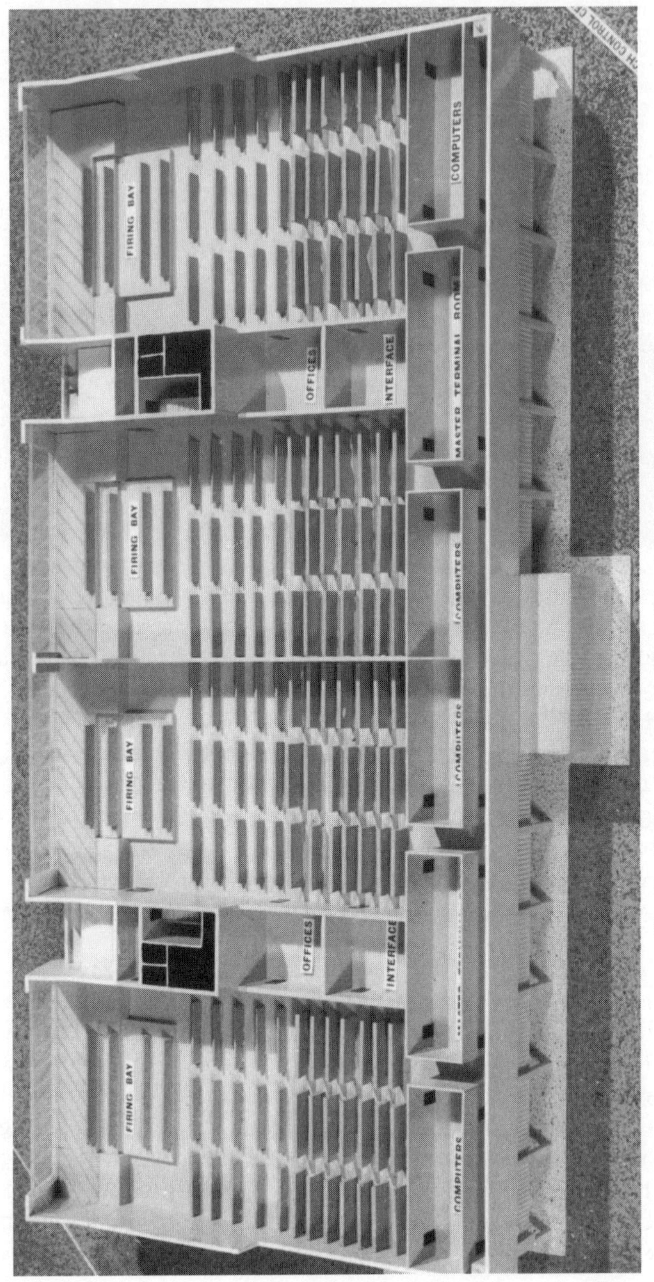

Fig. 59. Model of the proposed launch control center, February 1963.

While URSAM was designing the buildings, LOC engineers were determining what equipment would go inside. The control center's display stations were initially projected at a 13 September 1962 meeting between LOC representatives and Huntsville's Astrionics Division. Display consoles would monitor such things as propulsion, navigation, measuring, ordnance, propellant loading, ground support equipment, and emergency detection. Rack requirements included countdown clock, TV and communications, and discrete recorders. Criteria for the ten consoles located in each firing room for the control, test, and monitoring of the mobile launcher's electrical support equipment were furnished by the Launch Support Equipment Engineering Division. The design and fabrication of the panels were the responsibility of the Astrionics Laboratory at Marshall.[28]

During the following year, the emphasis shifted from a systems-oriented firing room to one organized by flight stage or hardware. Nearly 450 consoles would be operated by representatives from the stage contractors and Radio Corporation of America, General Electric, Saunders, Symetrics, International Telephone and Telegraph, NASA, and the Eastern Test Range. The consoles were arranged to permit the Boeing, North American, and other teams of engineers to sit together in their respective stage groupings. Responsibility for designing the consoles rested primarily with the various companies, but the designs were coordinated by LOC. W. O. Chandler, Jr., Deputy Chief of the Electrical Systems Branch, recalled making at least 25 trips to Houston and other Apollo offices to make certain that design change information for the consoles was current.

Design of the Crawlerway

While men had moved two- and three-story houses often enough—even some from Merritt Island to the mainland of Florida—no one had ever before moved a skyscraper. Yet that is what the mobile concept called for—or at least an Apollo-Saturn vehicle the size of a skyscraper. The problem was compounded by Merritt Island's marshy terrain and high winds. The combined weight of the crawler-transporter carrying the mobile launcher would exceed 7700 tons. No one knew what effect such a load would have upon the subsoil of Merritt Island. C. Q. Stewart of the Mechanical Engineering Division had commented on this problem in a memorandum of 1 August 1962 and suggested exploratory borings. He also spurned any type of rigid surface for the crawlerways as too prone to cracking, and urged instead a topping of gravel or crushed stone.[29]

On 1 February 1963, two months after the signing of the URSAM contract, the Detroit firm of Giffels and Rossetti agreed to design the crawlerway and the pads. Three weeks later 19 individuals representing NASA, the Corps of Engineers, and various contracting firms gathered at the LOC for a crawlerway conference. URSAM proposed building the crawlerway of layers with a total thickness of 1.4 meters, topped by crushed stone and a soft grade of asphalt. The Corps of Engineers agreed to compact sand to a depth of 7.6 meters below the pavement by vibroflotation in the areas adjacent to the VAB.[30]

At another meeting at the Cape on 27 March 1963, representatives of Giffels and Rossetti discussed the crawlerway with representatives of NASA, URSAM, the Corps of Engineers, Marion Power Shovel, and Brown Engineering Company. Donald Buchanan of LOC and one of the Marion representatives objected to the proposed use of an asphalt surface for the crawlerway. They feared that the asphalt would adhere to the treads of the crawler-transporter and cause severe wear of the road surface. The conferees then established two criteria for selecting materials: the surface material should not adhere to the crawler's treads and the coefficient of friction of the materials should not exceed 0.3 under the expected operating temperature range.[31]

During the next few months there were more meetings, one in Jacksonville on 27 June, another in Detroit on 14 August. At the former the Marion Power Shovel representative discussed the limits of friction. The conferees determined that the crawler would break up any type of hard surface, and the best surface would be crushed stone—as Stewart had suggested

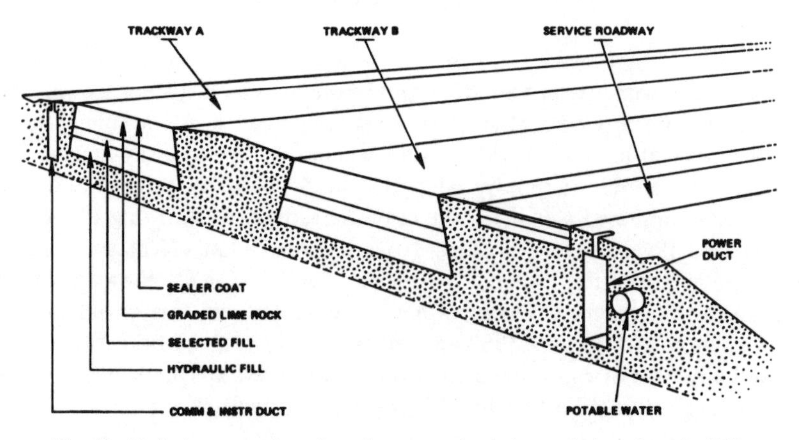

Fig. 60. Typical cross section of crawlerway, as the design took shape in early 1963.

a year before. After the latter meeting, J. B. Bing of the LOC programming office reported that there had been absolutely no coordination between Giffels and Rossetti and URSAM, even though their respective areas of design had an obviously close relationship.[32]

In the fall of 1963, eight representatives from LOC and the Corps of Engineers formed the "Construction Coordination Group for Complex 39"; the group's purpose was to manage the details of construction at the launch complex. The chairman briefed members on the construction status, problems, delivery of materials, and the impact of each change on critical construction schedules and contract costs. The scope of the group's work included scheduling, processing of changes, quality requirements, and funding. The Construction Coordination Group commenced operation immediately and was to continue until the completion of major Corps construction on complex 39.[33]

Flame Deflector and Launch Pads

If any one item virtually dictated the design of the launch pad, it was the flame deflector. This device would send the fiery exhaust of the five first-stage engines along the flame trench. The LOC designers who established criteria for the pads had wanted to keep the Saturn vehicle as close to the ground as possible in order to lessen wind stresses. They settled on a two-way, wedge-type flame deflector similar in design to those used on pads 34 and 37. The deflector, 13 meters in height and 15 meters in width, would weigh 317 tons. Since the water table was close to the surface of the ground, the criteria group wanted the bottom of the flame trench at ground level. The flame deflector and trench determined the height and width of the octagonal shaped launch pad; this in turn set the width of the space between the crawler treads, because the crawler straddled the pad.

During the last week of June and throughout July and August 1962, tests on 1:58 scale-model flame deflectors were conducted by the Test Division and Aeroballistics Division at Huntsville. They found that the launch complex 37 deflector, a copper, water-cooled, ridged model, suffered serious erosion from the concentration of heat and high gas velocities. By March 1964, the preliminary designs for a steel deflector and for a reinforced concrete deflector had been completed. By means of instrument readings and motion pictures, the aerothermodynamic flow characteristics were determined, and the flame deflector and trench designs were refined. In designing the deflectors for launch complex 39 pads, it was necessary to have a replaceable leading edge which eroded but was insulated. Four types of

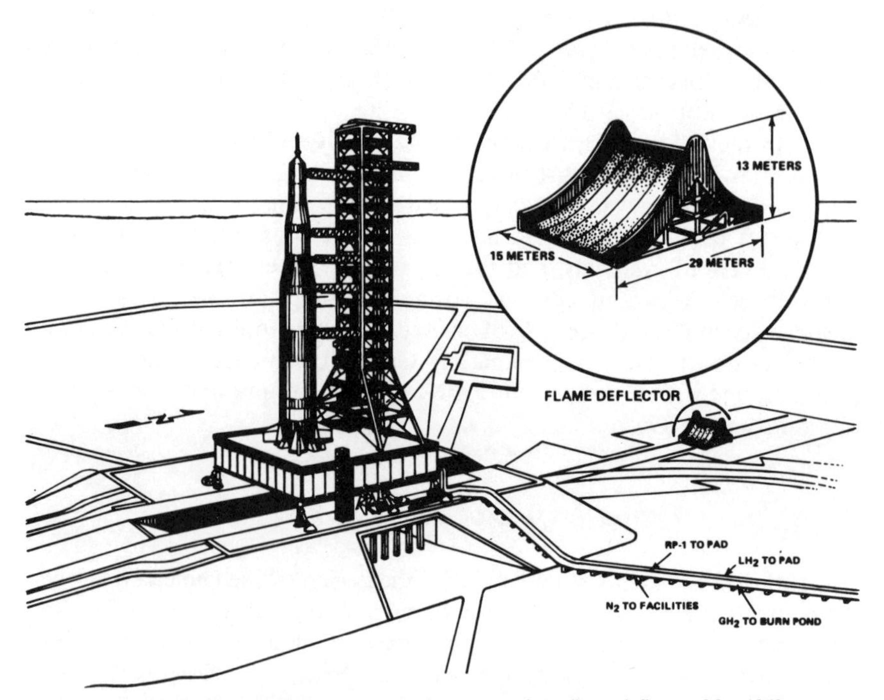

Fig. 61. Sketch of LC-39 pad, with enlargement of the flame deflector, May 1963.

deflector ridges were tested, using information gained in the study of heat-resistant shapes and materials for the Jupiter-C nose cone. When sufficient evidence was gathered, the design of the deflector proceeded with dispatch.[34]

The design of launch pads A and B presented further difficulties, many of which concerned the slab that covered the pads. This hardstand had to support the crawler-transporter and its pressure of 50 tons per square meter. Under a new proposal, a cellular construction, something like orange crates set in two rows, would support and protect the area adjacent to the flame deflector trench and beneath the crawler. The cellular construction, extending the full length of the flame deflector pit on either side, would provide an explosion buffer to the main launch facilities, reduce the pressure on the launch pad foundations, and offer additional space for service items.[35]

The selection of a refractory surface for the walls, floor, and an area outside of the flame trench was exacting. Such a surface had to withstand temperatures of 1922 kelvins and flame velocities four times the speed of sound. Special refractory fire bricks were held to the walls by interlocks, mechanical anchors, and a modified epoxy cement. All concrete surfaces

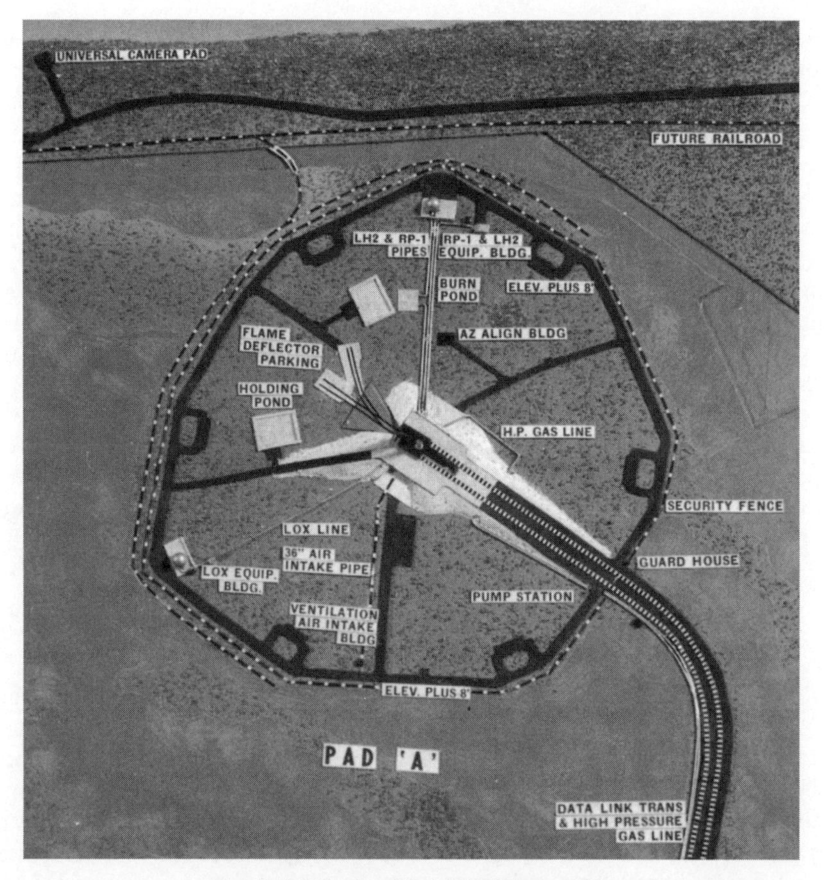

Fig. 62. Model of pad A, LC-39, February 1963.

protected by the brick had to have a smoothness tolerance of 0.3 centimeters in 3 meters to provide a bonding surface. This careful work was to limit the maximum temperature in the adjacent concrete structure during launch to 310 kelvins (37° C).[36]

Other components of the launch pad that required detailed design studies included the terminal connection room, the environmental control system, the high-pressure-gas storage facility, and the emergency egress system. The connection room, which contained extensive instrumentation facilities for testing during prelaunch and launch phases, and the environmental control system, which maintained the temperatures of the vehicle and compartments prior to launch, were designed to withstand concentrated pressures at any point. These rooms would protect ground support equipment located at the launch pad from heat, vibration, and shock during launch.[37]

By 1 June 1962, the design concept for the Saturn V propellant loading, high-pressure gases, and associated systems had been established. To use a reliable automated system and eliminate the cost of developing a new one, a modified version of the automatic propellant loading and associated systems used for LC-37 was selected for LC-39. Propellant servicing was controlled from the launch control center. Most of the hardware for the propellant-loading system was located in the terminal connection room at the pad. This room contained separate areas for each propellant and its associated systems. The remote command and display equipment in the control center was connected by an independent digital data transmission link to the hardware at the pad, which in turn was connected to the transporter-launcher and the storage facilities by electrical lines. Consequently, control commands could be initiated from the transporter-launcher, the launch control center, or the terminal connection room at the pad during servicing and checkout.[38]

Plans for the Industrial Area

The site plan for the industrial area, eight kilometers south of the VAB, was prepared by a joint Manned Lunar Landing Program Master Planning Board made up of NASA and Air Force personnel, and its subordinate joint planning committees for facilities, instrumentation, and communications. The committees had to plan and design facilities during a period when much of the equipment that would go into them was still under development. Yet a comparison of two site plans, one prepared in March 1963, the other in October 1965 after a more careful definition of program requirements, reveals few major changes. Most of the facilities remained as originally planned.[39] Some of the credit for this successful planning goes to the Air Force's contractor, the Guided Missile Range Division of Pan American World Airways. Back in December 1962 Pan American had completed a preliminary master plan for Merritt Island. The projection contained three sections: general plans for the launch area, a description of the Merritt Island industrial area, and detailed plans for the launch area. The Joint Facilities Planning Group, one of the several committees the Air Force and NASA set up, organized a task force to assist in preparation, correction, and development of this preliminary master plan. After the Webb-McNamara agreement, NASA used volume III of the Pan American master plan as a basis for its first plan, published in October 1963.[40]

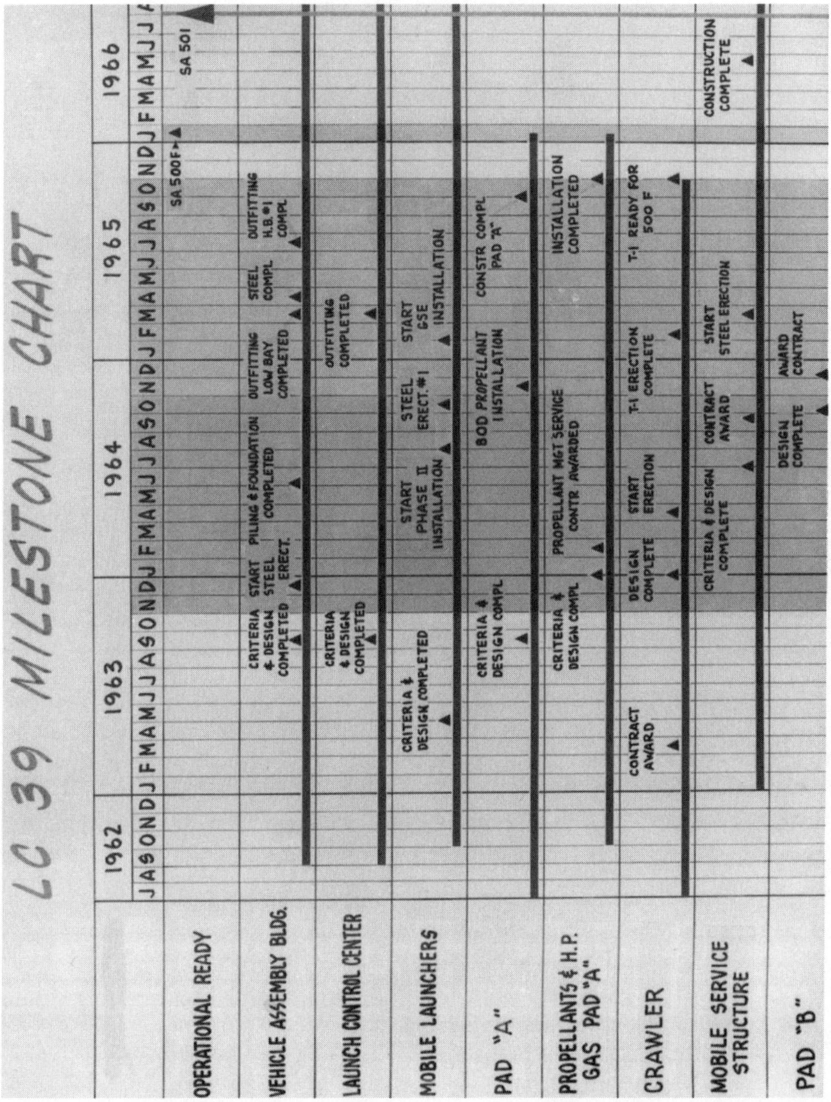

Fig. 63. LC-39 milestone chart, December 1964.

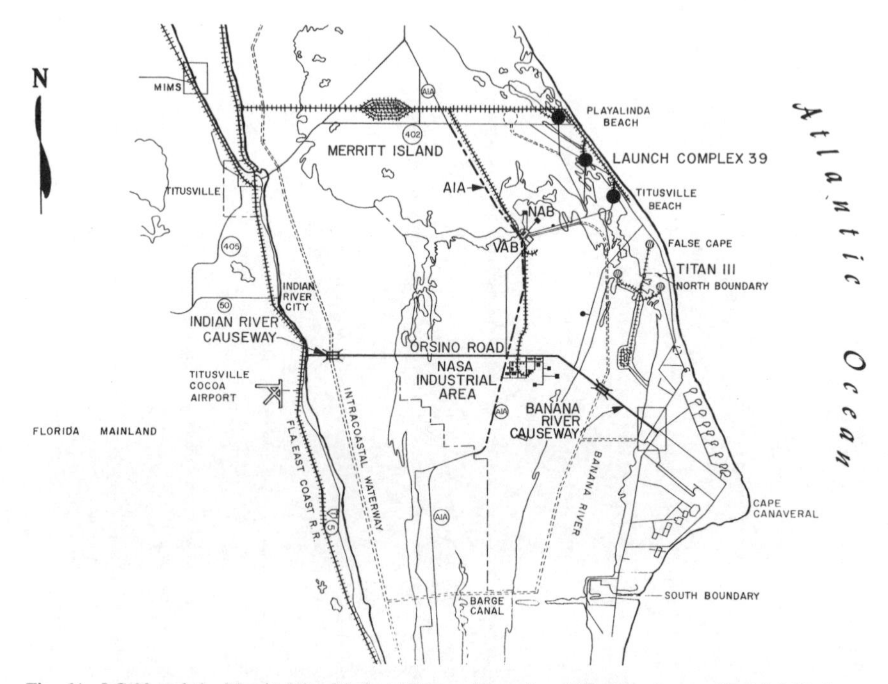

Fig. 64. LC-39 and the Merritt Island industrial area, December 1963. The three pads of LC-39, from south to north, were designated A, B, and C; C was never built. The NAB (nuclear assembly building) also was not built. Only two of the three Titan III sites were built.

Spacecraft support facilities took up the eastern half of the industrial area. Although the requirements for these facilities and equipment originated with the Manned Spacecraft Center in Houston or with its Florida Operations launch team, the responsibility for planning, siting, funding, and construction rested with the Launch Operations Center. Included among the Apollo mission support facilities were the following:

- Operations and checkout building
- Supply, shipping, and receiving building
- Weight and balance building
- Ordnance storage facility
- Fluid test complex, consisting of:
  - Hypergolic test building
  - Cryogenic test building
  - Environmental control systems building
  - Support building

The operations and checkout building was as essential to the spacecraft as the vertical assembly building was to the launch vehicle. The operations building would be used for the checkout of all non-hazardous systems in manned spacecraft. It would also provide space for the inspection of the spacecraft modules upon arrival at KSC, and for the mating and final integrated tests of the Apollo before it traveled to the VAB. Accommodations for astronaut preflight activities (living quarters, a technical and briefing area, a crew preparation area, and a bio-medical area) were included in the building. The floor area, 27 900 square meters, was divided into four functional areas: an administrative and engineering office area with an auditorium and cafeteria; a laboratory and checkout area with automated checkout equipment and data reduction and display facilities; a high-bay assembly and test area having a bridge crane hook height of 36 meters and a contiguous low-bay area with a crane hook height of 15 meters; and a service area containing shop space, a tool room, spare parts room, and space for electrical, mechanical, and vacuum equipment. The building was air conditioned and, where operationally necessary, humidity controlled and dust free.

Although it was always intended that the spacecraft modules (and the launch vehicle stages) would arrive at LOC in a flight-ready condition, the mechanics of shipping and the checkout process itself required that certain spacecraft parts be packed separately. Inevitably in the course of testing, some components had to be replaced, and those removed had to be returned to their makers for repair or modification. Also, various items of ground support equipment associated with the checkout and assembly processes were shipped with the spacecraft. The supply, shipping, and receiving building would provide the space for these functions. It was a one-story, L-shaped building of standard construction, with approximately 3720 square meters of floor area that included a machine room, a roofed-over loading dock, shipping and receiving and supply departments with a humidity-controlled storage area, a ground support equipment area, and cleaning, painting, carpentry, maintenance, and plastic shops.

The Planning Board isolated the facilities with hazardous operations in the southeast corner of the industrial area. KSC personnel frequently referred to the area located over a kilometer from the operations and checkout building as the "south 40." The facilities there included the weight and balance building, the ordnance storage building, and the fluid test complex. At the north end of the area was a 300-meter range for testing the rendezvous radar on the lunar module.

After checkout in the operations and checkout building, the spacecraft was to be moved to the weight and balance building, where the launch team would install solid-propellant motors, the launch escape tower, and various pyrotechnic items. After weighing and balancing the assembled spacecraft to determine its center of gravity, technicians would optically align critical components. Here, also, the spacecraft would receive its final servicing prior to departure for the VAB. The building would consist of a high-bay area having an overhead crane with a hook height of 30 meters and two adjacent low-bay areas. A door, of sufficient height and width to accommodate the assembled spacecraft in a vertical position, would take up almost all of one side of the building.

Contained under pressure in the spacecraft were environmental control, hypergolic, and cryogenic systems, all of which used corrosive and highly combustible fluids. Careful handling was obviously required. Because of the hazards in testing, adjusting, and verifying the proper operation of these systems and their component parts, and because the test methods were somewhat similar, the several buildings where these operations would be performed were grouped in the fluid test complex. The test buildings differed in size, but were similar in form. Each contained one or two test cells equipped with high-capacity exhaust systems, a floor system for collecting and diverting spilled fluids, and fire extinguishing systems. Adjacent to the test cells were control rooms designed and constructed to protect test operators from explosions or toxic fumes. An equipment storage room, dressing or locker rooms, and machinery rooms were included in each test building. Nearby was a support building containing offices, shops, and laboratories. It was air conditioned and equipped with a special filtering system to provide clean conditions in the laboratories. Miscellaneous service facilities for the test complex included stations for the parking of mobile fluid transfer tanks, and a dilution system and disposal dump for spilled fluids.

The ordnance storage building, slightly less than 370 square meters in floor area, would provide an environmentally controlled storage area for solid-fuel motors and aligned escape towers. This was designed to prevent any deterioration of explosives that could result in a misfire in space.[41]

Design of the Central Instrumentation Facility

Under the terms of the Webb-McNamara agreement, LOC was given certain instrumentation responsibilities on Merritt Island. Debus assigned these to Karl Sendler, the Director of Information Systems; and FY 1964 construction of facilities budget estimates for launch instrumentation reflected the new management.[42] Subsequent agreements concluded by Debus and the Missile Test Center clarified the instrumentation program and established a Joint Instrumentation Planning Group as the local coordinating body.

Fig. 65. Hypergolic building, fluid test complex, Merritt Island industrial area, under construction. The large doors were raised in sections, giving access to two large test chambers, which were served by common facilities in the lower, center section. Liquid and gas lines would enter the building in the concrete conduits beneath the fans.

The systems planned for installation in the central instrumentation facility were based upon those developed during Saturn I operations at complexes 34 and 37. The instrumentation systems criteria group held numerous meetings with design and operations personnel to determine what measurements were needed. Experience with LC-34 and LC-37 was of limited value however, because the distance from LC-39 to the control center was more than 14 times as long. After the criteria had been established, fixed-price contracts were negotiated. The digital acquisition equipment was designed by Scientific Data Systems; the computer was the GE 635. Since the number of on-board measurements for the vehicle had increased from 200 to 3000, it was necessary to procure equipment that produced accurate data in real time. For this type of instrumentation, there was no in-house design, but the specifications were assembled and bids were solicited from industry.[43]

By May 1963 the design criteria for the central instrumentation facility were available. The building—a three-story structure of approximately 12 480 square meters just west of the headquarters building—would house computers and other electronic equipment for reduction of telemetry data, analysis, and transmission to other NASA centers. A smaller building, later known as the CIF antenna site, was placed 2.5 kilometers north of the industrial area, to be free of radio-frequency interference and have clear lines of vision to the NASA launch complexes.[44]

The central instrumentation facility reflected the desire of Karl Sendler and his planners to centralize the handling of NASA data and provide housing for general instrumentation activities that served more than one complex. LOC coordinated the planning with the other NASA centers and with the Atlantic Missile Range. It was necessary to ground all metal in the structure and to ground separately the commercial power and the instrumentation power systems. Fluorescent lights were not permitted—they cause electromagnetic interference. When completed, the central instrumentation facility, with disc-shaped antennas adorning the roof, would be the most distinctive building in the industrial area.

Selection of MILA Support Contractors

While the design of LC-39 and the industrial area was still under way, LOC sought contractors who would operate and maintain the Merritt Island facilities. On 22 April 1963 LOC suggested four possibilities to Albert Siepert, NASA's Director of Administration in Washington, who would soon join LOC's management:

- Extension of the current Air Force contract with Pan American Airways to provide services for Merritt Island similar to those being provided for Cape Canaveral's Missile Test Area.
- Employment of a single NASA support contractor to provide all services.
- Employment of several contractors, each to perform a major function under direction of NASA staffs.
- Expansion of the LOC civil service staff.

LOC recommended the third solution, with 12 functional contractors.[45]

Earlier that month, Siepert and Brainerd Holmes had discussed the launch center's need for support services with Robert Seamans, NASA's Associate Administrator. Seamans turned down an Air Force proposal to handle the entire service support through its range contractor, Pan

American. Seamans wanted to spread contracts and did not want to increase civil service hirings. He favored the use of four or five prime contractors.

One week after the LOC staff report, Debus and Siepert met with Holmes in Washington. LOC recommended seven contracts with a separate food service contract handled by the NASA Employees' Exchange. Holmes thought seven contracts were too many for effective management and directed that LOC find a way to compress these into four. He did agree, after some discussion, that the food service could be a separate contract under the Exchange. LOC submitted a revised proposal on 7 May including a request that "LOC be authorized to initiate appropriate action to obtain contractor services . . . , grouped into four prime areas of activity. . . ." Holmes approved it two weeks later.[46]

As a temporary measure, NASA asked the Air Force for limited Merritt Island support services on a reimbursable basis. This was formalized as part of an interim agreement on management responsibilities signed by Dr. Debus and General Davis on 10 May 1963. Two weeks later, NASA Headquarters announced contract plans for more than 20 support functions in the four areas of management, communications, base support, and launch support. The prime contractors would be required to subcontract a substantial portion of the work to small firms.[47]

LOC's procurement office started work on one contract a week before the formal announcement. A request for proposals on the operation and maintenance of the communications system was issued 17 May. Contractor interest was heavy, but a disagreement between LOC and Southern Bell Telephone Company about interconnection points between the internal communication system and the Bell circuits delayed contract negotiations for over a month. Finally Southern Bell agreed to provide normal internal business and administrative telephone service (excepting service in hazardous or operationally critical areas). NASA Headquarters decided that source evaluation boards were necessary. Fourteen companies responded to LOC's second request for proposals. Administrator Webb narrowed the field to three firms, following an evaluation board presentation on 3 October. R.C.A. Service Company won and began work in December, although execution of the cost-plus-award-fee contract was delayed until mid-January 1964.[48]

Procurement action on the other three support contracts proceeded concurrently. A January 1964 cost-plus-incentive-fee contract gave Ling-Temco-Vought responsibility for photographic support, technical information, a field printing plant, and administrative automatic data processing. In February 1964 Trans World Airlines won the contract for supply, general maintenance, and utilities. In April, Bendix Field Engineering Corporation signed a contract for a variety of functions that included propellant services, precision shops, high-pressure-gas converter and compressor operations, cryogenic-equipment cleaning, spacecraft servicing facilities, and the operation of the crawler.[49]

12

FROM DESIGNS TO STRUCTURES

Making Big Sandpiles

Before thousands of skilled craftsmen could begin turning these unique designs into structures, the Launch Operations Center had to prepare the sites, dredge access channels, and fight mosquitoes. The Gahagan Dredging Corporation of Tampa won the contract to clear the land and dredge an access channel to the site of the vertical, or as it was soon to be called, the vehicle assembly building (VAB). The sand, dredged up by the Gahagan crews, was deposited on the projected sites of the VAB, the crawlerway, launch pad 39-C (shortly to be redesignated pad A*), and the causeway over the Banana River.

Gahagan started work on 31 October 1962 by clearing the land in the VAB area. Dredging began a week later. By the time the contract was awarded, Gahagan had three dredges at work and had already moved 20 600 cubic meters of fill. One part of the work involved the clearing of surface growth, ranging from palmetto scrub to orange trees, and the stripping away of undesirable surface material from the construction sites. Specialized equipment helped to speed this job. One device, a palmetto plow, pulled up trees by their roots, shook off the dirt, and piled them for burning. Bulldozers with heavy teeth on the blades knocked down whole rows of trees and brush, pushing them into piles to dry before burning. The bulldozers cleared some 2.5 square kilometers of land in this manner, while other earth-moving equipment removed 89 400 cubic meters of soft sand and muck.[1]

The second, and perhaps larger, part of Gahagan's job was dredging a barge canal 38 meters wide, 3 meters deep, and 20 kilometers long from the original Saturn barge channel in the Banana River to a turning basin near the

*At the time of the original siting of launch complex 39, the three projected launch pads were designated in accordance with standard Missile Test Center practice from north to south as pads A, B, and C. In January 1963, to bring the identification system in line with construction and operational use schedules, the pad designations were reversed, the southernmost becoming pad A. Early documentation carries the original designations; the revised designations are used hereafter in the text. C. Bidgood, Chief, Facilities Off., "Reidentification of Launch Complex 39 Launch Pads," 7 Jan. 1963.

Fig. 66. Dredging hydraulic fill from the Banana River and pumping it onto a construction site, December 1962. The water is draining back into the river above and right of the barge.

Fig. 67. The barge canal and turning basin, with assembly building site in lower left, August 1963.

Fig. 68. The wharf under construction at the turning basin, with assembly building site in background. August 1963.

Fig. 69. One of the big sandpiles: hydraulic fill piled on the site of LC-39, pad A, June 1963.

VAB. The canal would serve barges bringing in the first and second stages of the Saturn V. Gahagan dredged a channel to pad A so that barges could deliver material directly to the LC-39 construction site.

During the dredging operations, the powerful hydraulic pumps coughed up 6 876 000 cubic meters of sand and shell for fill. A major portion of it went into the 57-meter-wide, 2-meter-high crawlerway, which would stretch more than 4.8 kilometers from the VAB to pad A.

Out at the pad site, the pumps piled a pyramid of sand and shell 24.4 meters high, one of the highest recorded pumping operations. All the while, draglines, bulldozers, and other earth-moving equipment molded the mound into the approximate shape of the launch pad. Subsequent measurements revealed that this outsized sandpile had settled 1.2 meters and properly compressed the soil beneath. Bulldozers then removed part of the pile to bring the fill to the proper elevation.[2]

A final inspection of the land clearing, channel dredging, and fill in early September 1963 showed that Gahagan had completed all work on the contract. About six months after the completion of the fill for pad A, Gahagan began pumping and piling hydraulic fill for launch pad B and for the causeway from Cape Canaveral to Merritt Island east of the industrial area.[3]

NASA Declares War—On Mosquitoes

The *Spaceport News* had its share of sensational headlines during 1963, especially in May when Astronaut Gordon Cooper took Faith 7 into an earth orbit on a Mercury-Atlas launch vehicle from pad 14. But none quite reached the unique quality of a headline in the 8 August issue: "Peaceful NASA Declares War—On Mosquitoes."[4] It may well have been the most necessary, well-executed, and successful war in American history. For reasons of health and comfort, the mosquito population had to be reduced before workers could begin any sustained outdoor work during the prime mosquito season from April to late October. In the past, epidemics of malaria, yellow fever, and dengue (an infectious fever prevalent in warm climates)—all spread by mosquitoes—had periodically retarded the development of Florida. The discovery and application of successful methods of mosquito control had been one of the factors responsible for the state's rapid development in relatively recent years.

Almost from the outset, the mosquito figured prominently in NASA's operations. LVOD's Deputy Chief of the Mechanical, Structural, and Propulsion Office, Robert Gorman, spoke of the early days: "The mosquitoes were so bad . . . . Everyone wore long shirt sleeves and gloves, even in the summer . . . . In fact, one fellow with sensitive skin really got chewed up. He stayed in Huntsville after that." In recalling the first Redstone launch from the Cape, Gorman remarked: "You couldn't wear a white shirt. The mosquitoes would be so thick they'd turn it black." In an interview two years later, James Finn, who had come to the Joint Long Range Proving Ground in 1951 and joined the original Debus team in May 1954, said that "the mosquitoes were a hazard—but so was the mosquito repellant. . . . If any got on our badges, it rubbed our pictures off."[5]

The problem was at first almost unbelievable to all but former residents of the area. One acre of salt marsh was easily capable of producing 50 000 000 adult mosquitoes within a week after a heavy rainfall. The "landing rate" in bad areas was often more than 500 mosquitoes on a person in one minute. In 1962, two scientists from the Florida Entomological Research Center in Vero Beach collected with hand nets 1.6 kilograms of live mosquitoes in just one hour.

By April 1963, the Subcommittee on Mosquito Control of the Joint Community Impact Coordination Committee (see page 92) agreed upon a cooperative program using the services of the county, state, Air Force Missile Test Center, and the Launch Operations Center. The program sought both temporary and permanent control. At the time, the main breeding grounds of the salt-marsh mosquito included 57.7 square kilometers in northern Brevard County and 4.3 square kilometers in southern Volusia County. Within the Merritt Island Launch Area were also hundreds of acres capable of producing fresh-water mosquitoes.

The temporary control measures consisted of ground and aerial spraying of insecticide. The most effective permanent control on the Merritt Island Launch Area consisted of the construction of dikes to flood breeding areas during the peak summer months. With the flooding of marshes, the minnow population increased and mosquito eggs and larvae declined.

The Brevard County Mosquito Control District also agreed to continue work at the Merritt Island Launch Area. The county provided four draglines, two spray planes, and a helicopter for inspection purposes. The Launch Operations Center and the Air Force Missile Test Center provided two draglines and one bulldozer to accelerate the permanent control work that the county was doing. The Launch Operations Center supplied the insecticide and operated the ground fogging equipment. The State of Florida provided direct financial aid and scientific research. The master plan had originally estimated six years to accomplish reasonable mosquito control in the Merritt Island Launch Area. Fortunately the program moved much faster than that.[6]

Contracting for the VAB and the LCC

Four of the world's most unique buildings were to go up on Merritt Island during the succeeding years, two at one end of launch complex 39, two in the industrial area five miles south. While other structures, such as the more traditionally designed headquarters, were to be known at the center by their full titles, these four shortly became known by acronyms: the vehicle assembly building as the VAB, the launch control center as the LCC, the central instrumentation facility as the CIF, and the operations and checkout building as the O & C building. This last building was also called the manned spacecraft operations building.

The day after his visit to Florida in September 1962, President Kennedy stated at Rice University Stadium in Houston:

> In the last 24 hours, we have seen facilities now being created for the greatest and most complex exploration in man's history. We have felt the ground shake and the air shattered by the testing of a Saturn C-1 booster rocket . . . . We have seen the site where five F-1 rocket engines . . . will be clustered together to make the advanced Saturn missile, assembled in a new building to be built at Cape Canaveral as tall as a 48-story structure, as wide as a city block, and as long as two lengths of this field.[7]

No doubt many of the Rice engineers and students appreciated the remarks of the President. The concept, however, still stretched beyond the imagination of the average American. He could not picture a building so huge that the Rose Bowl or the Yankee Stadium would fit on the roof. Yet this was what URSAM planned for the vehicle assembly building.

During the first half of 1963, the Corps of Engineers was still acquiring land for the spaceport and simultaneously awarding contracts for continued site preparation and utility installations. Dredging operations to provide fill for the VAB, one launch pad, and the Banana River causeway were proceeding on schedule. In the industrial area, ground-breaking ceremonies were held in January on the site of the operations and checkout building, and the Corps of Engineers awarded a contract for the construction of primary utilities to provide for a water distribution system, sewer lines, an electrical system, a central heating plant, streets, and hydraulic fill for the Indian River causeway to connect the industrial area on Merritt Island with the Florida mainland. During this same period, the Launch Operations Center began awarding the first construction contracts for structures in the industrial and LC-39 areas.

The national goal of accomplishing the manned lunar landing "before this decade is out" dramatically affected the entire building program. With a deadline, scheduling became critical. At the beginning of 1963, the Office of Manned Space Flight's "official flight schedule" called for the launch of the first developmental Saturn V in March 1966 and of the first manned Saturn V in June 1967.[8] Meeting these dates was contingent upon the concurrent development of the Saturn V launch vehicle, the Apollo spacecraft, and launch facilities, and more particularly on the timely delivery of flight hardware to the launch center.

Of more immediate concern was the construction of launch facilities and checking them out many months before the first Saturn V launch. The first of December, 1965, was the most important date—the date when the launch complexes had to be ready for use. This in turn required that a number of facilities be ready by May 1965 to provide time for checking out and testing the launch complexes. Working backward from this date, LOC developed, and periodically revised, detailed schedules for completion of the construction and testing of each facility on launch complex 39 and in the industrial area. The demands of this tight schedule influenced construction as much as the development of the launch vehicle and the spacecraft.

On 31 May 1963, the Corps of Engineers advertised for bids on the structural steel and the erection of the VAB framework. On 9 July Col. G. A. Finley, District Engineer of the newly established Canaveral District of the Corps of Engineers, acting as agent for NASA, and officials of the American Bridge Division of U.S. Steel Corporation, Atlanta, signed the largest single contract NASA had yet awarded for work in the Cape area. This contract, in the original amount of $23 534 000, called for furnishing more than 45 000 metric tons of structural steel and the erection of the skeleton framework of the VAB, with completion by 1 December 1964. Workmen were busy at the site the day the contract was signed. The Blount Brothers Corporation of Montgomery, Alabama, signed an $8 000 000 contract on 11 July 1963 to provide the steel and concrete foundations and flooring of the VAB, with completion by 1 May 1964. The Blount firm also started work on the day of the signing.[9]

Laying the Foundations

Providing a firm foundation for construction on sandy soil had been one of the early design problems. Max O. Urbahn, head of one of the four firms in the design consortium, spoke of a second problem: "We were faced with the fascinating possibility that the shape of the building might make it react like an immense box kite; it could blow away in a high wind . . . ."[10] The solution to both problems was to drive thousands of piles, steel pipes 41 centimeters in diameter, through the subsoil until they rested on bedrock. These served to anchor the building as well as to prevent the structure from sinking into the ground.[11]

The building stood only a few feet above sea level and near the ocean. Salt water, saturating the subsoil, reacted with steel piling to create an electrical current. To prevent this electrolytic process from gradually eating away the steel pipe, workmen grounded the piling by welding thick copper wire to each pile and connecting the wires to the steel reinforcing bars in the concrete floor slab. Until this was done, the VAB could lay claim to being the world's largest wet cell battery.[12]

Blount Brothers moved rapidly to assemble pile-driving equipment, steel piling, and workmen at the work site. They drove the first piles on 2 August 1963 and by 15 August had driven 9144 meters of piling in the low bay area. The pipe for the piling came in 16.8-meter lengths, and welders had to join three and sometimes four lengths of pipe together to make up a single pile. To speed the work, Blount Brothers had their workmen weld at night and drive piles, which required better visibility, during the day. At the peak of activity, ten pile drivers were in action. Three of them were new, electrically driven, vibratory drivers that literally jigged the pilings into the ground. When the piles reached the first thin stratum of limestone at about 36 meters, steam- or diesel-driven pile drivers took over and pounded the piles into the bedrock, which ranged from 46 to 52 meters below the sandy surface. Although there were minor delays due to inclement weather—a week of unrelenting high winds and torrential rains brought all construction to a standstill in mid-September—the work on the foundations moved steadily ahead. The last of the piling was down on 3 January 1964, just five months after drivers pointed the first pile into the bedrock.[13]

As the pile drivers moved on, another group of Blount Brothers workmen moved in to erect the forms and place the reinforcing bars for the concrete pile caps and to bond the piles electrically to the reinforcing bars. To an observer in a helicopter, the VAB foundation site began to resemble a huge honeycomb with the concrete pile caps rapidly dividing the area into series of cells or boxes.[14] As soon as the concrete had set in a series of the pile caps, workmen removed the forms and replaced any fill removed in the course of work. Then they poured a layer of crushed aggregate into the boxes and poured the asphalt and concrete floor slab on top of the aggregate. All told, Blount Brothers poured 38 200 cubic meters of concrete for pile caps and floor slab before the foundation was completed in May 1964.

Driving piles for the assembly building

Figure 70

Figure 71

Fig. 70. August 1963. Fig. 71. September 1963.

Fig. 72. Pouring the floor of the assembly building, July 1964.

Even before sinking the first pile of the foundation and beginning the steel framework of the VAB, the Corps of Engineers had taken the initial step toward awarding a contract for three large bridge cranes in the VAB.[15] A 175-ton crane with a hook height of about 50 meters would run the length of the building and would traverse both the low bay and high bay areas above the transfer aisle. Two other cranes, with a 250-ton capacity and hook height of approximately 140 meters, would be capable of movement from an assembly bay on the opposite side. Although bridge cranes of this capacity are not unusual in heavy industry, the invitation for bids spelled out unique requirements for precision, smoothness, and control of their vertical and horizontal movement. The cranes would cost about $2 000 000. Colby Cranes Manufacturing Company of Seattle won the contract and agreed to have the cranes ready for final test on 1 September 1965.[16]

Structural Steel and General Construction

While work on the foundations and floor slab of the VAB was progressing rapidly during the latter half of 1963, there was little on-site activity on the part of the structural steel contractor. However, as the American Bridge Division began mobilizing its work force and assembling its equipment on Merritt Island, United States Steel plants throughout the country were fabricating the carbon steel plate and structural shapes required for the building's framework.[17]

On 4 October 1963, the Corps of Engineers advertised for bids on general construction and outfitting in the VAB area. In addition to completion of the VAB (including outfitting only high bays 1 and 3), the work covered general site preparation, roads and utility installations in the area, the construction and outfitting of the VAB utility annex and the launch control center—both good-sized buildings in their own right—and of two other support buildings, one for high-pressure-gas storage and the other for paint and chemical storage. Estimators set the price of the VAB alone at $52 000 000. The Corps of Engineers scheduled completion for 1 January 1966.[18]

A combine of three South Gate, California, construction-engineering companies—Morrison-Knudsen Company, Inc., Perini Corporation, and Paul Hardeman, Inc.—won the contract on 16 January 1964 with a bid of $63 366 378.[19] It was not long, however, before the contract grew considerably. On 9 March the South Gate combine assumed administration of the American Bridge Division's contract for structural steel work on the VAB and the Colby Cranes contract for fabrication, installation, and testing the three large bridge cranes, as the latter two companies became subcontractors to the South Gate group. With the absorption of these two earlier contracts, the Morrison-Knudsen, Perini and Hardeman general construction and outfitting contract reached a value of $88 743 386.[20] Although both American Bridge and Colby continued at their respective jobs until their completion, all VAB area brick-and-mortar construction was now under the direction of a single contractor.

With steel column sections and other structural steel arriving at the job site, erection of the framework began in January 1964 in the low bay area. By this time the original contract date for completing the structural steel (1 December 1964) had given way to a completion date of 7 March 1965. The job was a rather straightforward one although, because of the building's unique requirements, it appeared that the structure was being built wrong-side out. Because of the height of the assembly bay door openings—two on

The assembly building takes shape

Figure 73

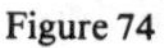

Figure 74

Fig. 73. The low-bay framework, with high bays rising in background, May 1964. Fig. 74. High bays to the right; the launch control center is the separate building on left, September 1964.

Figure 75

Figure 76

Fig. 75. In this view to the west in October 1964, the openings into the high bays are most distinct.
Fig. 76. Mobile launchers are visible to the right; the launch control center, seen end-on in the center, seems to be part of the assembly building. The view is northwest across the turning basin, January 1965.

each side of the building—the horizontal stiffening structure had to be installed on the interior of the building, parallel to the transfer aisle, rather than along the exterior sides.[21]

Less than a month after steel erection began, the general construction contractor, Morrison-Knudsen, Perini and Hardeman, started work in the VAB area. After setting up a temporary office, warehouse, and concrete mixing plant on the job site, the contractors began compacting and stabilizing the crawler erection area, preparing the erection site for one mobile launcher, excavating for the crawlerway base, and excavating for the foundation floor slab of the launch control center. Contractors completed the crawler assembly area by 11 March and the mobile launcher area by 1 June.[22]

Meanwhile, Morrison-Knudsen, Perini and Hardeman had also begun work in February on the high-pressure-gas storage building, the road system in the VAB area, the instrumentation and communication duct banks and tunnels from the launch control center to the crawlerway, and the foundation work for the control center itself. In March work started on the water distribution and storage system, on the sewage plant and sewer system, and on the electrical distribution system. In April construction began on the VAB utility annex, on the paint and chemical storage building, and on the VAB area crawlerways.[23] Thus, by the time the ironworkers of the American Bridge Division had progressed far enough with erection of the VAB framework to allow the Morrison-Knudsen, Perini and Hardeman workmen access to the building, almost all of the other construction in the general area was moving ahead. The combine started work in the northwest corner of the low bay in April.[24]

From this time on, employees of the two contractors worked jointly in the building, with the general construction men following closely behind the ironworkers. Joint occupancy was necessary if the building and related facilities were to be completed on schedule. Since contractors in widely scattered parts of the country worked on different parts of the total job, construction chiefs on Merritt Island had to test components regularly to see if they fitted and worked together. These so-called "fit tests" became important procedures in the early stages of construction. The installing of many pieces of vehicle-related ground support equipment—a necessity for facility checkout—had to await completion of most of the general construction.

The same combine of California construction-engineering companies built the launch control center as part of the VAB contract. URSAM had decided on a distinctively shaped four-story building adjoining the VAB on the southeast and connected with it by an enclosed bridge. The ground floor contained offices, cafeteria, and dispensary, the second floor telemetry and radio equipment. Firing rooms occupied the third floor, and the fourth floor had conference rooms and displays.[25]

Fig. 77. The launch control center under construction, September 1964. The assembly building shows to the left.

The original plan called for four rectangular firing rooms, 28 by 46 meters; one was never to be equipped. When completed, the firing rooms contained similar equipment set up on four levels. The first level took up over two-thirds of the room and would ultimately contain computers and five rows of 30 consoles each. Two rows of consoles (27 in one, 25 in the other) would fill the second level. The third level would contain the consoles of the Kennedy Space Center Director and other major officials. To the left of these consoles, two diagonal rows of seats with telephones and listening devices, but no control equipment, would provide a close-up view of operations for technical experts not directly involved in the launch. On the top level, a glassed-in triangular room would give visiting dignitaries a like view. They could either watch activities in the firing room or look out the windows at the launch pads. These double-paned windows extended the full width of the rear of the firing room and contained a special heat- and shock-resistant glass. Outside, large vertical louvers, resembling huge venetian blinds, could be closed in a few moments for further protection.

Cleo and Dora Visit the Cape

The nearby passage of hurricanes Cleo in late August and Dora in early September 1964 caused an estimated $35 000 worth of damage, but a

Fig. 78. Firing room 1 under construction, August 1965. The VIP viewing-area is behind the glass wall, left. The windows on the extreme left looked toward the pads. The triangular extension into the room above the VIP area was intended for the most distinguished guests.

Fig. 79. Firing room 1 ready for equipping, November 1965. The four large overhead screens, here reflecting ceiling lights, would display major milestones in the countdown.

delay of only three days. The editor's "Spotlight" in the 3 September edition of the *Spaceport News* reported the lack of major damage to NASA facilities and dispensed credit widely—from the people who drew up the storm plans to the man who laid the last sandbag in place a few hours before Cleo swept by. That everything went off without a hitch reflected favorably on the advance planning. "It was a team that got the job done," the editor wrote. "Everyone involved directly in securing operations had his work to do, and did it with the minimum of hubbub."[26]

The editor singled out "Hurricane" Jones. A KSC engineer with the Instrumentation Division's acoustic and meteorological section, bachelor Jones had volunteered to ride out the storm in the huge launch control center at complex 37, gathering weather data. From 10 a.m. on Thursday until relief came at 7:30 the next morning, he recorded winds that peaked at 112.6 kilometers per hour. The reluctant hero admitted that he had misgivings during his lonely vigil, even though he had thought the launch control center looked like the safest place in the vicinity.[27]

VAB Nears Completion

At the beginning of October 1964, a survey revealed that the construction force on all contracts at the new Merritt Island spaceport had reached a total of 4300, with about 500 more equipment installers at work. At that time, KSC had 1670 federal employees, 1902 support services contractor employees, and 863 employees of launch vehicle contractors. The Florida Operations Division of the Manned Spacecraft Center (deeply involved in Project Gemini, which would launch its first manned orbital flight the following March, and just becoming concerned with the activation of Apollo spacecraft facilities) had a force of 502 federal employees and 1042 persons in the employ of contractors. Overall employment at Cape Kennedy and Merritt Island was expected to exceed 15 000 by 1 January 1965.[28]

By Christmas of 1964, the ironworkers had erected nearly 38 000 metric tons of structural steel in the VAB, reaching the 128-meter level in all towers. The LCC building was nearing completion, although interior mechanical work and the installation of electrical fixtures continued on all four floors. The VAB utility annex was also nearing completion, with boiler stacks and skylights completed and installation of mechanical and electrical equipment continuing. Workers had finished the high-pressure-gas storage building on 2 October. The rest of the area facilities were all nearing the end of brick-and-mortar construction, although much installation and outfitting remained.[29]

Structural parts for the first of the extensible work platforms in the high bays (five pairs of platforms in each high bay) had arrived at the VAB site. Workmen assembled these platforms outside the VAB because of their size, approximately 18 meters square and up to three stories tall, and then moved them inside for mounting on the framework of the VAB. They would be vertically adjustable. Since they were of cantilever design, they could extend horizontally about 9 meters from the main framework of the building to surround the launch vehicle in the high bay.[30]

As construction and outfitting continued into 1965, the vertical assembly building got a new name but not a new acronym. It was still the VAB, but now officially the *vehicle* assembly building, as of 3 February 1965. The new name, it was felt, would more readily encompass future as well as current programs and would not be tied to the Saturn booster. The Office of Manned Space Flight formally approved the change in September

Fig. 80. An extensible work platform being prepared for installation in the assembly building, August 1965.

1965, but individuals at the facility continued to use both names interchangeably.[31]

The Colby Cranes Manufacturing Company had completed shop testing all three bridge cranes in Seattle and had shipped the 175-ton crane to the VAB site. By the end of January 1965, the two 250-ton bridge cranes had followed. They would soon be ready to install. In fact, countless details of the largest building in the world were approaching completion.[32]

Erection of the VAB's structural steel framework reached the top level of 160 meters at the end of March, and preparations began for the traditional topping-out ceremony. A 3600-kilogram, 11.6-meter-long steel I-beam, painted white and bearing the NASA symbol and the insignia of the American Bridge Division of the United States Steel Corporation, stood in front of several of the NASA buildings at KSC during early April to allow NASA and contractor employees to sign their names on it. The signed beam then went under the roof of the VAB over the transfer aisle.[33]

Fig. 81. The topping-out ceremony with the signed beam, April 1965.

Construction in the Industrial Area

While the vehicle assembly building and other facilities moved steadily toward completion at LC-39, the industrial area began to take shape to the south. Preliminary to any actual construction, the Azzarelli Construction Company had completed ground work for the operations and checkout building in early November 1962.[34] Azzarelli had used the surcharging method in preparing the soil by piling sand on the construction site until its weight was approximately equivalent to the weight of the proposed structure. The heavy surcharge compressed the underground layers of clay and coral, squeezing out liquids. The contractors used piling under later parts of the building, as well as for all other buildings on Merritt Island.

On 16 January 1963, the Paul Hardeman and Morrison-Knudsen combine, which would also construct the basic utility systems in the industrial area, signed a contract in the amount of $7 691 624 for the construction of the operations and checkout building. By the end of February, men were clearing the construction site and the right of way to it and removing excess surcharge material.[35] The O & C building was to undergo continuous addition, modification, and alteration during the succeeding five years. Some contractual changes reflected planned phasing of construction over several fiscal years' funding; others, the evolving design of the spacecraft; some were intended to improve the original design of the building.

Early criteria for the building had envisioned flight crew training equipment among the astronaut facilities. Early in 1964, however, the Manned Spacecraft Center's Flight Crew Support Division in Houston decided on a separate building for crew training. To assist in the preparation of the separate structure, it forwarded criteria and sketches of a similar facility located at the Manned Spacecraft Center. Other changes from the initial O & C building design included additions to the administrative and engineering area, to the four-story laboratory and checkout section, and to the assembly and test areas. Three firms in joint contract, Donovan Construction, Power Engineering, and Leslie Miller, Inc., completed these additions.[36]

In September 1964, designers began drawings for a clean room, or white room, for the Gemini program. This was a dust-free room with high quality temperature and humidity controls to prevent contamination of the space vehicle. The air intakes would have special filters. All persons who entered the room would wear clothing resembling surgical uniforms. To be located in the O & C building's assembly and test area, the room was built by S. I. Goldman of Winter Park, Florida.[37]

The O & C building, a multi-storied structure of approximately 17 200 square meters, contained as much flexibility as the Apollo spacecraft that it would test. A high bay, 68.2 meters long and 30.5 meters high, and an adjacent 76.5-meter-long low bay accommodated the three-man Apollo capsule. Two altitude chambers were prominent fixtures in the high bay. In these tanks, each 17 meters high and 10 meters in diameter, KSC engineers would check out the command and service modules and the lunar module. After pumps had evacuated the air from the chambers, the Apollo modules were checked out in a near vacuum. Two airlocks, measuring 2.6 meters in height and width, provided access to each chamber. They also housed the rescue teams. Should a loss of oxygen occur in the spacecraft, the physiological effects on the crewmen would be the same as in space. The rescue teams

Fig. 82. Interior of operations and checkout building, August 1965. The two partly visible silos at left are altitude chambers.

would have to move fast, after rapidly pressurizing the chamber to a simulated altitude of 7600 meters.[38] After testing, the mated spacecraft components—the command module, the lunar excursion module, and the service module—would be moved from the integrated test area to the VAB in a vertical attitude, ready for stacking on top of the launch vehicle.[39]

The headquarters building, just west of the O & C building and a much less complicated structure, went up in two phases: the central structure, measuring 80 by 72 meters, first, and the east and west wings later. The building stood three stories high, except in the front where a fourth floor contained top administrative offices. The main section of the building extended east and west. The original plan called for four arms stretching to the south. Later on, the east and west additions brought with them two other southward extensions. The Franchi Construction Company of Newton, Massachusetts, which had begun work on the fluid test complex in mid-April 1963, started on the headquarters building in February 1964. On the day that the headquarters building got under way, the Blount Brothers Construction Corporation began the two buildings that comprised the central instrumentation facility.[40]

The Florida Operations team of the Manned Spacecraft Center was the first organization to occupy NASA's new facilities on Merritt Island. Most of the 1270 MSC and contractor employees who moved over from the Cape in September and October 1964 took new offices in the O & C building. Although heavily committed on the Gemini program, representatives of Florida Operations coordinated with contractor personnel and KSC in determining Apollo checkout requirements. By the following year the launch team had formulated a ground operations requirements plan. Some of the requirements, anticipated in early 1963 when the facilities were designed, no longer appeared necessary, e.g., Houston had decided to pack the Apollo parachutes at the factory rather than in Florida. Houston officials were coming to believe that, because of the spacecraft's complexity, it was undesirable to postpone major operations until the prelaunch checkout. As much as possible should be accomplished at the factory. This view would alter considerably the scope of Apollo operations in Florida.[41]

Ceremonies at Completion

With construction nearing completion, Kennedy Space Center celebrated two formal dedications in the spring of 1965. On 14 April, 30 dignitaries came for the topping-out ceremonies at the vehicle assembly building: officials of KSC, the Corps of Engineers, the newly renamed

Eastern Test Range, U.S. Steel, the Morrison-Knudsen, Perini, and Hardeman consortium of contractors, and the design team of Urbahn-Roberts-Seelye-Moran. In a brief address, Debus stated:

> This building is not a monument—it is a tool if you will, capable of accommodating heavy launch vehicles. So if people are impressed by its bigness, they should be mindful that bigness in this case is a factor of the rocket-powered transportation systems necessary to provide the United States with a broad capability to do whatever is the national purpose in outer space . . . .[42]

In a less formal, but equally effective way, Ben Putney summed up the workers' feelings when he quipped: "This is the biggest project we've ever worked on. There just ain't nothing bigger!" American Bridge's senior construction superintendent, John Pendry, said of the VAB: "You can't call it a high-rise building, it's more like building a bridge straight up."

Although workers had topped out the structural steel in the VAB, the work was far from finished. Steven Harris, VAB project manager, noted that one of the biggest tasks was keeping up with evolving equipment as the work went along. He remarked: "The VAB was designed and is being constructed concurrently with the development of the Saturn V vehicle, and any changes made on the vehicle or its support equipment may require changes in the building."[43] At the time he was speaking, designers had already incorporated some 200 changes into the VAB since construction began, the most recent being modification of the extensible platforms as required by the final design of the mobile launcher.

The formal opening of KSC headquarters on 26 May provided another opportunity for ceremonies. Prior to the formalities, a 40-piece Air Force band entertained the guests. Maj. Gen. Vincent G. Huston, Commander of the Air Force Eastern Test Range; Maj. Gen. A. C. Welling, head of the Corps of Engineers, South Atlantic Division; and Col. W. L. Starnes, Canaveral District Engineer, shared the podium with Debus, who thanked the Administration, the Congress, NASA, and the American people for the faith they had placed in the KSC team. Then he handed American and NASA flags to members of the security patrol who raised them to the top of the pole in front of the new headquarters.[44]

At the same time, the people who were going to support, maintain, and operate these facilities and their equipment had begun to move in. By mid-September "Operation Big Move" had brought 7000 of KSC's civil service and contractor employees from scattered sites at Cocoa Beach, the Cape, and Huntsville to Merritt Island, mostly to the industrial area; 4500 more would move to Merritt Island during the following months, mostly into the VAB. During 1965 the civil service personnel at KSC rose from 1180 to more than 2500, chiefly through the addition of the Manned Spacecraft Center's Florida Operations and the Goddard Space Flight Center's Launch Operations Division; the latter specialized in unmanned launches.[45] Even more significant for many than the physical move was the psychological move from the "pads where they had their hands in the operation" to desks where they directed the actions of others.

This description of the spaceport's construction has emphasized the material and the contractual. A later chapter will discuss the intermittent walkouts that made some wonder if the contractors would ever finish. This chapter has dealt little with the human side of the workmen who slaved and sweated and suffered—and in a few instances died as a result of accidents. On 4 June 1964, workers were stacking concrete forms for the third floor deck in the low bay area of the VAB. Apparently, the forms became overloaded and collapsed. Five men fell and were injured, two seriously. A month later, on 2 July 1964, Oscar Simmons, an employee of American Bridge and Iron Company, died in an accidental fall from the 46th level of the VAB. On 3 August 1965, lightning killed Albert J. Treib on pad B of launch complex 39.[46]

To some, construction at KSC was just another job. Others, however, were keenly aware of the contribution they were making to the task of sending the first man to the moon and bringing him back safely.

13

NEW DEVICES FOR NEW DEEDS

The Crawler-Transporter

The four unique structures going up on Merritt Island—the vehicle assembly building, the launch control center, the central instrumentation facility, and the operations and checkout building—had their match for distinctiveness in a group of devices being designed and built at the same time: the crawler-transporter, the mobile launcher, the mobile service structure, and the service arms. These novel mechanisms almost defy verbal description, and the reader should refer frequently to the illustrations in this chapter.

Fig. 83. Sketch of Saturn V and mobile launcher on a crawler-transporter, November 1963.

Something like the crawler-transporters that would eventually move the Apollo-Saturns from the VAB to the launch pads of LC-39 had been used for surface coal mining in Paradise, Kentucky (page 118). These huge vehicles ran on four tank-like treads, much like bulldozer treads in shape, but considerably larger. One of these double-track mechanisms supported the vehicle at each corner. Since Bucyrus-Erie of Milwaukee was the only firm that had built such giant contrivances, the Launch Operations Center (LOC) sent engineers to inspect them in Kentucky and the Bucyrus-Erie plant in Wisconsin.[1] LOC moved toward closing a contract with the Milwaukee firm, as the only company that could build the crawlers in the allotted time. Approval by NASA Headquarters seemed assured.[2]

Negotiations, however, did not prove so simple. A Bucyrus-Erie employee, Barrett Schlenk, had first interested LOC in using the crawler to carry the spaceship from the VAB to the pad. But when it appeared that Bucyrus-Erie would get the contract under sole-source procurement, William C. Dwyer, Vice President of Marion Power Shovel Co. of Ohio, protested to NASA. Brainerd Holmes urged Debus to use competitive bidding. Twenty-two industrial firms sent representatives to a procurement conference, but only two submitted proposals—Marion for 8 million dollars, Bucyrus-Erie for 11 million dollars.[3]

Now Senator William Proxmire (D., Wis.) protested. Webb met with him and other members of Congress to discuss the matter. Previously, Proxmire had tried in vain to amend the NASA Authorization Bill for fiscal 1963 to require competitive bidding to the "maximum possible extent" (page 162); but now he advanced the cause of a Wisconsin firm, even though it had lost out in competitive bidding. He questioned the validity of Marion's estimate of an 8-million-dollar cost for the crawler-transporter. Congressman Henry Reuss (D., Wis.) next urged a fixed-price contract to hold Marion to its estimate. But Webb countered that continual modifications would come during construction and insisted on the cost-plus contract.[4]

A second major factor in Marion's favor, besides its considerably lower bid, was its announced intention of choosing a project manager from its own personnel, thus saving considerable time in building a team.[5] Bucyrus had said it would bring in one from outside. Having received the contract, Marion selected a competent manager, Philip Koehring, not from its own company, but from—of all firms—Bucyrus-Erie. When Marion finally completed the contract two years later, the price had risen above 11 million dollars.

By contract, Marion undertook to assemble the first complete crawler-transporter on Merritt Island by 1 November 1964 and finish road

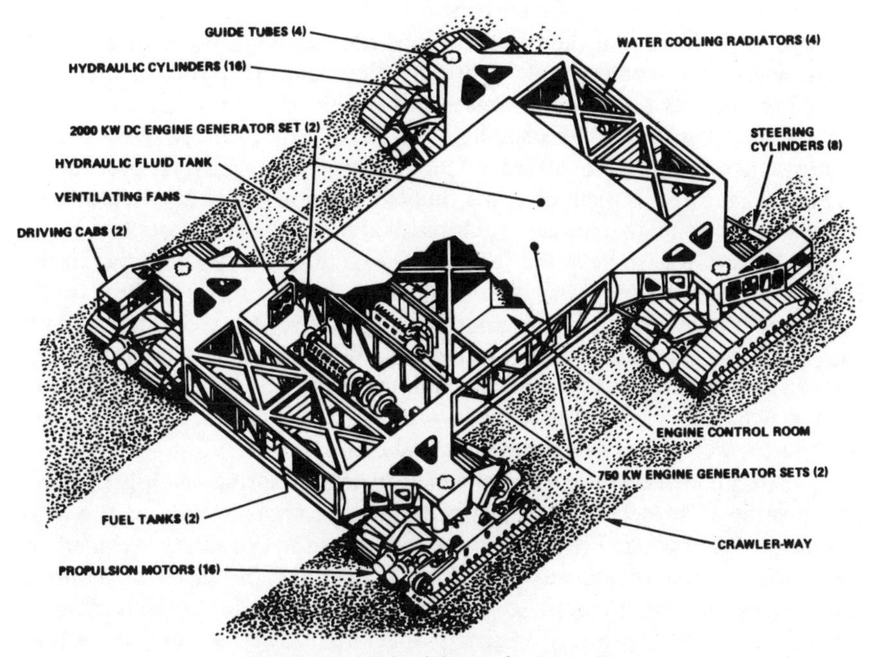

Fig. 84. Schematic of the crawler-transporter.

tests, mating, and modifications by 1 March 1965.[6] By early December 1963, Marion had completed 90% of the design and promised that parts of the vehicle would begin to arrive at the launch area—now the Kennedy Space Center—in March 1964.[7]

In the meantime, Marion had run into trouble with one of its subcontractors, American Machine & Foundry, over the hydraulic system for steering and levelling the crawler.[8] Balancing a 5400-metric-ton load called for precision. The motion of the transporter, the height of the load, variations in the level of the roadway, the wind—all would combine to throw the cargo off balance. Marion hired Bendix Corporation to check the levelling and equalization systems on the crawler. The Bendix study, made by mathematician-inventor Edward Kolesa, criticized the levelling systems as too quick and sensitive in their actions. The difficulties between Marion and its subcontractor were not the direct responsibility of KSC. Nevertheless, KSC sent the Bendix study to General Electric for analysis. The GE experts agreed with Kolesa's calculations.[9] As a result, Marion had to adjust the designs. Among other things, a separate power system, distinct from the diesel engines that powered the treads, was added for load-levelling, jacking, steering, and ventilating.

"Prophetically," said an article in *Aviation Week* sometime later, "NASA early identified the transporter as the one item which would most likely encounter trouble and whose development, therefore, should be started as soon as possible."[10] One set of problems arose from a factor in Marion's background. The company had previously held few government contracts and its management lacked familiarity with the intricate procedures and tests that a government contract entailed. Marion had to hire new men to carry out the new procedures, which resulted in unexpected costs.[11]

Marion was to have the first of the two units ready for testing in the late fall of 1964, although it was to make its initial trip in April 1965. In the meantime, the monster ran into another snag: someone noted that it had no fire alarm system or fire detection devices. With flammable materials and extensive electronics and mechanical equipment aboard, an alarm system was needed when the crawler was not in use.[12] As first designed, the crawler-transporter would carry only dry chemical extinguishers.[13]

After considerable correspondence during the spring of 1965, the Factory Insurance Association of Hartford made a complete study of fire protection on the crawler. Fifteen recommendations for fire safety included an automatic carbon dioxide extinguishing system for the electrical control room, the entire engine room, and the hydraulic equipment compartment; a limited, automatic sprinkler system as backup protection for the carbon dioxide system; an automatic, total-flooding foam system; flexible water connections to the sprinkler and hose systems in the transporter parking area; meticulous housekeeping and cleanliness inside the crawler-transporter; and the use of 100% noncombustible materials in all future construction and modifications to the crawler-transporter.[14] The contractor set about putting in a satisfactory fire prevention system.

When finally assembled, the crawler-transporter would not have won any awards for beauty. From a distance it looked like a steel sandwich held up at the corners by World War I tanks. Each crawler-transporter was larger than a baseball infield and weighed about 2700 metric tons. Two 2750-horsepower diesel engines powered 16 traction motors, which moved the four double-tracked treads. Each tread had 57 "shoes." Each shoe, 0.3×2.3 meters, weighed close to 900 kilograms. Quite naturally, a great deal of experiment and readjustment preceded the final success of such treads. Because of their importance and cost, they were nicknamed, "Them Golden Slippers." Many people recalled the next line of that song: "Golden shoes I'm going to wear, to walk that golden street."[15] The crawlerway would be such a street.

Fig. 85. Crawler-transporters under construction, April 1965.

Fig. 86. Richard L. Drollinger, Director of Engineering, Marion Power Shovel Co.; Theodor A. Poppel and Donald D. Buchanan, both of KSC; S. J. Fruin, Executive Vice President, Marion; Phillip Koehring, Project Engineer, Marion; and Kurt H. Debus, KSC, in front of a crawler truck at Marion, Ohio, in July 1964. The group observed the first test of the vehicle.

Fig. 87. The first crawler-transporter ready for service, January 1966.

Building a New Kind of Road

For the safe movement of the crawler-transporter, mobile launcher, Saturn V, and Apollo (a load exceeding 8400 metric tons), engineers would have to design a unique roadbed. The completed crawlerway would look something like the many interstate highways under construction throughout the nation in the 1960s. Beyond surface appearance, however, the resemblance ended. The crawlerway would support loads never envisioned for a public road—loads in excess of 58 000 kilograms per square meter.[16]

Gahagan Dredging Company had already begun preliminary site preparation. After excavating softer, unsuitable surface material, Gahagan had pumped nearly 2.3 million cubic meters of hydraulic sand fill into place on the crawlerway route. Vibratory rollers had compacted this fill under the trackways, and then a 90 600 kilogram vehicle proof-rolled them.

Each of the dual trackways, separated by a median strip, would consist of slightly over a meter of selected sub-base material, topped by a meter of graded crushed aggregate, with a blacktop sealer over all. A service road would border the south side of the crawlerway from the VAB to pad A. Underground ducts for communication and instrumentation lines to link the control and assembly areas with the launch pads would parallel the north side of the crawlerway; power line ducts and a pipeline for drinking water would go along the south side. Where any of the ducts or pipes had to pass beneath the crawlerway, the access tunnels had to be capable of withstanding the load conditions. The completed crawlerway would be level with the terrain, 2.3 meters above sea level.

Two firms, the Blount Brothers Construction Company of Montgomery, Alabama, and the M. M. Sundt Construction Company of Tucson, Arizona, acting jointly, agreed to build pad A and the crawlerway for $19 138 000, somewhat under the estimated cost of $20 000 000. Blount-Sundt started work on 19 November 1963. The contract called for the construction of about 5500 meters of crawlerway from the VAB to launch pad A, the elevated pad, several related facilities in the pad area, and the parking site for the tower. Subsequently a high-pressure-gases converter-compressor facility was added to the contract, at a cost of $155 000. The converter-compressor facility was to be complete on 1 May 1964, the arming tower (mobile service structure) parking site by mid-May, the crawlerway ready for test by 1 November, and the overall project by 1 June of the following year. The George A. Fuller Company of Los Angeles signed a contract on 30 November 1964 to construct pad B and extend the crawlerway 2100 meters. Using experience gained by Blount-Sundt, the Fuller personnel were well on their way with their work by the middle of 1965.[17]

Crawlerway under construction

Figure 88

Fig. 88. The terrain was not the best for supporting heavy vehicles (February 1964). Fig. 89. The site of the assembly building is top, center; pad A is off to the right (April 1964). Fig. 90. The communication and instrumentation duct is open in the right foreground (July 1964).

Figure 89

Figure 90

The converter-compressor facility was built just north of the crawlerway, about one-third of the distance from the VAB to pad A. It consisted of a one-story equipment building and a 1 892 000-liter spherical tank for storing liquid nitrogen, together with an access road and paved parking areas. A railroad spur brought tank-car loads of helium and nitrogen to the facility. Its evaporators, compressors, and pumps, in turn, supplied high-pressure gaseous nitrogen and helium to storage and distribution facilities at the VAB and the launch area.

Since plans called for the construction of the mobile service structure on the parking site, this facility would have to support considerable loads. The service structure would weigh 4763 metric tons. When the crawler-transporter moved beneath it, the total load on the parking position would be nearly 7500 metric tons, heavier than the USS *Halsey,* a guided missile frigate. In addition to this, calculations showed that, should wind velocities reach 200 kilometers per hour, the service structure, standing by itself on its four support legs in the parked position, with side struts and hold-down arms for each leg, could exert about 6300 metric tons of force. To withstand these anticipated forces, the parking site had to have a heavily reinforced base.

The Swing-Arm Controversy

The most difficult of all launch mechanisms to describe verbally is the mobile launcher, at times called the launch umbilical tower. It consisted of three main features: a two-story platform 49 meters long by 40 meters wide, on which the launch vehicle stood both on the crawler-transporter during its journey from the VAB to the pad, and on the pad itself, held erect by four hold-down arms; a tower that resembled the Apollo-Saturn in shape and

Fig. 91. The three mobile launchers under construction, October 1964.

Fig. 92. The mobile launcher: the platform and base of the tower. Note the four hold-down arms around the square opening for the rocket exhaust. The crawler-transporter is moving beneath the mobile launcher, May 1965.

Fig. 93. Mobile launcher on a crawler-transporter in front of the assembly building, July 1965.

size, and stood beside it surmounted by a hammerhead crane; and, attached to the tower, nine swing arms of various sizes that carried electric, propellant, and pneumatic lines to the space vehicle. These swing arms would automatically move away from the vehicle between the time of ignition and liftoff. The two-story launch platform housed computers that were connected to the launch control center. The platform also had a 14-meter square opening in the center for the rocket exhaust. Two high-speed elevators were centrally located in the tower. Besides their ordinary function of bringing personnel and equipment to various levels, they formed part of an emergency egress system.[18]

The Jacksonville-based firm of Reynolds, Smith, and Hills designed the tall mobile launcher that replaced the umbilical towers previously used at the Cape. Ingalls Iron Works of Birmingham, the prime contractor for steel erection, began work on the first launcher in December 1963. Nine months later workers hoisted the last major piece of steel, a 19-ton crane boom, into place on the first mobile launcher. The crews proceeded to outfit the finished tower with ground support equipment and electrical apparatus. They expected to have the giant completed in another 12 months.[19] In February 1965, Ingalls topped out the second tower, and on the afternoon of 1 March of the same year topped out the third (and last) with the hoisting of its huge hammerhead crane to the top of the 136-meter structure.

In planning and building the mobile launcher, the most difficult features were the nine swing arms, or service arms, as they were also called. The Brown Engineering Company of Huntsville, Alabama, designed the service arms in conjunction with Theodor A. Poppel's Launch Support Engineering Division. Brown faced unusual problems: the equipment was novel—no one had built such large access and umbilical devices in combination before; the vehicle for which the swing arms were being designed was developing so fast that the criteria changed continually, even after NASA had let the contract for construction; and the service arms were to be amazingly complex pieces of equipment. By way of example, as many as 24 electric cables, each 50 millimeters in diameter, and about 44 fluid service lines,

Fig. 94. One of the nine swing (service) arms; when installed, this one would connect to the second stage 43 meters above the base of the rocket.

ranging from 12 to 25 millimeters thick, went into a single umbilical carrier. Each arm would be wide enough for a jeep to drive across—though none ever was to do so. Their length varied with the configuration of the vehicle; they would average over 22 metric tons in weight.[20]

Employing 250 people on the project, Brown Engineering made 5000 drawings and 11 000 pages of specifications, but NASA designers found many unsatisfactory features. The company admitted errors in the drawings—but not in numbers or significance out of proportion to the average error rate for such a complicated enterprise. In retrospect, Cliff Boylston, the design project engineer for Brown at the time, was to agree with individuals at KSC that "one typical arm should have been totally tested before going into production." Boylston concluded that the "design was started before the criteria were set . . . . The developmental effort was not complete before the production started . . . . [In spite of this] we gave the customer the best effort that he could have gotten anywhere in the time, and within the limitations we had on us . . . ."[21]

Boylston was correct in saying that NASA had not developed a prototype of an entire service arm. As early as 30 July 1963, however, William T. Clearman, head of the Apollo–Saturn V Systems Office, had authorized a prototype within the allowable funds and schedules.[22] Before the end of June 1964, NASA had built and tested a partial prototype of arm 6—a typical one that included all critical aspects. Contemporary photographs of the prototype compare favorably with the final version of service arm 6.[23]

When NASA opened bids on the service arms on 31 July 1964, the low bidder was Hayes International Corporation of Birmingham, Alabama. A pre-award service team made an on-site inspection of Hayes. The following day Raymond Clark, in charge of the team, reported: "Past experience with sub-standard quality from Hayes, under previous contracts, along with the results of this survey, may dictate an evaluation of Martin-Baltimore." Debus penned a note on the bottom of this statement: "This is in conflict with what you told me."[24] The service team concluded that Hayes had the personnel to do the job, but needed additional facilities and tools and would have to incorporate into their plans further recommendations of the survey team. It seemed that Hayes had built up a good team in earlier years, but had lost many of its better men during a time it had fewer contracts. On 25 August the Launch Support Equipment Engineering Division expressed serious reservations about Hayes's technical capacity to perform the task. Yet during the previous week, the Division had changed several hundred drawings, which would have strained the capacities of any bidder.[25]

In spite of these misgivings, the contract went to Hayes International on 10 September 1964, with a fixed price of $11 480 113 and a completion date of 26 April 1966.[26] At this time, incidentally, KSC was under pressure from the Office of Manned Space Flight to contract every possible item by fixed price competitive bidding. The fixed price contract, however, soon proved untenable. The details of the service arms were in a state of constant change, and a fixed price contract is valid only for a fixed design. The arms were a new design, more complex and mechanically larger than those used on the Saturn I.[27]

Shortly after Hayes International started work, it "uncovered innumerable discrepancies in the design and bills of material." Hayes notified the KSC Procurement Division that although it had believed it had a complete document package, many drawings were missing or tentative. On 27 October, Procurement delivered to Hayes less than 100 missing drawings. Ultimately, however, drawing changes for all reasons, including research and development, went above 4000.[28]

To alleviate these problems, in late November 1964 the Apollo-Saturn V Test and Systems Engineering Office of KSC's Apollo Program Management Office concurred with the recommendation of Procurement Division's Launch Support Equipment Section to change the contract so as to incorporate revised lists of drawings. These lists would supersede all documentation previously incorporated, including documents attached to the original invitation for bids and subsequent change orders that revised, added, or deleted drawings.[29] In an effort to maintain control over the many changes and revisions, a change review board made up of representatives of the Apollo Program Management Office, Procurement, Quality Assurance, and Launch Support Equipment Engineering began to meet in late September. By early November, the board had promulgated a formal procedure for handling engineering changes in the Hayes contract. Petrone, KSC Apollo Program Manager, approved this procedure on 20 November 1964.[30]

By the time the review board approved complete drawing documentation, another problem surfaced. On 2 December Hayes could not buy 190 items specified in the contract for sole-source procurement. It seems that, in designing the swing arms, Brown engineers had changed specifications on components without the knowledge or approval of the manufacturers of these components. Further, Brown engineers had accepted sales representatives' promises that their respective companies could meet specifications or proposed changes. Some companies, however, did not back their salesmen's promises and refused to deliver. Hayes then took the position that since the items were listed as sole source, the government was required to specify alternate sources. Hayes would do no engineering or expend any effort to supply items from sources other than those specified in the contract. Neither would

Hayes rework substitute items to make them meet specifications unless NASA furnished detailed rework designs, nor would it provide the necessary engineering design to facilitate rework without a contract modification.[31]

The review board evaluated all design changes for their impact on contract time, costs, and delivery schedules. Hayes and KSC revised the contract whenever necessary. In addition KSC in mid-1965 established a Resident Apollo Program Office, headed by Willard L. Halcomb, at the Hayes plant in Birmingham to reduce the time involved in approving decisions. KSC also set up a so-called tiger team, an ad hoc team that went to Birmingham every Monday to review and identify problems in design and production and returned to KSC to report on progress each Friday. In addition Hayes employed the consultant firm of Booz, Allen, and Hamilton to recommend managerial improvements.[32] By late summer of 1965, however, the situation had reached the point that Hayes International management felt it necessary to approach the KSC Procurement Division with a formal recommendation to change the contract from fixed-cost to some form of cost-reimbursable method.

Debus, wishing to have time to assess contract progress objectively following the establishment of the resident program office, waited until early November before replying personally to Hayes's management. In his reply, Debus said:

> There is no doubt in my mind that we both entered into the fixed price contract (NAS10-1751) in good faith. . . . It is indeed unfortunate that it was necessary for a considerable number of design changes to be introduced subsequent to the award of the contract. . . . I am fully aware of your recommendation that the method of contracting should be converted from fixed price to a cost reimbursable type. . . . You have been briefed in detail on the reasons why a conversion is neither feasible nor satisfactory to us since it would, in all probability, generate more complications than it would solve. I do, however, have the utmost confidence that so long as a proper spirit is evidenced by our respective representatives at all levels, then we each will be able to achieve our joint objectives—delivery of the highest quality arms, in an acceptable time frame, at a fair and reasonable price. I trust that you too are now encouraged that continuation on a fixed price basis does not present an unworkable contract relationship.[33]

In spite of Debus's hopes, a subsequent reappraisal converted the contract into a cost-plus arrangement.[34]

It had originally been planned to transport completed service arms from the Hayes plant in Birmingham to the Marshall Space Flight Center in Huntsville for testing, calibration, and acceptance. When it became clear that such testing would take a great deal of time and the deliveries were already late, NASA decided to have Hayes International deliver untested arms directly to the Kennedy Space Center, for installation on the first mobile launcher. After validation of the overall complex, KSC could then remove the arms and transport them to Marshall for a thorough testing of the service arms themselves. The arms that were not needed at KSC for validating the complex went directly to Marshall.

Eventually the total cost of the contract tripled from the initial $11.5 million. Major mixups had occurred, but none of them was deliberate and, given the press of time, none may have been avoidable. In the end these unprecedented devices performed with astounding reliability and majestic smoothness.[35]

Hold-Down Arms and Tail Service Masts

Four hold-down arms had to secure the Saturn V firmly on the mobile launcher during assembly, transportation to the launch site, and its stay on the launch pad in all kinds of weather. These devices also had to have the strength to hold down the launch vehicle after ignition, until all engines registered full thrust. Then they automatically and simultaneously released the Apollo-Saturn for liftoff. They did not, of course, have to overcome the full power of all the engines; the great weight of the fueled vehicle counteracted much of the thrust. As an indication of the unusual design requirement, James D. Phillips of KSC Launch Support Equipment Engineering Division won the 1965 steel-casting design contest sponsored by the Steel Founders Society of America for the design of the casting forming the base for the hold-down arms.[36] The arms would weigh over 18 metric tons each; the base was to be just under two meters wide, and not quite three meters long. They would stand 3.35 meters high. Nevertheless, in contrast to the huge Saturn vehicle, the hold-down arms seemed much too small to anchor—even momentarily—the huge rocket. On 17 February 1964 the KSC Procurement Division issued a contract to Space Corporation, Dallas, for the manufacture of 16 hold-down arms for the mobile launchers. The cost of the fixed price contract was $676 320, with completion date set for 25 July 1965.[37]

The first hold-down arm arrived at Huntsville on 31 October 1964, and testing began on 20 November. Due to a strike at a subcontractor's

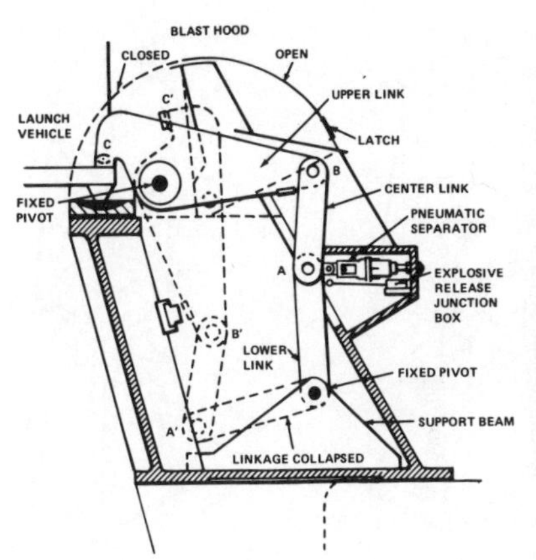

Fig. 95. Schematic of hold-down arm. The leverage produced 350 metric tons of force at C. The solid lines show the arm at work; the dotted lines represent the condition following release of the Saturn, when the linkage has collapsed and the blast hood closed.

Fig. 96. A hold-down arm ready for installation on a mobile launcher, November 1964.

plant, the second arm, scheduled for delivery on 19 November, came on the 28th. On 17 May 1965, engineers tested the ability of the first hold-down arm to sustain a vertical thrust of 725 747 kilograms. After the successful completion of all other tests on this arm on 25 May, workers installed and aligned an operational set of hold-down arms on launcher 3 at KSC. The other hold-down arms were ready by the end of the year.[38]

In addition to the four hold-down arms, three tail service masts would also stand on the base of each launcher. These provided support for electrical cables, propellant loading lines, hydraulic lines, and pneumatic lines servicing the first (S-IC) stage of the Saturn V. At liftoff a sequencer would hydraulically retract them, swinging them up and away from the Saturn V. A protective hood would fold over the umbilical connections on the end of each mast, protecting the connectors from the rocket engine's exhaust. After constructing and testing a prototype of these devices, the American Machine & Foundry Company of York, Pennsylvania, built the tail service masts.[39]

Launch Pads

The launch pads at complex 39 were more than just raised, hardened areas for the launching of the Saturn V. There would be no permanently

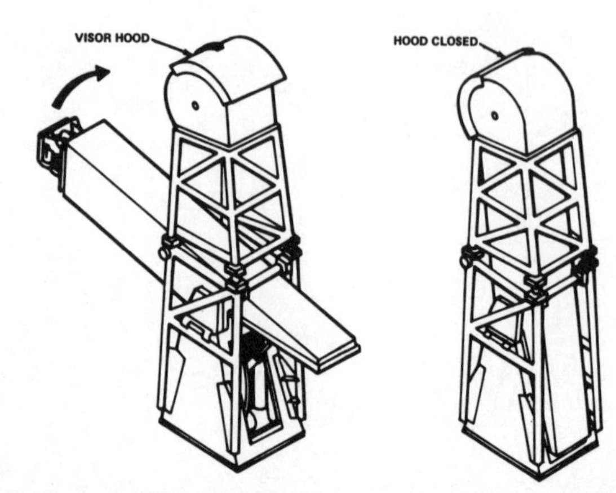

Fig. 97. Tail service mast for delivering propellants and electrical connections to the first stage.

emplaced launch stands, umbilical towers, and service structures as previously associated with a complete launch complex. At LC-39 these structures would be mobile, and the pad had to be of sufficient strength to support their weight and that of the crawler-transporter. But the pad would have many other appurtenances common to its predecessors.

The site of launch pad A, approximately 0.7 square kilometers, was roughly octagonal. A contract with Blount-Sundt called for the construction of the pad proper, roads, camera mounts, utilities, and several other small facilities. The elevated launch pad, which would rise 12 meters above ground level, lay in a north-south direction. This orientation required the crawlerway to make a near right-angle turn before approaching the ramp sloping 5° upward to the top of the pad. A flame trench, level with the surrounding area at its base, 18 meters wide and 137 meters long, would bisect the pad. On each side of this flame trench a cellular structure would support a thick surface, called a hardstand. The crawler-transporter would place the mobile launcher and the Apollo-Saturn vehicle on top of this reinforced slab.

The two-story pad terminal connection room and the single-story environmental control systems room would be within the western side of the pad. The former would house the electronic equipment that would connect communication and digital data link transmission lines from the launch control center to the mobile launcher when it was on the pad. The environmental control systems room would serve as the distribution point for air conditioning and water systems. The high-pressure-gas storage facility, to store and distribute nitrogen and helium gases piped from the converter-compressor facility, would lie beneath the top of the pad on the east side.

Should a hazardous condition arise that allowed safe egress from the spacecraft, the astronauts could cross over to the mobile launcher on a swing arm and then ride one of the high-speed elevators from the 104-meter level to level A, thirty stories down at 183 meters per minute. From there they would slide down an escape tube to a thickly padded rubber deceleration ramp. Steel doors, much like those of a bank vault, allowed access to a blast room, which could withstand an on-the-pad explosion of the entire space vehicle. Those inside could stay alive for at least 24 hours to allow rescue crews time to dig them out. The emergency egress system was part of the pad A contract.[40]

From Arming Tower to Mobile Service Structure

Originally conceived as a stationary arming tower, the mobile service structure went through many design changes before arriving at its final form. The structure, 125 meters high, nearly matched the mobile launcher in height as it stood on the opposite side of the Apollo-Saturn on the launch pad. The tall steel framework included five work platforms—the two lower ones vertically adjustable—that provided access to the space vehicle, and a base that contained several rooms. Shortly before launch, the crawler-transporter would move the mobile service structure along the crawlerway to a safe distance from the pad. The changing operational requirements during the construction of facilities made the mobile service structure one of the last essential facilities at launch complex 39 to get under way.

When the Rust Engineering Company of Birmingham undertook the design of the arming tower in February 1963, it faced a difficult task—designing a structure that would satisfy Apollo requirements but not exceed the load-carrying capability of the crawler. Initial meetings between NASA and Rust engineers concentrated on requirements for installing vehicle ordnance. Discussions on 21 March disclosed a major problem with the installation of linear-shaped charges—charges that would separate the stages during flight and, if necessary, destroy the vehicle. Their placement on the Apollo-Saturn required access from the arming tower at the interstage sections of the S-IC, S-II, and S-IVB, as well as at each stage that required a destruct package. As the tentative tower design with these features exceeded the load capability of the crawler's front end, NASA engineers agreed to see if the shaped charges could be installed in the VAB. The tower would still serve to arm the various destruct charges and install the Saturn's retro and ullage rockets. After reviewing the matter the following week, Gruene supported the use of the VAB; he also recommended a hazards study to confirm

Construction of pad A, LC-39

Figure 98

Figure 99

Fig. 98. Looking north, July 1964. The large sandpile has been removed and the first concrete poured. The rectangle (upper right) is the hydrogen burn pond. Fig. 99. The cellular construction of the hardstand, either side of the flame trench, is evident by September. The view is to the west.

Figure 100

Figure 101

Fig. 100. Looking southwest, November. Most of the concrete has been poured. Fig. 101. Same view two months later; pad 39A essentially complete. The flame deflector would move along the tracks in the foreground. The crawlerway enters the picture in the upper right, passes the service structure (née arming tower) in its parking position, and makes a near-90° turn in the upper left to approach the pad.

the safety of the proposed change. Meanwhile the Rust engineers dropped from their design the requirement to install shaped destruct charges.[41]

Wind loads were a second major concern for the tower's designers. On 28 March representatives of the Marion Power Shovel Company, the Corps of Engineers, LOC, and Rust agreed to design for a maximum wind velocity of 100 kilometers per hour. When resting on its supports at the launch pad, however, the arming tower was to be able to sustain considerably higher winds. NASA officials cancelled the latter requirement two weeks later: in the event of a hurricane, the tower would be removed from the pad area.[42]

Rust Engineering completed the criteria for the mobile arming tower on 1 May and began the design work two days later. At a design review in September, the Corps of Engineers asked Rust for a thorough analysis of the tower's weight and wind-load factors on the crawler. The review showed that the arming tower was overweight. During the next two months, numerous changes were made to bring the weight down to the crawler's capability, but the efforts met with little success.[43]

On 3 December 1963 Debus asked the Corps to reexamine the arming tower. Within three weeks, the Corps submitted the results of several studies. In the first, Strobel and Rengved, consulting engineers, retained the basic Rust configuration but reduced the size of the work platforms and eliminated the air conditioning equipment. In contrast, Rust recommended reaching the weight limit by eliminating one of the five service platforms, one of the three elevators, and the air conditioning. While the Rust proposal got rid of more weight than the Strobel and Rengved study, it also reduced the operational flexibility of the tower. The Corps then asked the two firms for another study, this time a completely new design. Rust's study employed fixed platforms, Strobel and Rengved's movable platforms. Both resulted in overall weight reductions but no significant reduction of wind loads on the crawler.[44]

While the studies were in progress, a KSC decision rendered the work superfluous. At a Huntsville meeting on 10 December, KSC's representative announced a new policy for the installation of ordnance at LC-39. Ullage rockets, retrorockets, the separation charges for the Apollo escape system, and other small ordnance items would be installed in the VAB. Detonators would be installed at the pad after the arming tower had been removed, by technicians using the mobile launcher's swing arms for access. The arming tower, no longer required for ordnance installation, thereby became the mobile service structure.[45]

That change made Rust's job much easier. Despite delays in receipt of spacecraft data from the Manned Spacecraft Center (page 183), Rust completed the redesign work by July 1964. The combine of Morrison-Knudsen,

Fig. 102. Model of the service structure, July 1964.

Fig. 103. The crawler-transporter ready to pick up the service structure, August 1966.

Perini and Hardeman won a construction contract on 21 September in the amount of $11 587 000. Steel fabrication started at once, and actual construction got under way on 21 February 1965. Because of the late start, minor labor delays, and the late delivery of material, the framework of the mobile service structure had only reached the 13-meter level by mid-1965. From then on, however, work moved rapidly ahead. By the end of September, steel erection had reached the 68-meter level, and the workers topped out the structure at 122.5 meters on 19 November 1965, only four months late.[46]

Work on the crawlerway progressed steadily, and by the end of 1964 it was 83% complete. The converter-compressor facility was complete, and mechanics were installing equipment. The concrete paving, the supports for the high-pressure-gas lines, the foundation for the mobile service structure at its parking position—all were ready. Interior architectural, electrical, and mechanical work moved forward in the pad terminal connection room, while joint occupancy of the environmental control system room began on 28 December 1964.[47]

At pad A, activities moved ahead of schedule, with the completion of all major concrete work. As the middle of 1965 approached, the launch pad lacked only the paving of aprons and road, the placement of refractory brick, the digging of ditches, and the testing of components and systems.[48]

Lightning Protection for Apollo Launch Operations

KSC officials had been concerned about lightning strikes since the start of the Apollo program. The Cape Canaveral area averaged more than

70 thunderstorms per year, twice the national average. Although there had been little lightning damage to missiles during the 1950s, the height of the Saturn vehicle greatly increased the chances of a strike. Studies made in 1962 pointed up the hazard to LC-39. General Electric engineers predicted that the VAB would receive five lightning strikes per year, the mobile service structure and mobile launcher four strikes per year. The potential for lightning damage had prompted a Marshall–LOC meeting in August 1962. The group recommended contracting with General Electric's High Voltage Laboratory for a lightning protection study and appointed LOC the technical supervisor. In Februry 1963, Petrone set up a committee on lightning protection, under Hayward D. Brewster, to review the GE proposals.[49]

The GE study served as the basis of the committee report submitted to Debus in July 1963, which concentrated on four problems, all during prelaunch operations: protection of the vehicle against a direct strike, the induced effects of a strike, mobile launcher grounding in transit, and corona. The placement of a lightning mast atop the mobile launcher was the straightforward solution to the first problem. Some authorities, however, did not think the traditional "cone of protection" applied to a structure as tall as the mobile launcher.* Electrical engineers also differed as to whether bonding and shielding on the mobile launcher would lower induced voltages to an acceptable level. Unless KSC could protect the launcher from both a direct strike and the secondary induced voltage, some other lightning diverter, such as balloons, would have to be used. GE module tests demonstrated that the cone-of-protection theory did apply to the mobile launcher and that practical measures would protect the vehicle and support equipment circuits from induced voltages. Besides a retractable mast for the launcher,† the committee recommended "general grounding, shielding, and bonding techniques . . . throughout the LC-39 area in order to keep the high voltage imposed by lightning strokes anywhere on the complex to a safe level."[50] An extensive underground counterpoise of rods and interconnecting conductors was eventually built into LC-39. Thousands of ground rods, driven deep enough to achieve a one-ohm resistance, tied together the crawlerway, service structure parking area, perimeter fence, and pad. Similar counterpoises protected cross-country cabling.[51]

GE engineers recommended certain precautionary measures when the mobile launcher was in transit. If a threat of lightning existed, personnel

---

*In the early 1960s experts disagreed about the generation and incidence of lightning and about its behavior and effects. The cone-of-protection theory held that all strokes would terminate on a tall structure in preference to a shorter structure located within the conical volume whose apex was the height of the tallest structure and whose base radius was equal to the apex height. Evidence from lightning strikes on skyscrapers and church steeples indicated the theory applied to the top half of the cone; the disagreement concerned the protection provided to the lower half.

†The mast retracted so that the mobile launcher could get into the VAB.

would stay inside the mobile launcher, or at least six meters from the crawler. Insulated ladders would be used for movement on or off the crawler. The committee proposed a backup warning system to alert personnel of approaching storms. The actual grounding of the crawler was simple—it would drag a chain along a conductor buried in the crawlerway.

A bluish electrical discharge, sometimes called St. Elmo's fire, occurs frequently when storm clouds pass over tall structures. GE investigated the possibility of this phenomenon igniting a hydrogen explosion, but found that the corona would likely appear on the top outer edges of the mobile launcher and mobile service structure. This posed no threat, since the S-IVB lines ran 30 meters below the top of the launcher. The hydrogen lines to the Apollo service structure would shield the spacecraft connections during loading. The GE team rated the corona hazard a "negligible risk."[52]

During the next three years, Brewster's committee implemented the safety features on LC-39 while KSC's Instrumentation Division set up a system to collect more data. A GE study of LC-34's and LC-37's needs led to a second set of committee proposals approved by Debus in November 1964. At a September 1965 meeting of the Lightning Protection Committee, R. H. Jones, an Instrumentation Division engineer, reported thirteen measured strikes during the previous year. One bolt had killed a construction worker on LC-39, pad B. Another strike on the Cape side had delayed Gemini II operations at LC-19 by several weeks (the lightning had damaged a number of electrical components in the spacecraft and supporting equipment). E. R. Uhlig of GE's High Voltage Laboratory pointed out the correspondence between the measured incidents and GE's earlier predictions.[53]

When Apollo launch operations began in 1966, KSC applied strict safety rules for lightning protection. All launch personnel evacuated the mobile launcher, mobile service structure, and space vehicle when lightning was detected within five miles of the pad. A half dozen lightning storms delayed operations but never for more than a few hours. KSC relaxed its provisions somewhat in 1970 as experience demonstrated the safety of the mobile launcher and service structure. Thereafter operations on the tall structures, excepting electrical work, continued in the face of an approaching storm.[54]

Flame Deflectors

The last of the major facilities for launch complex 39 to reach the contract stage was the 635-metric-ton flame deflector. It would protect the lower section of the Saturn launch vehicle and the launch stand from high pressures and flame during ignition and liftoff. It would move on rails in the flame trench to a position beneath the Saturn V's massive booster engines. The reflector, shaped like an inverted V, would send the flames down each side of the trench. It would be constructed of structural steel beams and trusses, supporting a steel skin. The skin was covered by 10 centimeters of ceramic material capable of withstanding the direct flame and pressure effects of the Saturn first stage engines. On 5 November 1965, Heyl and Patterson, Inc., signed a contract in the fixed amount of $1 465 075 for the manufacture, installation, and erection of three deflectors.[55] Ultimately there would be a fourth, with one in use and another in reserve at each pad.

Without doubt, the many amazing structures under way on complex 39—the world's largest building, the crawler-transporters, the hold-down arms, the mobile launchers—constituted one of the most awesome building programs in the world. After the American Society of Civil Engineers considered engineering projects from every part of the country in 1966—the Astrodome in Houston, the North California Flood Rehabilitation work, the Trans-Sierra Freeway from Sacramento to the Nevada line, and the hurricane barrier at New Bedford, Massachusetts, among others—it recognized launch complex 39 as the outstanding civil engineering achievement of the year.[56]

14

Socio-Economic Problems on the Space Coast

Labor Problems at the Missile Center

Socio-economic problems went hand in hand with the engineering problems encountered in sending men to the moon. The relocation of a large number of people—many of them from urban centers—to the small towns of Florida's east coast where newcomers were not always welcome, the tenfold increase in population in Brevard County within 20 years, and the construction of many buildings and the assembling of highly complicated machinery in a previously quiet corner of a nonindustrial state brought about dramatic changes in the quality of life.

Many factors complicated the relations of labor, management, and government at the Kennedy Space Center, especially during the construction years, chiefly 1963 through 1965. Disputes of various kinds held up work on the assembly building, on other phases of LC-39 construction, and in the industrial area. The major labor issues will be discussed here.

First, Florida had an open-shop law, called by its supporters a "Right to Work Law." Such laws tend to create a climate of suspicion for union workers and are accompanied by strife between union and nonunion workers. At KSC, the unions were wary of any increase in contracts with nonunion contractors or subcontractors.

Second, Florida was not an industrialized state. In central Florida, the Cape–KSC area was at once the largest industrial center and the area where labor relations most closely paralleled the practices of more industrially developed states. As a result, some labor leaders did not hesitate to use the KSC arrangements as a possible club over contractors in nearby areas. One of the building trades unions, for instance, jockeyed for advantage with an Orlando contractor by using KSC arrangements as a lever.

Third, many contractors failed to enter serious contract negotiations until workers actually went on strike. Most of these strikes were short, and the contractors could have avoided them had they settled with the union one day before the strike, instead of agreeing to union demands after a one-day walkout.

Fourth, jurisdictional disputes caused endless problems. To understand the worker's point of view in this regard, one should remember that the welfare of an entire trade often depended on the protection of certain tasks that came within its jurisdiction. If a trade lost a particular type of work, the union simply found its members unemployed. Further, precedent so influenced jurisdictional assignments that unions zealously and carefully protected their existing areas. Sometimes, however, these jurisdictional disputes went beyond common sense and outraged everyone concerned. Carpenters walked out in a dispute: ironworkers were installing aluminum door frames. Labor leaders on occasion acted in the "public-be-damned" spirit of the 19th century industrial "Robber Barons."

Fifth, certain attitudes of construction workers, such as carpenters and plumbers, differed from those held by industrial workers, such as steelworkers. The more highly centralized industrial unions tended to heed decisions made on a nationwide basis or at national headquarters. The loosely bound construction locals, on the other hand, enjoyed greater autonomy. The construction worker never felt the same loyalty to his employer that the industrial worker felt. His term of employment was relatively short and his job security came from the union hiring hall, not from the company. It did not really matter a great deal to a plumber whether he was putting pipes in a motel, an industrial plant, or a missile site. He had little emotional involvement with the work itself or with the company he worked for at the moment. When he finished a job, he looked to the union for another. The construction worker thus tended to identify himself with his craft and his union, not with his employer or even with a major purpose such as sending a man to the moon.

Many construction workers were transient by background. Accustomed to moving where the work happened to be, oftentimes they did not put down roots. Some men came in for only a few days, sometimes sleeping in their own cars, then moving on. With the increase of work at the Cape and at KSC—the only diversified construction activity in Florida at the time—so many new workers came in with permits from other locals that they swamped the local unions and made their business agents edgy. At one time, for instance, between 600 and 700 electricians worked at KSC with permits from locals outside the region. The building trades thought they saw a lack of consistent policy and felt they had to scrap for everything they could get. These factors often made dealing with construction workers more difficult than dealing with industrial workers, as several officials at Kennedy Space Center were to comment.[1]

Labor troubles at missile sites, especially the Cape, had grown acute even before President Kennedy issued his lunar landing challenge to the nation.

On eight days from 25 April to 5 May 1961, the permanent Subcommittee on Investigations of the Senate Committee on Government Operations had held hearings in Washington. Senator John L. McClellan (D., Ark.) chaired this subcommittee, whose prestigious membership included Senators Ervin, Muskie, Jackson, Mundt, and Curtis. They took testimony from 38 individuals. The witnesses showed that work stoppages and slowdowns were commonplace at missile sites.[2]

The hearing brought to light many abuses including excessive overtime, exorbitant wages, low productivity of workers, improper classification of work, and inefficiency by contractors. The subcommittee criticized both labor and management. Work stoppages resulted in a total loss of 87 374 man-days at Cape Canaveral during a 4½-year period in the late 1950s and early 1960s. Wildcat strikes, slowdowns, and a deliberate policy of low productivity further delayed progress. Workers gouged the taxpayer with unnecessary and exorbitant overtime costs. The international unions did nothing to discipline the locals. Some contractors, operating under a cost-plus-fixed-fee contract, did nothing to stop skyrocketing costs in excessive overtime payments. They overmanned jobs and did not properly supervise.

The subcommittee insisted that the military and civilian officials on construction sites try to rectify unsatisfactory labor conditions. It pointed out that while Congress had passed the Davis-Bacon Act of 1931 to keep construction wages on government contracts consistent with the wages prevailing in a given area, some labor leaders improperly used it as a device for settling jurisdictional disputes. To conclude its findings, the subcommittee pointed out that work conditions at the missile sites improved for a time after the subcommittee began its hearings, then deteriorated.[3]

The Center's Labor Policy

Such was the industrial climate at the Cape shortly before NASA was challenged to send men to the moon. Only four days before President Kennedy gave that call on 21 May 1961, he signed Executive Order 10946, establishing the Missile Sites Labor Commission, with Secretary of Labor Arthur J. Goldberg as chairman. He and three representatives of management were to establish policies and procedures that were intended to improve labor relations within the missile and space industry. Section 2 of the order provided for the establishment of local on-site committees to anticipate problems and to prevent their becoming acute. The Missile Site Labor Relations Committee at KSC included one representative of each of the following: the Defense Department, NASA, building contractors, the Building and Construction

Trades Department of the AFL-CIO, the industrial contractors, the industrial unions, and the Federal Mediation and Conciliation Service.[4]

The work of the committee, coupled with other factors, resulted in a marked decrease in man-days lost at the Cape. The threat of further action by the McClellan Committee weighed heavily. McClellan introduced Senate bill 2361, which would have outlawed strikes and called for compulsory arbitration at strategic defense facilities.[5] One of the most important achievements of the Missile Sites Labor Commission at the Cape stemmed from a series of meetings between representatives of the Department of Defense, NASA, building and construction contractors, and international and local building trades unions. On 20 February 1962, they agreed to the Project Stabilization Agreement that standardized local arrangements between various unions and contractors. Two years later all parties were to accept a slightly revised agreement for three years more.[6]

A major dispute between NASA and certain of the building trades unions concerned the point where construction work ended and installation of equipment began. Further, the Air Force and NASA took different views on this question. Contractors working for the Air Force early reached an understanding with the construction unions and established an unwritten range policy to allow construction trades to install almost all ground support equipment. NASA never really accepted this policy.

Because of the research and development nature of its work, NASA maintained that each missile firing was essentially a laboratory experiment for the purpose of gathering data, testing feasibility of design concepts, operational techniques, and future development; and, therefore, all ground equipment, including launch controls, plumbing, and instrumentation that connected directly with the missile formed an integral part of the missile system. Thus, all such equipment should come under the direct control, from installation to final use, of the NASA missile teams. NASA saw many advantages to this viewpoint. It ensured quality control, increased reliability, reduced cost, and rendered unnecessary elaborate contract specifications for installation of launch facilities. At times, too, KSC saw the advisability of having the firm that built a piece of equipment bring its own workers to Florida to assemble it. The next chapter will discuss this issue with regard to the crawler-transporter—and the union disapproval that resulted. In line with NASA's attitude, and in spite of the Air Force's unwritten policy differing from NASA's, some Air Force missile contractors would have preferred to have their own personnel do the entire job. This had come up in at least one significant case with Convair before the Senate hearing on work stoppages at missile bases.[7]

The Air Force had also drawn up ground rules that allowed the use of nonunion contractors, but never on the same specific job as a union contractor, such as inside the same blockhouse at the same time. The Air Force, further, won an agreement that disallowed picketing on the Cape itself. Although the commanding general readily listened to the complaints of labor leaders, the Air Force rarely intruded in disputes that arose between contractors and their workers.[8]

NASA did not duplicate all these policies. As a result, many unions had one set of rules east of the Banana River and another on the west bank, and the difference showed from time to time. On one occasion, construction unions walked off their jobs, causing a loss of 491 man-days to NASA contracts and 3867 man-days to NASA-financed Corps of Engineers contracts. At the same time, Air Force contracts and Air Force–financed Corps of Engineers contracts of about the same size did not lose a single man-day.[9]

As the Launch Operations Center moved toward the period of construction, its Industrial Relations Office increased in importance. In June 1963, Oliver E. Kearns, who had worked with the Federal Mediation and Conciliation Service in Toledo, and before that had been field examiner with the National Labor Relations Board in Seattle, became Industrial Relations Officer. Later in the year, John Miraglia, who had worked in industrial relations for NASA at the Cape, returned to the Space Center as Industrial Relations Chief, with Kearns as his deputy. In the NASA-wide administrative reorganization of early 1964, Paul Styles became Labor Relations Director, with Miraglia his deputy. With this new office added to his duties at KSC, Miraglia served as trouble-shooter at NASA centers throughout the country.

Miraglia had experience both as a textile worker and a representative of the textile workers' union and had worked with the National Labor Relations Board. He understood that many labor problems were emotional as well as economic and that the first essential was proper communications.[10] He and Kearns would have plenty of opportunity to develop the art of communication and to extend their patience to the limit during 1964, an especially trying year. But all the construction years at Kennedy Space Center would prove exasperating.

A Spring and Summer of Strikes

In early February 1964, KSC signed an agreement with the Florida East Coast Railroad for the operation of a spur line on NASA property. Eleven nonoperating unions, such as telegraphers and maintenance-of-way workers, had been on strike against the railroad for 13 months in an effort to bring their pay up to the national scale as accepted by 190 railroads in 1962. Violence during the strike had caused suspension of passenger traffic. But the Florida East Coast continued to move freight, and during the week before the agreement two trains had been blown up.[11]

NASA Administrator Webb had warned board chairman Edward Ball that a paralyzing strike might endanger the nation's space and security program. Vice President of the railroad W. L. Thornton believed that the unions would not shut down the Cape operations because such action would constitute an illegal secondary boycott. Thornton had refused President Kennedy's recommendation for "final and binding arbitration" the previous year. Thornton did not seem to take seriously the pledge of the almost 12 000 spaceport union employees to honor picket lines.[12] The railroad, in fact, had tried to operate a train on NASA property before the agreement. A confrontation with NASA security personnel had prevented unloading of the train.[13]

The nonoperating unions placed pickets at all entrances to the space center and to Cape Kennedy on 10 February, halting construction on the Cape and Merritt Island.[14] The National Labor Relations Board obtained a temporary restraining order from the Federal District Court of Orlando on the grounds that in halting space construction, the pickets violated a ban on secondary boycotts.[15] The unions removed the pickets on 12 February and the workers returned to their jobs, even though the attorney for the union contended that the Florida East Coast came under the purview of the Railway Labor Act, and thus the National Labor Relations Board had no jurisdiction.[16] In his weekly report to Debus, Miraglia correctly assumed that one or two months would elapse before pickets reappeared.[17]

The meetings that followed between Assistant Secretary of Labor James J. Reynolds and the officials of the railroad transcended the local situation at the spur line to KSC. Reynolds suggested that the President's Missile Sites Labor Commission arbitrate the strike—a proceeding that Ball had steadily opposed for 13 months. When President Johnson spoke at Palatka, Florida, later in the month, a blast blew up a Florida East Coast train 25 kilometers away.[18] Ball continued to oppose compulsory arbitration and the dispute dragged on. But wider aspects of the battle did not affect the situation at KSC.

Paul Styles represented NASA at a meeting of the Missile Sites Labor Relations Committee on 20 April 1964. In the previous year, jurisdictional disputes between building trades unions and disagreement over working conditions had caused 33 work stoppages. Styles stressed the need for a new dedication by labor organizations and contractors to adjust jurisdictional disputes without work stoppages. The representatives of the contractors and the union pledged greater efforts to follow the prescribed methods of settling such disputes. Government, labor, and management all felt the meeting successful.[19] Actualities were to betray their hopes.

The Missile Sites Labor Relations Committee held a special meeting to avert picketing of KSC and the Cape by members of Steelworkers Local 6020 of Tampa. This union had been on strike against the Florida Steel Company of Tampa for 12 weeks. KSC used steel from this company, and the union felt that placing pickets at the spaceport would bring the dispute to the attention of the public. KSC prevailed upon the union to postpone action until a committee had studied the situation. The committee suggested that a reduction or possibly total elimination of the use of steel from this company would remove the threat of picketing.[20] This was obviously a case of a union using KSC as a lever to win a strike against a particular firm.

So many work stoppages occurred during the next few months one might well have thought that the building of the space center would stagger on forever. In late May and early June the ironworkers refused to work for the American Bridge Company in the assembly building, alleging unsafe practices; 736 man-days were lost. Since workers left their jobs contrary to the orders of union representatives, the walkout indicated a loss of control by the union. At the same time, 20 pipe-fitters left their jobs on complex 36B in the cable terminal building. When Akwa Construction Company sent several nonunion workers, the carpenters' business agent pulled out the remaining union workers. The firing of 5 men for allegedly drinking and gambling on the job provoked 129 laborers in the assembly building and 29 cement masons in the industrial area to stay off the job beginning 3 June. Conciliation brought about the rehiring of three of the men on the basis of inconclusive evidence and termination slips for milder reasons for the other two, so as not to impair their chances of future employment. Eight laborers and 9 carpenters walked off the job on 1 June at the cable terminal building and at the site of the communications ducts to protest the hiring of 4 nonunion carpenters. Nonunion men then took over.[21] Twenty-five operating engineers left their jobs on 5 June to protest the discharge of one member; 11 man-days were lost. The business agent ordered the men back to work at the direction of the Corps of Engineers.

On the morning of 8 June, Locals 2020 and 717 of the Brotherhood of Maintenance of Ways placed pickets at all entrances to Merritt Island and the Cape at 5 a.m. without giving prior notice. Members of the building trades honored the picket lines, closing down nearly all construction work at KSC and at the Cape. About 4000 of 4500 workers stayed away. The railroad trouble had surfaced again.

At a meeting of the Missile Sites Labor Relations Committee on the following day, Paul Styles admonished the building trades unions for violating the no-strike clause—Article 6 of the Project Stabilization Agreement. The committee insisted that the unions needed more effective leadership and that the contractors had to discipline violators of the agreement. Styles urged the heads of 14 building trades unions to get the men back to work. The union officials responded that the workers had refused to cross the picket lines spontaneously and not under orders from the union leadership.[22]

On the same day (9 June 1964), Styles notified all employees of the Florida East Coast Railroad, its subcontractors, and its suppliers that they had to use one entrance to the Merritt Island area. If unions wanted to picket, they could do so only at the one gate. This decision of the Director of NASA's Office of Labor Relations followed a procedure established at many multiemployer work sites throughout the country and repeatedly upheld by the National Labor Relations Board. At this juncture, Federal District Judge George C. Young ordered the maintenance-of-way unions to cease picketing the railroad at Kennedy Space Center. His temporary injunction would last until the following Monday. Early the following week he extended the injunction until Friday the 19th. In the meantime the National Labor Relations Board issued an opinion that the railroad unions involved, principally the telegraphers and the maintenance-of-way men, fell under its jurisdiction. Judge Young extended the injunction indefinitely.[23]

And the month of June had barely passed the midway point!

Representatives of the unions and contractors who had signed the Project Stabilization Agreement met in Orlando on 18 June to find out if the unions intended to adhere to the no-strike provision. Representatives of NASA and the Department of Defense attended. The meeting failed to produce any change in attitude of union representatives toward the Project Stabilization Agreement. Basically, the locals resented this restriction agreed to by the international unions and tended to ignore it. International unions, in turn, were not insisting on compliance by the locals.[24]

Strikes and work stoppages piled one on top of another with such frequency that Debus penned these words at the bottom of Miraglia's weekly notes: "John: The continuation of the 'little' walkouts precipitated by sometimes unknown causes is *very alarming*. What can be done about it?"[25] Jurisdictional strikes especially galled. At one time several jurisdictional disputes took place simultaneously and were to drag on through much of the summer of 1964. Carpenters walked off the job at the assembly building following a dispute with the contractor, Morrison-Knudsen, Perini, and Hardeman, over the assigning of aluminum door frames to the ironworkers.[26]

In the third week in July, Kearns, who gradually assumed more of Miraglia's duties at KSC, thought it noteworthy to record that no jurisdictional disputes had caused work stoppages during the past week, although three previous disputes were still pending. Now a new area of dispute took center stage. Five plumbers left the operations and checkout building in the industrial area protesting the award of a contract to a nonunion prime contractor who had subcontracted the mechanical work to another nonunion contractor. The strike lasted one day.

Unions began to show concern over the number of contracts that went to open-shop employers. The Brevard Building and Construction Trades Council asked for information on the number of nonunion contractors winning contracts from local government agencies—even though many open-shop contractors did use union workers or subcontracted to firms that had union workers. A cursory check by NASA during late August showed that 94% of the workers on KSC contracts were union men. This represented a rise in nonunion workers from 1.7% in June to 5.8% in August. The percentage of contracts let to nonunion contractors was between 15 and 20%. By dollar volume, however, it was only 5%.[27]

The *Orlando Sentinel* for 8 September 1964 depicted NASA's relations with labor as being in decay. To the Industrial Relations Office at KSC, it appeared that Clifford Baxley, the coordinator of the Brevard Building and Trades Council, had given false information to the newspaper, and Kearns recommended boycotting informal, off-the-record discussions whenever Baxley represented labor. In his report to Debus, Kearns mentioned that Baxley did not have to support all unions and that his conduct completely destroyed the purpose of meetings, particularly when the information Baxley gave to the press was not accurate. On Kearns's report, Debus wrote an emphatic "No!" and underlined the word twice. "We cannot take this attitude," he insisted. "Discuss this with Mr. Siepert."[28]

In line with the insistence of Debus, Kearns wrote the following week:

> NASA will continue to attend these informal labor management meetings if they are resumed. Other Government agencies that have participated in these meetings agree that certain rules be established to retain the trust and confidence the attendees must have towards each other in order to assure the success of such meetings. No date has been set for another meeting.[29]

The long hot summer of 1964 proved frustrating for Miraglia and Kearns; indeed, labor relations were not to improve during the construction period at KSC. One of the most significant strikes came in mid-September 1965, when construction neared its conclusion throughout Merritt Island.

Most other strikes had been purely local, or at most regional, such as the strike against the Florida East Coast Railroad. This one was part of a nationwide walkout of Boeing Company employees. The strike directly affected only about 50 members of the International Association of Machinists and Aerospace Workers on KSC's Saturn program, and about 225 on the Air Force Minuteman program at Cape Kennedy.[30] Revolving around a new contract, it hinged on such issues as the grading of employees, insurance coverage for dependents, and the union shop.

When contract negotiations broke down, the union struck Cape Kennedy and KSC on 16 September 1965. W. J. Usery, regional representative of the machinists, made considerable but fruitless efforts to prevent the walkout of the nonstriking machinists (those who worked for firms other than Boeing). The striking machinists, in general, honored the one-gate picketing procedure that Paul Styles had set down in the railroad strike of the previous year. A large number of construction workers walked off the job for a time in support of the machinists. All the workers from the Marion Power Shovel Co., who had come south to assemble the crawler-transporter, went home.[31]

Boeing would not grant the union shop request. But the negotiations eventually resulted in a new contract that satisfied the international leadership of the union, and the spaceport machinists voted on 4 October to end the 19-day strike.[32]

The Spaceport's Impact on the Local Communities

During the years that Merritt Island changed from citrus groves, sand bars, and swamps to a major launch site, the local communities reflected dramatic growth. The area had no major city like Houston; further, no one community dominated the Cape area as Huntsville did the environs of Marshall. Instead, the newcomers dispersed over a wide area.

A short distance south of Cape Canaveral, Cocoa Beach early assumed a central role in the space program. Many industrial contractors located there. Numerous motels and an excellent beach imparted a holiday atmosphere and made the town popular with tourists. The area's night life centered there. The nation came to identify the space program with Cocoa Beach rather than with other communities in the vicinity. *Time* magazine carried a lurid picture of activities at Cocoa Beach night clubs on weekends and especially at launchings and splashdowns.[33] Cocoa Beach, however, had no television station—there was none in Brevard County. As a result, the cities of Orlando and Daytona Beach influenced the region through their television facilities, even though they were 64 and 80 kilometers distant, respectively.

In 1963 NASA funded three studies of the social and economic development of the area. A regional planning commission looked at roads and water systems, a Florida State University team dealt with community affairs, and a University of Florida research group studied population and economics. The study groups were to finish their reports within two years. The three principal investigators met profitably with NASA's local officials and delegates of NASA headquarters. They further got in touch with representatives of the various Brevard County communities. Florida State University set up an urban research center in the area and published materials developed by the three studies.[34]

Between 1950 and 1960, the population of Brevard County, 106 kilometers long and 32 kilometers wide, had grown faster than any other county in the country—from 23 653 to 111 435—an increase of 371%, in contrast to the 79% increase for the state of Florida and 19% for the entire nation. Most of the people settled in four towns: Titusville, the county seat, in the north, Cocoa in the center, Eau Gallie and Melbourne in the south. Titusville reached only half the population of each of the other three in the 1960 census.[35]

In 1950 Brevard County's 13 schools had an average daily attendance of 4163; by the school year 1963–64 there were 46 schools with an average daily attendance of 39 873. Classrooms grew from 117 to 1473 in the 14-year period.[36]

An infinitesimal percentage of the residents of the four main communities of Brevard County had been born there. Roughly one-fourth of the newcomers came from each of these categories: villages of less than 5000, towns between 5000 and 25 000, cities of 25 000 to 100 000, and cities over 100 000. Industrial firms transferred 13% of the newcomers from plants in other areas; 25% freely accepted Florida jobs with a firm they already worked for; and slightly over 25% sought better economic opportunities by coming to the area on their own to seek employment. Some 35–40% came from southern states other than Florida; close to 20% from other counties of Florida; and 15–20% from both the northwest and the midwest. Thus over half were southerners.[37]

In community involvement, the churches and PTAs led the way. Recreational and hobby clubs grew faster than economic and service-related institutions. Not surprisingly, women tended to involve themselves more in community participation than men. Melbourne, Cocoa, and Cocoa Beach developed active theatre and musical groups, including the Brevard Light

Opera Association in Melbourne and the Brevard Civic Symphony in Cocoa. The Surfside Players at Cocoa Beach presented six plays a year.[38] Recreationally, Titusville suffered in a way the more southerly areas did not. Its nearest beach, the rough but challenging Playalinda, was so close to the new launching pads that it would remain closed during many months each year.

A Florida State University survey showed slight participation of the newcomers in the political activities of the community or even of the nation. While only 23% of the old-timers, for instance, had failed to vote in the 1960 presidential election, 43% of the early migrants and 52% of the most recent arrivals did not go to the polls.[39] Registration requirements naturally influenced voting patterns. The newcomers, in general, willingly lent a hand in such activities as the United Fund; but they did not in any noticeable degree seek political control within the community. The few who did hold office often found older residents suspicious and uncooperative. The main loyalty of the newcomers lay with the space program, with their particular firm, and sometimes with a particular project of that firm, so that many did not feel Brevard County their permanent home, but merely a temporary assignment, as a soldier might look at a tour of duty.

The 1960 census gave Brevard County 111 435 residents. In May 1963, the Florida Power and Light Company estimated the booming population at 156 688. In the estimates for three years after that, the company expected the 72 650 people in southern Brevard County to admit over 52 000 newcomers; the 54 940 in central Brevard to grow to 100 000; and the 25 760 in northern areas at least to double.[40]

A month later (June 1963) Paul Siebeneichen and his staff at KSC's Community Development Office presented more detailed statistics on the population of the county. By that time, 42 new residents were arriving every day. Nine out of 10 homes were single-family units, and each housed an average of 3.4 people—the statistic the Florida Power and Light Company had used the previous month. The number of men approximated the number of women. Three out of every four men over 14 were married. More than one-third of the women over 13 had jobs. The median income per family was $6123—far and away the highest in the state. Consistent with this, the median value of homes was $13 000, compared to the state's average of $11 800.[41]

In May 1964, NASA and the Air Force took a residential survey by questionnaire of more than 28 000 military, civil service, and contractor employees in the area. This study, tabulated by a team from Florida State University, showed that up to that time residents tended to remain where they had located in the late 1950s. South Brevard had 42.1% of the population, with 20.8% on the mainland and 21.3% in the beach areas. Central

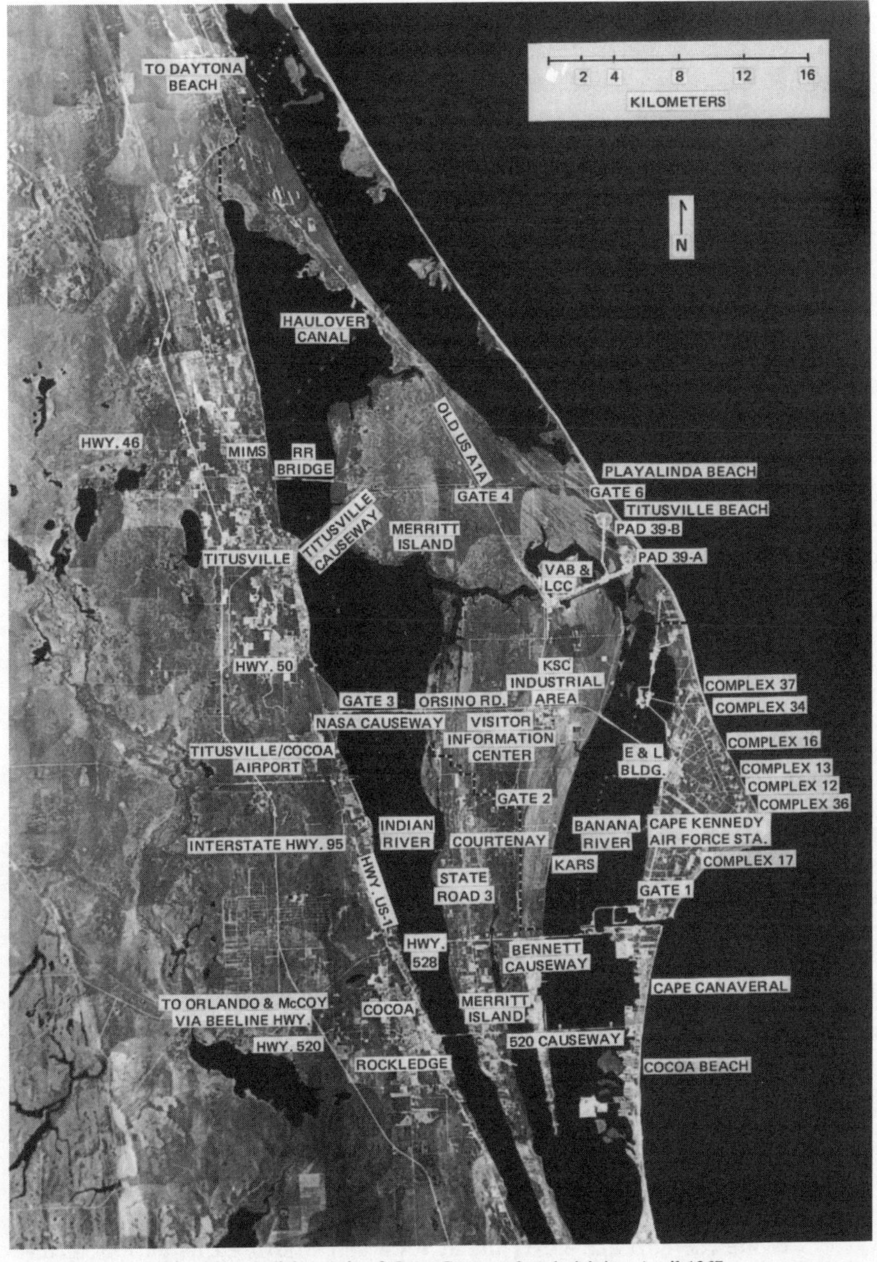

Fig. 104. Aerial mosaic of Cape Canaveral and vicinity, April 1967.

Brevard had 40.4%, with 12.7% on the mainland, 15.6% on the north beach area, and 12.1% on Merritt Island. North Brevard (the general area of Titusville) had 12.4%. Orange County had 2.4%; Volusia 1.6%.[42]

As the population of the area continued to grow, the automobile remained the only significant means of local transportation. Two roads that figured prominently in KSC plans were the north-south Merritt Island road (U.S. A1A) and the Orsino Road, an east-west street that deadended near the Indian River. The industrial area was southeast of the junction of these two roads. KSC improved the Merritt Island road as the main north-south artery within NASA property. A four-lane divided highway extended from 1.7 kilometers south of the industrial area to the Titusville Beach road, about 8 kilometers north of the assembly building. Studies by the Joint Community Impact–Coordinating Committee, which antedated the Regional Planning Commission, gave no indication of the tremendous growth ahead for the residential area on Merritt Island, about 16 kilometers south of the KSC industrial area.[43] As a result, the State of Florida did not widen the two-lane road (Florida Highway 3) that ran south from the KSC area through Courtenay to the Bennett toll road (Highway 528). After 1964, four FHA-backed apartment complexes were to spur extensive residential growth in that area of Merritt Island. As a result, Florida Highway 3 became a bottleneck during peak traffic hours.

East-west traffic was never to present a problem. The four-lane divided highway (the old Orsino road), a few blocks north of the industrial area, ran east a kilometer, then turned southeastward to a two-lane causeway over the Banana River to the Air Force Missile Test Center industrial area on the Cape; there it connected with the four-lane traffic artery to the Cocoa Beach area and south. The building of a five-kilometer long, four-lane causeway across the Indian River to the west connected the Orsino road with U.S. Highway 1 on the mainland a few kilometers south of Titusville. Originally intended as a limited-access road for KSC-badged personnel only, this road became a public highway a few years later with the opening of the Visitors Information Center several kilometers west of the KSC industrial area. On the west, beyond U.S. Highway 1 on the mainland, state road-builders were ultimately to continue the east-west road as a four-lane divided highway just north of Ti-Co Airport to its junction with Florida 50 near the intersection of Interstate 95. Thus traffic could move rapidly west from the industrial area across the Indian River and on to Titusville to the north, Cocoa, Rockledge, Eau Gallie, and Melbourne to the south, and the suburbs of Orlando to the west.

While the national government took steps during this period to increase the opportunities for employment of members of minority races, aerospace employers had few openings for blacks. Black engineers were few.

Black applicants in other categories of work often lacked the necessary background, training, or union membership. Thus while the white community multiplied, the black population of Brevard County remained the same, declining noticeably as a percentage—from 25% in 1950 to 11% in 1960.[44]

KSC and many contractors tried to improve the situation. An Equal Employment Opportunity meeting of 21 April 1964, with most contractors represented, planned a program to draw on the local black population rather than recruit from outside sources. The meeting set up two committees: one for job development and employment, the other for education and youth incentive. There was an obvious need to develop jobs suitable to available blacks, and to include in the local high school curriculum such courses as shorthand, typing, and the like. A month later NASA representatives attended a luncheon meeting sponsored by the contractors' Equal Employment Opportunity committee. That organization set up a program for employing local black teachers during summer vacation to give them first-hand knowledge of the academic skills necessary for employment at the space center, so that they could better counsel their students. Principals of three local black high schools, a representative of the National Association for the Advancement of Colored People, and an Air Force Equal Employment Opportunity coordinator attended this meeting.[45]

Harry W. Smith, Chief of KSC's Recruitment and Placement Branch, attended a meeting on 30 March 1965 of Governor Claude Kirk of Florida, his cabinet, and black leaders of the State. Smith participated at the request of the black leaders and explained KSC's Equal Employment Opportunity Program. The black leaders commented favorably on the program and hoped that the State government would adopt at least a part of it.[46]

In order to give wives and children a better understanding of the activities of their husbands and fathers, Kennedy Space Center's Protocol Office began to hold Saturday tours of Merritt Island and launch complexes 34 and 39. The Air Force had begun such a program in 1963 and KSC followed in the summer of 1964. On each of the first two Saturday trips, more than 200 wives and children made the trip.[47]

By late 1964 other visitors besides the families of employees wanted to see the growing wonders of Merritt Island. As a result, on the first Sunday of 1965, KSC began a Sunday tour. Guards handed out brochures and a letter of welcome from Director Debus as the cars passed through the gate. More than 1900 visitors came the first Sunday, some from as far away as Nebraska and Ontario. As the Sunday tours grew more popular, KSC laid plans for a permanent Visitors Information Center. In late June 1965, a group of architects met with Debus and other KSC officials to discuss design possibilities, while the National Park Service estimated the potential visitor attendance by 1967 to be in the millions.[48]

Familial and Personal Tensions

The move of Hans Gruene's launch vehicle team and Theodor Poppel's design group in 1964 and 1965 brought about 1000 families from Alabama to Brevard County. Except for 40 Boeing families, newly arrived in Alabama, most had lived for some time in the Huntsville area. In spite of the best efforts of the Community Impact Committee to provide information about Florida's east coast, relocation proved difficult for many of the newcomers. The families settling in the Titusville area found no large shopping center closer than Orlando. Titusville had only one small department store. Sears and Penney's would arrive three years later, in response to the rapid population growth.[49]

To provide a place where all could come together on occasion for relaxation, a group of employees developed a recreation area five kilometers east of Highway 3 on KSC, halfway between headquarters and the residential area farther south on Merritt Island. Situated on the west bank of the Banana River, with 762 meters of shore line and a boat basin, the tract, one kilometer square, boasted a setting of live oak, palm, persimmon, and pine trees, and provided playgrounds, picnic areas, and a swimming area.[50] The Spaceport Travel Club also organized a year-round series of trips that specialized in Caribbean cruises and air journeys to Europe, Hawaii, and the Orient. In spite of these efforts, the KSC employees remained segmentized, close to their own division or contractor, united only in the purpose of sending men to the moon and bringing them back.

Mobility was a major factor in the lives of many on the Apollo project. Military men had grown accustomed to it and accepted it as part of their lives. Engineers who worked for a particular contractor expected a change of residence when a contract was completed. Some saw the east coast of Florida as only a temporary home and did not sell their residences near the Douglas or Boeing central plant. Others viewed it as their permanent home and intended to find permanent employment when their work at KSC ended. Still others lived in constant uncertainty—a factor that influenced their entire family life.

These tensions made family life difficult in many ways. Articles in the local newspapers and national magazines regularly carried features on the domestic strain in the space communities. As *Time* magazine was to state:

> The technicians who assemble and service the rockets have chosen a tense career, and it has taken its toll on their personalities, their marriages and their community. . . . The rhythms of life at Cape Kennedy are set not so much by the clock or the

> seasons as by the irregular flights of the missiles. Bouts of furious activity and 14-hour days may be followed by periods of idleness.[51]

The *Time* article saw some difficulties stemming directly from the nature, training, and background of the engineering profession. Many engineers were perfectionist males, surrounded all day by scientific precision, who could not brook the sight of an unwashed coffee cup in the sink on their return home. Many carried their work home with them, spending the evening hours not with their families but in reading technical material. Intelligent, but not liberally educated, their interests focused primarily on the technical world.

Debus told an interviewer:

> There is so much tension, so much anxiety in putting men into space. Yes, we've lost men because of family problems. When a man is so dedicated that the NASA program becomes his personal life, it takes much time away from wife and children. We need a great many understanding wives here . . . in the end we usually have to tell them their husbands will be working even harder next year.
>
> Such exposure to stress is rare elsewhere. We live with it constantly. In fact, it is so much with us that we are studying it—how it is affecting our hearts, our nerves, our functions, our aging processes. We don't know yet.[52]

Putting men into space caused grave family problems. But readjusting to the decline in employment that followed was to cause even greater problems, especially to children. A prominent pediatrician of the region, Dr. Ronald C. Erbs of Titusville, noted a high incidence of ulcers in children, especially during the last half of the Apollo program. "Before coming to this area," he stated, "I did not see ulcers in children, except for rare examples."

> It is my opinion that the life generated by the Space Program was basically unhealthy for the families of space personnel. . . . With the decline of the Space Program, these highly trained men became very insecure regarding their futures. It is extremely difficult to keep the emotions of work away from the emotions of the family, hence increased family tensions. These tensions then were felt by the children, and since the problems were not usually discussed, the children had no outlet for these emotions, leading to the development of ulcers.[53]

Dr. Erbs had recommendations for future space programs, but they came too late for Apollo.

One compensating social attitude was the almost total lack of snobbishness among the space workers in the neighboring communities. No doubt it stemmed partially from most of them being newcomers trying to set up homes on Florida's east coast. A major contributing factor was the sense of the importance of each member of the Apollo team to the success of the mission. The most brilliant design engineer knew that the man who bolted on the hatch hinges did an important piece of work. All saw the unheralded contributions of countless persons around them. This appreciation of the worth of the individual carried over into the communities beyond KSC. One technician asked: "Where else in America would my closest friends be two men who make twice as much money as I do?"[54]

Source Notes

Notes, bibliography, and index encompass both *Gateway to the Moon* and its companion volume, *Moon Launch!*

Chapter 1: The First Steps

1. Army Ballistic Missile Agency (hereafter ABMA), *Juno V Space Vehicle Development Program (Phase I), Booster Feasibility Demonstration*, by H. H. Koelle et al., report DSP-TM-10-58 (Redstone Arsenal, AL, 13 Oct. 1958), p. 1; Oswald Lange, "Development of the Saturn Space Vehicle," in *From Peenemünde to Outer Space*, ed. Ernst Stuhlinger et al. (Huntsville, AL: Marshall Space Flight Center, 1962), p. 6. Probably the best source for an understanding of the complex developments of the American space program during the late 1950s is *The History of Rocket Technology*, ed. Eugene Emme (Detroit: Wayne State Press, 1964). Maj. Gen. John B. Medaris gives an interesting, albeit one-sided, account of ABMA's activities during this period in *Countdown for Decision* (New York: G. P. Putnam's Sons, 1960).
2. R. Cargill Hall, *Project Ranger: A Chronology* (Pasadena: Jet Propulsion Laboratory, California Institute of Technology, 1971), pp. 48–52; ABMA, *Juno V Development*, pp. 1–2.
3. ABMA, *Juno V Development*, pp. 1–2; Lange, "Saturn Space Vehicle," p. 6; Medaris, *Countdown*, pp. 151–241, passim.
4. Memo of agreement, Advanced Research Projects Agency and Army Ordnance Missile Command, "High Thrust Booster Program Using Clustered Engines," 23 Sept. 1958, printed in ABMA, *Juno V Development*, Appendix A; NASA, *Historical Pocket Statistics*, July 1972 (Washington, 1972), p. E-4. The tenfold increase in the cost of the Saturn I program can be explained in large part by the changing purposes of the program. Initially the Defense Department viewed it as a four-vehicle test series relying extensively on available engines, fuel tanks, and tooling machinery. The program evolved into something quite different, requiring much unanticipated construction for launch vehicles and facilities. Warren G. Hunter, ARPA Coordinator, SSEL, to Hans Hueter, Dir., SSEL, "Juno V (Saturn) Program," 3 Oct. 1958. Unless specified otherwise, manuscript sources are in KSC Archives.
5. ABMA, *Juno V Development*, pp. 7–11, 19–20, 25–27, 47–51; ABMA, *Juno V Transportation Feasibility Study*, by J. S. Hamilton, J. L. Fuller, and P. F. Keyes, report DLMT-TM-58-58 (Redstone Arsenal, AL, 5 Jan. 1959), pp. 1–4; ABMA, *Juno V Space Vehicle Development Program* (Status Report—15 Nov. 1958), by H. H. Koelle et al., report DSP-TM-11-58 (Redstone Arsenal, AL, 15 Nov. 1958), pp. 2–3, 19–20.
6. NASA Special Committee on Space Technology, *Recommendations Regarding a National Civil Space Program* (Stever Committee Report), Washington, 28 Oct. 1958; ABMA, *Juno V Development*, pp. 19–20, 65; Army Ordnance Missile Command (hereafter cited as AOMC), *Saturn Systems Study*, by H. H. Koelle, F. L. Williams, and W. C. Huber, report DSP-TM-1-59 (Redstone Arsenal, AL, 13 Mar. 1959), pp. 16–19, 61–63, 183–89; House Committee on Science and Astronautics, *Equatorial Launch Sites—Mobile Sea Launch Capability*, report 710, 87th Cong., 1st sess., 12 July 1961, pp. 1–5 (see hearings of same committee and topic, 15–16 May 1961, for fuller discussion); Mrazek interview. The debate over the merits of an equatorial launch site or a mobile sea launch capability continued for several years with congressional hearings in the spring of 1961. Vice Adm. John T. Hayward was a leading advocate of shipboard launches.
7. Missile Firing Laboratory, "Project Saturn, Facilities for Launch Site," n.d.

8. "Champagne Flight," *Spaceport News* 2 (18 July 1963): 3. For other details of this first attempt, see L. B. Taylor, *Liftoff: The Story of America's Spaceport* (New York: E. P. Dutton & Co., 1968), pp. 42–44.
9. House Committee on Science and Astronautics, *Management and Operation of the Atlantic Missile Range*, 86th Cong., 2d sess., 5 July 1960, pp. 1–2.
10. Zeiler interview, 24 Aug. 1972.
11. H. H. Koelle, ed., *Handbook of Astronautical Engineering* (New York: McGraw-Hill Book Co., 1961), pp. 28-8 through 28-10.
12. Deese interview, 16 Mar. 1973.
13. E. R. Bramlitt, *History of Canaveral District, 1950–1971* (So. Atlantic Dist. U.S. Corps of Engineers, 1971), pp. 17–21.
14. AOMC, *Saturn System Study*, pp. 4–5, 21; AOMC, *Saturn System Study II*, report DSP-TM-13-59 (Redstone Arsenal, AL, 13 Nov. 1959), pp. 1–2.
15. Dept. of the Army, *Project Horizon, A U.S. Army Study for the Establishment of a Lunar Military Outpost*, I, *Summary* (Redstone Arsenal, AL, 8 June 1959).
16. Minutes, NASA Research Steering Committee on Manned Space Flight (the Goett Committee), 25–26 May 1959, pp. 2–10, NASA Hq. History Office. The authors wish to thank historian Thomas Ray of NASA Hq. for assistance on this subject.
17. NASA Hq. working draft, "Long Range Objectives," 1 June 1959, NASA Hq. History Office.
18. Medaris, *Countdown*, pp. 241–47; Eugene Emme, "Historical Perspectives on Apollo," NASA Historical Note 75 (Oct. 1967), pp. 14–17.
19. Medaris, *Countdown*, pp. 247–69.
20. Emme, "Historical Perspectives," p. 17.
21. Medaris, *Countdown*, pp. 262–66; ABMA, *Saturn System Study II*, pp. 1–2.
22. AOMC, *Saturn System Study II*, pp. 5–10; *Report to the Administrator, NASA, on Saturn Development Plan by Saturn Vehicle Team*, 15 Dec. 1959, p. 1.
23. *Report on Saturn*, pp. 4, 7, 8, and table III.
24. Emme, "Historical Perspectives," p. 18; Robert L. Rosholt, *An Administrative History of NASA, 1958–1963*, NASA SP-4101 (Washington, 1966), p. 114.

Chapter 2: Launch Complex 34

1. Chief, MFL, to Chief, Ops. Off., Guided Missile Development Div. (GMDD), "Manning Charts," 5 Jan. 1953; Chief, MFL, to Dep. Chief, GMDD, "Official List of Operations Personnel for Missile #1"; Launch Operations Directorate (hereafter cited as LOD), "Special Report on Support Operations at the AMR by LOD," 21 Dec. 1960, part 5.
2. ABMA, *Juno V Development*, p. 47; Georg von Tiesenhausen, "Saturn Ground Support and Operations," *Astronautics* 5 (Dec. 1960): 30.
3. Debus to C.O., Atlantic Missile Range (hereafter cited as AMR), "Juno V Program," 1 Oct. 1958; Robert F. Heiser, Technical Asst., Off. of the Dir., MFL, memo for record, "Juno V," 26 Sept. 1958.
4. Deese to Debus, priority TWX, "Feasibility Study and/or Criteria for a Launch Site at AMR for a Clustered First Stage of Juno V Project," 8 Oct. 1958; Koelle, ed., *Handbook of Astronautical Engineering*, pp. 28-1 to 28-10; MFL, *Juno V (Saturn) Heavy Missile Launch Facility, 1st Phase Request, 2d Phase Estimate*, by R. P. Dodd and J. H. Deese, 14 Feb. 1959, pp. 2–3.
5. Deese to Debus, "Feasibility Study for a Launch Site"; Warren G. Hunter, ARPA Coordinator, SSEL, memo, "Meeting at MFL, CCMTA on Juno V Launch Complex," 10 Nov. 1958; Pan American Aviation, "Juno V Program Siting Study," 24 Oct. 1958, pp. 1–3.

6. Glen W. Stover, Chief, Facilities Br., AMR, Army Field Off., memo for record, "Criteria Contract, Juno V Facilities," 10 Nov. 1958; Maurice H. Connell and Assoc., *Heavy Missile Launch Facility Criteria* (Miami, FL, 15 Mar. 1959).
7. ABMA, *Juno V Development*, p. 55; LOD, "Complex 34 Safety Plan for SA-1 Launch," 24 Oct. 1961, p. 2; Porcher interview.
8. MFL, *Juno V (Saturn) Facility*; Connell and Assoc., *Launch Facility Criteria*; Sparkman interview, 13 June 1974. For detailed descriptions of the Saturn C-1 Launch Complex with its ground support equipment, see Marshall Space Flight Center (hereafter cited as MSFC), *Saturn SA-1 Vehicle Data Book*, report MTP-M-S&M-E-61-3 (Huntsville, AL, 26 June 1961), pp. 133–65, and MSFC, *Project Saturn C-1, C-2 Comparison*, report M-MS-G-113-60 (Huntsville, AL, 16 Nov. 1960), pp. 33–47, 123–290.
9. Davis interview; Walter interview, 21 Sept. 1973.
10. Von Tiesenhausen interview, 20 July 1973; Buchanan interview, 22 Sept. 1972.
11. Davis interview; Koelle, ed., *Handbook of Astronautical Engineering*, p. 28-44.
12. MSFC, *C-1, C-2 Comparison*, pp. 167–83; Wasileski interview, 14 Dec. 1972.
13. Zeiler interview, 24 Aug. 1972; Connell and Assoc., *Launch Facility Criteria*, p. 2-9.
14. Zeiler, Chief of Mechanical Br., MFL, to Debus, "Servicing Equipment for Juno V on Launch Site," 24 Nov. 1958.
15. Connell and Assoc., *Heavy Missile Service Structure Criteria*; Army Engineer Dist., Jacksonville Corps of Engineers, "Minutes of Conference on Review of Criteria for Saturn Facilities," 7 Apr. 1959.
16. Robert E. Linstrom, DOD Saturn Project Engineer, memo for record, "Summary of Fifth Saturn Meeting," 1 Apr. 1959; Debus's Daily Journal (hereafter cited as DDJ), 13 Apr. 1959, KSC Director's Office; Debus, memo for record, "Meeting on Saturn Service Structure," 9 Apr. 1959.
17. Debus, memo for record, "Meeting on Saturn Service Structure," 9 Apr. 1959; Connell and Assoc., *Alternate "I" Heavy Missile Service Structure*, 24 Apr. 1959, R. P. Dodd's personal papers.
18. MSFC, *SA-1 Data Book*, pp. 131–33.
19. DDJ, 13 Apr., 22 June, 11 July 1959; "Corps of Engineers Contract Tabulations for LC-34, LC-37, and MILA Facilities," p. 3.
20. "Vibroflotation," Vibroflotation Foundation Co., Div. of Litton Industries, Pittsburgh, PA, undated pamphlet.
21. J. P. Claybourne to Robert Heiser, priority TWX, "Saturn Launch Facility Costs," 7 Sept. 1960.
22. J. P. Claybourne, memo for record, "Cost of Saturn Launch Facilities and Ground Support Equipment," 13 Sept. 1960.
23. C. C. Parker, Technical Program Dir., memo for record, "Saturn Launch Facilities at AMR," 3 Dec. 1959; Connell and Assoc., *Siting Study and Recommendation, Saturn Staging Building and Service Structure, Complex 34 AFMTC*, Jan. 1960, pp. 2–3; MSFC, *C-1, C-2 Comparison*, p. 35; Debus to Rees, "Additional Saturn Launch Complex," 29 Jan. 1960; ABMA, *A Committee Study of Blast Potentials at the Saturn Launch Site*, by Charles J. Hall, report DHM-TM-9-60 (Redstone Arsenal, AL, Feb. 1960).
24. MSFC, *C-1, C-2 Comparison*, pp. 100–102; Poppel interview, 12 Feb. 1973.
25. MSFC, *C-1, C-2 Comparison*, pp. 123–63, 206–20, 244–67; Poppel, memo, "Launch Equipment Installation at Complex 34, AMR," 27 May 1960; Sparkman interview, 15 Dec. 1972; Wasileski interview, 14 Dec. 1972.
26. Davis interview, 2 Feb. 1972; Poppel, memo for record, "Contract NAS8-46 Extension," 29 Dec. 1960; Poppel to Petrone, "Documentation of Major Facility and/or GSE Changes for Project Saturn," 23 Feb. 1961, p. 5.
27. C. C. Parker, Chief, Ops. Off., MSFC, "Saturn FY-61 LOD Budget," 29 July 1960, encl. 3.
28. Debus, memo, "Priority of Effort on Saturn Launch Facilities," 8 July 1960.

29. R. P. Dodd to C. C. Parker, 8 Aug. 1960.
30. Debus to Wernher von Braun, "Labor Situation," 21 Sept. 1960.
31. George V. Hanna, "Chronology of Work Stoppages and Related Events, KSC/NASA and AFETR through July 1965," KSC historical report (KSC, FL, Oct. 1965), pp. 26–27.
32. Ibid., pp. 30–35; DDJ, 15 and 28 Nov., 5 and 22 Dec. 1960.

Chapter 3: Launching the First Saturn I Booster

1. W. R. McMurran, ed., "The Evolution of Electronic Tracking, Optical, Telemetry, and Command Systems at the Kennedy Space Center," mimeographed paper (KSC, 17 Apr. 1973), fig. 2; MSFC, *Saturn SA-1 Flight Evaluation*, report MPR-SAT-WF-61-8 (Huntsville, AL, 14 Dec. 1961), p. 235. The Saturn Flight Evaluation Working Group at MSFC published reports on all the Saturn C-1 launches. See also MSFC, *Results of the First Saturn I Launch Vehicle Test Flight, SA-1*, report MPR-SAT-64-14 (Huntsville, AL, 27 Apr. 1964) which superseded the above report, and MSFC, *Results of the Saturn I Launch Vehicle Test Flights*, report MPR-SAT-FE-66-9 (Huntsville, AL, 9 Dec. 1966).
2. MSFC, *C-1, C-2 Comparison*, pp. 3–7. See pp. 68, 78–82 for Long-Range Program (Sloop Committee Report) of Sept. 1960.
3. Oswald H. Lange, "Saturn Program Review," 27 Jan. 1961; Akens, *Saturn Illustrated Chronology*, pp. 13, 19.
4. F. A. Speer, "Saturn I Flight Test Evaluation," American Institute of Aeronautics and Astronautics paper 64-322 given at Washington, D.C., 29 June–2 July 1964, p. 2; ARINC Research Corp., *Reliability Study of Saturn SA-3 Pre-Launch Operations*, by Arthur W. Green et al., publication 247-1-399 (Washington, 3 Jan. 1963), pp. 2-21 through 2-23.
5. ABMA, *Organizational Manual* (Redstone Arsenal, AL, 5 Mar. 1959), Sec. 530, pp. 13–15, copy available in Historical Div., Sec. of General Staff, Army Missile Div., Redstone Arsenal, AL; Russell interview.
6. Moser interview, 30 Mar. 1973.
7. Grady Williams interview.
8. Debus to Joseph Shea, "Principles of Operations of the MSFC Firing Team at Cape Canaveral," 22 Mar. 1962, Debus papers, KSC Archives.
9. Ibid.
10. Ibid.
11. Debus to Committee for LOD Scheduling and Test Procedures, "Day-by-Day Test Schedule for Saturn SA-1," 16 Mar. 1961, ibid.; DDJ, 14 Mar. 1961.
12. Moser interview, 30 Mar. 1973; DDJ, 17 Apr. 1961.
13. MSFC, *Catalog of Systems Tests for Saturn S-1 Stage*, pp. III-46–III-54 (Huntsville, AL, 1 Sept. 1961), Moser papers, Federal Archives and Records Center, East Point, GA, accession 68A1230, boxes 436257, 436259.
14. "Launch Facilities and Support Equipment Office [hereafter cited as LFSEO] Monthly Progress Report," 12 June 1961, p. 2; "LFSEO Monthly Progress Report," 13 July 1961, pp. 2–3.
15. DDJ, 11, 12 May 1961; "LFSEO Monthly Progress Report," 12 June 1961, p. 3.
16. Zeiler to Debus, "Work Statement," 14 July 1961.
17. "Saturn SA-1 Schedule," 15 Aug. 1961, Moser papers, Federal Archives and Records Center, East Point, GA, accession 68A1230, boxes 436257, 436259.
18. DDJ, 24 Mar., 26 Apr. 1961; Akens, *Saturn Illustrated Chronology*, p. 18; Georg von Tiesenhausen, "Ground Equipment to Support the Saturn Vehicle," paper 1425-60 presented at the 15th annual meeting of the American Rocket Society, Washington, D.C., 5–8 Dec. 1960, pp. 2–3.

19. "LFSEO Monthly Progress Report," 13 July 1961, p. 4; Akens, *Saturn Illustrated Chronology*, pp. 21, 26–27.
20. MSFC, *SA-1 Vehicle Data Book*, pp. 123–30.
21. Karl L. Heimburg to MSFC Dep. Dir. for R & D, "Water Route for NASA Vessels to Cape Canaveral," 9 Feb. 1962, attached to a response from Debus, 14 Feb. 1962, in Debus papers.
22. MSFC, *SA-1 Flight Evaluation*, p. 7; Akens, *Saturn Illustrated Chronology*, p. 26; Crunk interview; Zeiler interview, 23 July 1973; von Tiesenhausen, "Equipment to Support the Saturn," pp. 2–3.
23. MSFC, *SA-1 Flight Evaluation*, p. 7; "LOD Daily Journal," 27 July 1961; "Saturn SA-1 Schedule," 15 Aug. 1961.
24. MSFC, *SA-1 Flight Evaluation*, p. 8; MSFC, "Saturn Quarterly Progress Report," July–Sept. 1961, p. 1; "Saturn Schedule," 15 Aug. 1961; interviews with Newall, Marsh, and Humphrey.
25. Grady Williams interview.
26. MSFC, *SA-1 Vehicle Data Book*, pp. 74–81.
27. Interviews with Edwards and Glaser.
28. White interview; MSFC, *Consolidated Instrumentation Plan for Saturn Vehicle SA-1*, by Ralph T. Gwinn and Kenneth J. Dean, report MTP-LOD-61-36.2a (Huntsville, AL, 25 Oct. 1961).
29. White interview.
30. MSFC, *SA-1 Flight Evaluation*, p. 7.
31. Moser interview, 30 Mar. 1973.
32. Ibid.; "Saturn SA-1 Schedule," 15 Aug. 1961; MSFC, *SA-1 Flight Evaluation*, pp. 8, 200–202.
33. Moser interview, 30 Mar. 1973; "Saturn SA-1 Schedule"; MSFC, *SA-1 Flight Evaluation*, pp. 7–9; LOD, "Saturn Test Procedures, SA-1 G & C Overall Test #3," Moser papers, Federal Archives and Records Center, East Point, GA, accession 68A1230, boxes 436257, 436259.
34. Moser interview, 30 Mar. 1973; "Saturn SA-1 Schedule"; MSFC, "SA-1 Flight Evaluation," pp. 8–9; George Alexander, "Telemetry Data Confirms Saturn Success," *Aviation Week and Space Technology*, 6 Nov. 1961, pp. 30–32.
35. "Saturn Test Procedures: SA-1 Mechanical Office L – 1 Day Prelaunch Preparations," Moser papers; MSFC, *SA-1 Flight Evaluation*, pp. 9–10; interview with Chester Wasileski by Benson, 14 Dec. 1972; Pantoliano interview.
36. MSFC, *Launch Countdown Saturn Vehicle SA-1*, report MIP-LOD-61-35-2 (Huntsville, AL, 3 Oct. 1961), pp. 9–15, Moser papers.
37. MSFC, *SA-1 Flight Evaluation*, p. 10; MSFC, *Countdown SA-1*.
38. MSFC, *Countdown SA-1*, pp. 20–23.
39. MSFC, *SA-1 Flight Evaluation*, pp. 11–12; MSFC, *Countdown SA-1*, pp. 28–30; LOD, *Saturn Test Procedures: Set Up LO_2 Facility for Fast Fill (T – 100)*, procedure LOD-M-703; LOD, *Saturn Test Procedures: Fast Fill LO_2 Loading (T – 60)*, procedure LOD-M-704.
40. MSFC, *Countdown SA-1*, pp. 36–39; Alexander, "Telemetry Confirms Success," p. 31; "Emergency Procedures SA-1," LOD Networks Group, pp. 2–3, Moser papers.
41. Richard Austin Smith, "Canaveral, Industry's Trial by Fire," *Fortune*, June 1962, pp. 204, 206; "Saturnalia at Canaveral," *Newsweek*, 6 Nov. 1961, p. 64; *Miami Herald*, 28 Oct. 1961, p. 1; "Saturn's Success," *Time*, 3 Nov. 1961, p. 15.
42. *New York Times*, 28 Oct. 1961, pp. 1, 9; *Miami Herald*, 28 Oct. 1961, p. 1. The MSFC news release on the SA-1 launch, dated 1 Nov. 1961, included a paragraph on the sound effect.
43. *Miami Herald*, 28 Oct. 1961, p. 1 (UPI release).
44. *New York Times*, 27 Oct. 1961, p. 1; 28 Oct. 1961, pp. 1, 9.

Chapter 4: Origins of the Mobile Moonport

1. House Committee on Science and Astronautics, *Report on Cape Canaveral Inspection*, 86th Cong., 2d sess., 27 June 1960, p. 1.
2. House Committee on Science and Astronautics, *Management and Operation of the Atlantic Missile Range*, 86th Cong., 2d sess., 5 July 1960, p. 4.
3. Francis L. Williams interview.
4. David S. Akens, *Saturn Illustrated Chronology* (MSFC, Jan. 1971), pp. 7–8; J. P. Claybourne, Saturn Project Office, memo, "Saturn C-2 Configurations," 6 July 1960; NASA, "A Plan for Manned Lunar Landing" (Low Committee report), 7 Feb. 1961, pp. 7–13, figs. 4, 7, NASA Hq. History Office.
5. Interview with Debus by Benson, 16 May 1972; H. H. Koelle, "Missiles and Space Systems," *Astronautics* 7 (Nov. 1962): 29–37.
6. Claybourne, "Saturn C-2 Configurations," 6 July 1960.
7. DDJ, 24 Apr. 1961.
8. Livingston Wever, Support Instrumentation Div., to Porcher, Facilities Br., Army Test Off., AFMTC, "Addendum to Scheme for Offshore Launching Platform for Space Vehicles," Mar. 1960; Wyle Laboratories, *Sonic and Vibration Environments for Ground Facilities—A Design Manual*, by L. C. Sutherland, report WR68-2, 1968, pp. 5-21, 10-2.
9. Nelson M. Parry, Army Test Off., AFMTC, "Land Developments for Missile Range Installations (Preliminary Notes)," 30 Dec. 1958, p. 3; Nelson Parry to Porcher, "Offshore Launch Platform for Heavy Space Vehicles," 6 Apr. 1960.
10. Porcher interview.
11. Von Tiesenhausen interview, 29 Mar. 1972; Sparks interview; von Tiesenhausen, "Vorversuche für Project Schwimmiweste," Electromechanische Werke Peenemünde, 11 Sept. 1944, typescript, von Tiesenhausen's private papers.
12. Poppel to Debus, "Offshore Complex," 6 May 1960.
13. MSFC, *Preliminary Feasibility Study on Offshore Launch Facilities for Space Vehicles*, by O. L. Sparks, report IN-LOD-DL-1-60 interim (Huntsville, AL, 29 July 1960).
14. DDJ, 28 Feb., 2 Mar. 1961.
15. Debus to Col. Asa Gibbs, Chief, NASA Test Support Off., AFMTC, "Future Saturn Launch Sites at Cape Canaveral (SR 2953)," 14 Feb. 1961.
16. Parry to Charles J. Hall, "Future Launch Sites at Cape Canaveral," 9 Mar. 1961.
17. Debus to Gibbs, "Siting for Third Saturn Launch Complex at AMR (SR 2953)," n.d.; Debus to Gibbs, "Siting of Fourth and Fifth Saturn Launch Complexes at AMR," 6 Apr. 1961. The siting requests were canceled 22 Sept. 1961, after the MILA purchase had changed the situation.
18. Future Launch Systems Study Off., LOD, "Progress Report," Jan. 1961; DDJ, 6 Feb. 1961.
19. Debus to Poppel, "Offshore Launch Facility Study," 4 Apr. 1961.
20. Nelson M. Parry, "Land Development (Offshore and Semi-Offshore Launch Sites)," 14 Apr. 1961; Deese interview, 10 May 1972; DDJ, 12 May 1961.
21. Petrone interview.
22. Joint Air Force–NASA Hazards Analysis Board, AFMTC, *Safety and Design Considerations for Static Test and Launch of Large Space Vehicles*, 1 June 1961, p. I-B-1.
23. Debus to von Braun, "Offshore Facilities Studies," 24 May 1961.
24. "Death on Old Shaky," *Time*, 27 Jan. 1961, pp. 15–16; von Tiesenhausen interview, 29 Mar. 1972; Debus interview, 16 May 1972.
25. Zeiler interview, 11 Aug. 1970.
26. Memo for record, "Phoenix Study Program," 3 July 1961, p. 2; Aerospace Corp. News Release, "Titan III Management and Technology to Be Model for Future Systems" (Los Angeles, June 1965), p. 2.

27. KSC Public Affairs Office, *Kennedy Space Center Story* (Kennedy Space Center, FL, Dec. 1972), p. 5. No document is cited for the statement that the three men met 16 days after JFK's inaugural. This would place the meeting on Sunday, 5 Feb. The Daily Journal for 6 Feb. mentions a Saturday meeting of the three men.
28. Duren interview, 16 May 1972.
29. DDJ, 22, 30, 31 Mar., 10 Apr. 1961.
30. MSFC, *Interim Report on Future Saturn Launch Facility Study*, by Olin K. Duren, report MIN-LOD-DL-1-61 (Huntsville, AL, 10 May 1961).
31. DDJ, 17, 20, 26 Apr. 1961.
32. Douglas Aircraft Co., *Saturn C-2 Operational Requirement Study*, prepared by J. Simmons, report SM-38771 (Santa Monica, CA, July 1961), p. 188.
33. Ibid., pp. 171–205; The Martin Co., *Saturn C-2 Operational Modes Study, Summary Report*, report ER-11816 (Baltimore, MD, June 1961), pp. 19–25, 46–47, 66–68.
34. John M. Logsdon, "NASA's Implementation of the Lunar Landing Decision," NASA HHN-81, Aug. 1969, typescript, pp. 1–6; Lunar Landing Working Group (Low Committee), "A Plan for Manned Lunar Landing," 7 Feb. 1961.
35. House Committee on Science and Astronautics, *Hearings, 1962 NASA Authorization*, 87th Cong., 1st sess., 23 Mar. 1961, pt. 1, p. 177.
36. Ibid.; Akens, *Saturn Illustrated Chronology*, p. 4.
37. Akens, *Saturn Illustrated Chronology*, pp. 17, 19, 22; Logsdon, "NASA's Implementation," p. 6; House, *1962 NASA Authorization*, pp. 170–77.
38. DDJ; Debus to von Braun, "Offshore Launch Facilities," 24 May 1961; MSFC, Future Projects Off., "Procurement Request," 26 Apr. 1961.
39. Logsdon, "NASA's Implementation," pp. 8–18.
40. DDJ, 6, 26 June 1961. The Fleming Master Flight Plan called for 167 flights prior to the first lunar landing, but this included launchings of Atlas, Agena, Centaur, Saturn C-1, Saturn C-3, and Nova rockets. Fourteen C-1s and 24 C-3s were to be launched in 1965–1966.
41. DDJ, 6, 20 June 1961.
42. NASA, *A Feasible Approach for an Early Manned Lunar Landing* (Fleming Committee Report), 16 June 1961, p. 26.
43. Robert C. Seamans, Jr., to Maj. Gen. Leighton I. Davis and Debus, "National Space Program Range Facilities and Resources Planning," 23 June 1961.
44. Seamans to Davis and Debus, "National Space Program Range Facilities and Resources Planning," 30 June 1961.
45. MSFC, LFSEO, *Preliminary Concepts of Launch Facilities for Manned Lunar Landing Program*, report MIN-LOD-DL-3-61, 1 Aug. 1961, pp. 4–6.
46. NASA-DOD, *Joint Report on Facilities and Resources Required at Launch Site to Support NASA Manned Lunar Landing Program* (hereafter cited as Debus-Davis Report), 31 July 1961, p. 3; Owens interview, 12 Apr. 1972; Petrone interview, 25 May 1972; Clark interview.
47. Petrone interview; Clark interview.
48. Petrone and Leonard Shapiro, "Guideline for Preparation of NASA Manned Lunar Landing Project Report," 7 July 1961; KSC Biographies, in KSC Archives.
49. Debus-Davis Report, passim; Owens interview, 12 Apr. 1972.
50. MSFC, *Interim Report on Future Saturn Launch*, p. 16.
51. Zeiler interview, 11 July 1972; von Tiesenhausen interview, 29 Mar. 1972.
52. Debus-Davis Report, pp. B-1 through B-7.
53. Ibid., pp. B-9, B-10.
54. The authors are indebted to Rocco Petrone for this idea: interview of 25 May 1972 and remarks delivered by Petrone to Apollo History Workshop, NASA Hq., 19–21 May 1971.

Chapter 5: Acquiring a Launch Site

1. DDJ, 26 Apr. 1961.
2. Joint Air Force–NASA Hazards Analysis Bd., *Safety and Design Considerations for Static Test and Launch of Large Space Vehicles*, 1 June 1961, part I, "Hazards Analysis," p. I-A-1.
3. Ibid., part I, pp. I-D-3, I-B-1, and I-B-2.
4. AFMTC, NASA, & Pan American, *Preliminary Field Report, Cumberland Island & Vicinity for Nova Launch Facilities*, 13–14 June 1961, pp. 22–23; Hal Taylor, "Big Moon Booster Decisions Looming," *Missiles and Rockets*, 28 Aug. 1961, p. 14.
5. Charles J. Hall to R. P. Dodd, "Land Development for Future Launch Sites," 12 May 1961.
6. Debus, memo for the record, "Land Acquisition Book II," 11 June 1961.
7. NASA, *A Feasible Approach for an Early Manned Lunar Landing* (the Fleming Report), 16 June 1961.
8. Roswell Gilpatric to the Secretaries of the Army, Navy, and Air Force, "National Space Program Facilities Planning," 16 June 1961.
9. *Air Force Systems Command News Review*, Mar. 1963.
10. Debus-Davis Report, reference b.
11. Petrone and Shapiro, memorandum of understanding, "Guideline for Preparation of NASA Manned Lunar Landing Project Report," 7 July 1961.
12. Gibbs interview.
13. Debus-Davis Report, pp. 19, 20.
14. Ibid., p. 20.
15. Ibid.
16. Ibid., pp. D-70, D-71.
17. Ibid., pp. D-19, D-20.
18. Gordon E. Dunn and Banner I. Miller, *Atlantic Hurricanes*, rev. ed. (Baton Rouge: LSU Press, 1960), pp. 266–67.
19. Bramlitt, *History of the Canaveral District*, p. 34.
20. Debus interview, 22 Aug. 1969; Owens interview, 12 Apr. 1972.
21. Debus interview, 16 May 1972.
22. DDJ, 27 July 1961.
23. *Washington Post*, 3 Aug. 1961.
24. Milton Rosen to Hugh Dryden and James Webb, "Selection of a Launch Site for the Manned Lunar Program," attached to a letter from Dryden to Webb, same subj., 18 Aug. 1961.
25. Hugh L. Dryden to Webb, "Selection of a Launch Site for the Manned Lunar Program," 18 Aug. 1961.
26. NASA release 61-189, "Manned Lunar Launch Site Selected," 24 Aug. 1961.
27. Interview with Gibbs by James Covington, 7 Aug. 1969.
28. Petrone interview.
29. "Agreement between DOD and NASA Relating to the Launch Site for the Manned Lunar Landing Program," signed by James Webb and Roswell Gilpatric, 24 Aug. 1961.
30. Ibid., p. 2.
31. NASA-LOD proposal, "Integrated Master Planning, Atlantic Missile and Space Operations Range," 22 Nov. 1961; Minutes of the meeting prepared by Raymond L. Clark, Asst. to the Director.
32. House Committee on Science and Astronautics, Subcommittee on Manned Space Flight, *Hearings, 1963 NASA Authorization*, 87th Cong., 2d sess., pt. 2, pp. 643–55.
33. Lloyd L. Behrendt, comp., *Development and Operation of the Atlantic Missile Range* (Patrick A.F.B., FL, 1963), pp. 63–64, Air Force Eastern Test Range Archives.

34. "Background Information on Agreement between DOD and NASA re: Management of the AMR of DOD and the Merritt Island Launch Area of NASA," prepared by Paul T. Cooper, Brig. Gen., USAF, Mar. 1963, p. 4.
35. Behrendt, *Development of AMR*, pp. 66–67.
36. Debus interview, 22 Aug. 1969; reiterated in an interview 16 May 1972.
37. DDJ, 14 July 1961.
38. Senate, *Amending the National Aeronautics and Space Administration Authorization Act for the Fiscal Year 1962*, report 863 to accompany Senate Bill 2481, 87th Cong., 1st sess., 1 Sept. 1961.
39. Seamans to Lt. Gen. W. K. Wilson, Chief, Corps of Engineers, U.S. Army, 21 Sept. 1961.
40. Senate Committee on Appropriations, 87th Cong., 1st sess., *Hearing, Second Supplemental Appropriation Bill for 1962*, p. 154; Arthur G. Procher, memo for record, "Land Acquisition for NASA Lunar Launch Facility," 26 Sept. 1961.
41. NASA Audit Div., "Review of Management Controls over Contract Modifications Executed by the Corps of Engineers, Launch Operations Center, Cocoa Beach, Florida," report Bu/LO-W 64-7, Washington, DC, 23 Sept. 1963, p. 3.
42. Real Estate Div., U.S. Army Engineer Dist., Jacksonville, FL, "Progress and Status, Real Estate Acquisition and Manned Lunar Landing Program Project," Cape Canaveral, FL, 6 June 1962.
43. Sollohub to Debus, 8 June 1962.
44. *Titusville Star-Advocate*, 17 Feb. 1962.
45. Morgan T. Nealy, Jr., Proj. Mgr., Real Estate Project Off., Titusville, FL, to LOC, 21 Mar. 1963.
46. Telephone interview with Faherty, 1 May 1972. The lady preferred to remain anonymous.
47. "List of Buildings Retained for Interim Use, NASA–Merritt Island Launch Area," in R. J. Pollock's files, KSC Maintenance Div.
48. AFMTC, "Cape Canaveral Missile Test Annex Development Plan," 12 Mar. 1962; AFMTC, "Status Report, NASA–Merritt Island Launch Area Development," Patrick A.F.B., 27 Sept. 1962; James Trainor, "Titan III Plan Awaits DOD Approval," *Missiles and Rockets*, 14 May 1962, p. 35.
49. Shriever to Seamans, 14 Mar. 1962.
50. NASA, "Management Council Meeting of 29 May 1962," corrections, dated 7 Mar. 1962.
51. Debus, memo for record, "Holmes-Shriever-Davis-Debus Meeting, Saturday, 17 Mar. 1962, on Titan III Siting," 19 Mar. 1962.
52. NASA, "Minutes of the Management Council Meeting of 27 Mar. 1962."
53. *Hearings, 1963 NASA Authorization*, pp. 634–35.
54. Ibid., pp. 641–55.
55. Ibid., p. 652.
56. Behrendt, *Development of AMR*, p. 69.
57. Davis to District Engineer, "Real Property Accountability for MLLP Area," in "Background Information and Agreement between DOD and NASA: Re: Management of the AMR of DOD and the Merritt Island Launch Area of NASA," Mar. 1963. Brig. Gen. Paul T. Cooper, USAF, stated that when the question of title arose between the Launch Operations Center and the AF Missile Test Center, the two staff elements agreed to bring the matter to the attention of higher levels by this letter of Gen. Davis. We have seen no NASA document suggesting that this letter arose from a mutual decision, although several NASA communications refer to Davis's letter (Debus to Holmes, 3 Apr. and 4 Apr. 1962; Webb to Gilpatric, 23 May 1962).
58. Petrone interview.
59. Seamans to Webb and Dryden, 13 Apr. 1962.
60. Webb to McNamara, 17 Apr. 1962.

61. *Missiles and Rockets*, 30 Apr. 1962, p. 75.
62. Bell to McNamara, 3 May 1962.
63. Jackson to McNamara, 21 May 1962. The 18 June 1962 issue of *Aviation Week and Space Technology* credited Rep. George Miller with a major role in securing NASA's autonomy on Merritt Island.
64. McNamara to Jackson, 24 May 1962.
65. Debus to Davis, 14 June 1962.
66. Subcommittee of the Senate Committee on Appropriations, *Hearings on H.R. 12711*, 87th Cong., 2d sess., pp. 861–904.
67. P.L. 87-584, *NASA Authorization Act for 1963*, sec. 5.
68. Webb to Gilpatric, 14 Aug. 1962, with enclosure, "Relationships and Responsibilities of the DOD & NASA at the AMR and the Manned Lunar Landing Program Area."
69. Cmdr., AFMTC, to Sec. of Defense, 4 Jan. 1963.
70. "Agreement, the Department of Defense and the National Aeronautics and Space Administration regarding Management of the Atlantic Missile Range of DOD and the Merritt Island Launch Area of NASA," signed by Robert S. McNamara and James E. Webb, 17 Jan. 1963.
71. The interim agreement that implemented the Webb-McNamara Agreement was signed on 10 May 1963 by Davis and Debus. Addenda covered joint instrumentation planning procedures, calibration equipment and services, chemical analysis, security and law enforcement, facilities management, visitor control, and other topics. Under the terms of addendum 5, LC-34, LC-36, LC-37, the Saturn Barge Canal, and other facilities were transferred by AFMTC to NASA-LOC. The 11 separate agreements were understood to be consolidated into a single document. Debus, "AFMTC-LOC Agreements Implementing the Webb-McNamara Agreement of 17 January 1963," 5 June 1963.
72. John P. Lacy to Center Dir., "Status of KSC Land Acquisition," 7 Dec. 1967.
73. Policicchio interview.
74. "List of Buildings to Be Retained for AFMTC Use in the Expanded Cape Area," Col. Colie Houck, 29 June 1962, to Mr. Owens; Joseph Hester, memo for record, "Disposition of House Trailers, Area 1," 5 Mar. 1962.
75. Debus to Peterson, 5 Mar. 1962; Bidgood to Debus, "Real Estate Procurement Policy," 22 Dec. 1961.
76. Senate Committee on Appropriations, *Hearings on H. R. 11038*, 87th Cong., 2d sess., 4 Apr. 1962, pp. 155–56; *Spaceport News* 2 (10 Oct. 1963): 8.
77. "Audit of Merritt Island Purchase," Office, Chief of Engineers, 16 Nov. 1971, in Corps of Engineers Files, per telephone conversation with Joseph Hester, 18 May 1972.
78. *Spaceport News* 11 (1 June 1972): 1, 4; O'Conner interview.
79. "Agreement between National Aeronautics and Space Administration and Bureau of Sport Fisheries and Wildlife for Use of Property at John F. Kennedy Space Center, NASA," 2 June 1972, signed by Willis H. Shapley for NASA and Nathaniel P. Reed for the Dept. of Interior.

Chapter 6: LC-39 Plans Take Shape

1. Logsdon, "NASA's Implementation," p. 22; Ivan D. Ertel and Mary Louise Morse, *The Apollo Spacecraft, A Chronology*, vol. 1 (NASA SP-4009, 1969), pp. 95, 108–109.
2. Logsdon, "NASA's Implementation," p. 34.
3. Ibid., pp. 40–44; Shea interview; Rosen interview, 14 Nov. 1969; Ertel and Morse, *Apollo Chronology*, 1: 118–20, 134.
4. Ertel and Morse, *Apollo Chronology*, 1: 131–34; Akens, *Saturn Chronology*, pp. 33–35.
5. James Grimwood and Barton Hacker, with Peter Vorzimmer, *Project Gemini, A Chronology* (NASA SP-4002, 1969), pp. 2–20.

6. Ertel and Morse, *Apollo Chronology*, 1: 111.
7. Ibid., p. 101.
8. Ibid., pp. 101–104, 121, 128; NASA release 66-15, "Apollo Spacecraft Contract," 21 Jan. 1966.
9. Martin Marietta Corp., *Saturn C-3 Launch Facilities Study Final Report*, vol. 1, *Selection of Optimum Concept*, report ER 12125-1 (Baltimore, Dec. 1969), p. 1.
10. Ibid., pp. vii, 1–11, 70–85; Martin Co., *Special Study Saturn Launch Facilities*, report ER 11996 (Baltimore, 17 Oct. 1961), pp. II-1 through II-5; calendar and schedule of events in O. K. Duren's private papers.
11. LFSEO, LOD, *A Preliminary Study of Launch Facility Requirements for the C-4* (Huntsville, AL, 27 Oct. 1961), p. 38.
12. George von Tiesenhausen, memo for record, "Launch Complex 39," 11 Oct. 1961.
13. Harvey F. Pierce, Maurice H. Connell & Assoc., Inc., to Debus, 21 Nov. 1961, Debus papers.
14. Poppel to Petrone, memo, "Saturn C-3/C-4 Study," 9 Oct. 1961.
15. Biographies, KSC Archives; Owens interview, 21 Nov. 1972.
16. MSFC, *Saturn Mobile (Canal) Concept Flame Deflector and Launcher/Transporter Emplacement Evaluation*, by George Walter, report MIN-LOD-DH-2-62 (Huntsville, AL, Feb. 1962).
17. Poppel to E. House, "Temporary Employment of Naval Architecture Consultant," 22 Dec. 1961; "LFSEO Monthly Progress Report," 15 Feb. 1962, p. 8.
18. Martin, *Saturn C-3 Study*, vol. 3, *Design Criteria for Launch Facilities*, report ER 12125-3, Dec. 1961, pp. 57–84.
19. "LFSEO Monthly Progress Report," 15 Feb. 1962, p. 8.
20. Debus to Petrone, "Transportation Proposals for Complex 39," 30 Jan. 1962; DDJ, 30 Jan. 1962.
21. Duren interview, 29 Mar. 1972; Zeiler interview, 24 Mar. 1972; private papers of Duren.
22. "LOD Weekly Notes," Petrone, 8 Feb. 1962; W. T. Clearman, Acting Sec., Heavy Vehicle Systems Off., memo for record, "Complex 39 Staff Meeting," 12 Mar. 1962, Petrone papers.
23. MSFC, *Appraisal of Transfer Modes for Saturn C-5 Mobile Systems as of 11 June 1962*, by Donald D. Buchanan and George W. Walter, report MIN-LOD-DH-9-62 (Huntsville, AL, 11 June 1962), pp. 5–8; Buchanan interview, 7 Nov. 1972; Walter interview, 7 Nov. 1972.
24. MSFC, *Transporter for Nova Track Design and Stresses*, by William H. Griffith, report NASA-MFSC-LOD-D; MSFC, *Appraisal of Transfer Modes*, p. 5; MSFC, *Saturn Mobile (Rail) Concept: An Examination of Rail Transfer Systems for a Launcher/Transporter*, by George W. Walter, report MIN-LOD-DH-3-62 (Huntsville, AL, 3 Apr. 1962).
25. Maurice H. Connell & Assoc., Inc., *Saturn C-5 Launch Facilities Complex 39: Study of Rail Systems for Vertical Transporter/Launcher Concept* (Huntsville, AL: MSFC, May 1962), p. 7.
26. MSFC, *Appraisal of Transfer Modes*, pp. 9–11; MSFC Weekly Notes, Debus to von Braun, 28 May 1962; Buchanan, memo for record, "Analysis by H. Pierce, 15 May 1962," Buchanan's private papers.
27. LOD Weekly Notes, Zeiler, 15 Feb. 1962; Poppel, Zeiler, Buchanan, and Duren made up the team.
28. Donald Buchanan, memo for record, "Launcher/Transporter Crawler Version," 23 Mar. 1962; Buchanan interview, 28 Nov. 1972; Duren interview, 29 Mar. 1972; MSFC, *Appraisal of Transfer Modes*, pp. 5, 8–9.
29. LOD Weekly Notes, Poppel, 16 May 1962; MSFC, *Appraisal of Transfer Modes*, p. 9; Buchanan, memo for record, "TDY at Bucyrus-Erie, South Milwaukee, Wisconsin," 16 Apr. 1962, Buchanan's private papers; Buchanan interview, 28 Nov. 1972; Buchanan, memo for record, "Analysis by H. Pierce, 15 May 1962," Buchanan's papers.

30. Army Corps of Engineers, Jacksonville Off., "Summary of Opinions Developed by the Jacksonville District Engineering Staff on Mobile Launch Concepts for the Advanced Saturn C-5 Vehicle," June 1962, in Buchanan's papers.
31. MSFC, *Appraisal of Transfer Modes*, p. 11.
32. William T. Clearman, Jr., memo, "Launch Operations Directorate Complex 39 Review," 18 Sept. 1962; E. M. Briel's notes, 12–13 June 1962.
33. Biographies, in KSC Archives; Claybourne interview; Clearman interview, 5 Jan. 1973.
34. Launch Operations Center (hereafter cited as LOC), "Summary of Conference with Members of Manned Space Flight Sub-Committee of House Committee on Science and Astronautics at the NASA Launch Operations Center," 23 Mar. 1962, p. 27.
35. Martin, *Saturn C-3 Study*, 3: 46–52.
36. Clearman, "Complex 39 Staff Meeting," 12 Mar. 1962; Deese to Moser et al., "Preliminary Concepts, Vertical Building," 6 Mar. 1962; Brown Engineering Co., Inc., *An Evaluation of an Enclosed Concept for a C-5 Vertical Assembly Building (VAB)*," 2 Apr. 1962, pp. 7–8; Brown Engineering Co., Inc., *An Evaluation of an Open Concept for a C-5 Vertical Assembly Building*, 2 Apr. 1962, pp. 9–10.
37. Briel's notes, 12–13 June 1962; Brown Engineering Co., "Evolution of the Saturn C-5 Mobile System Vertical Assembly Building," a mimeographed report prepared by E. M. Briel, 7 Sept. 1962.
38. LOC, "Minutes of the Saturn C-5 Launch Operations Working Group Meeting, 18–19 July 1962," 8 Aug. 1962, pp. 2–6.
39. Briel's notes, 31 July 1962; DDJ, 15 Aug. 1962.
40. URSAM, "VAB-LC39: A Report of Meeting with Representatives of LOC, Corps of Engineers and Component Contractors" (Cape Canaveral, FL, 28 Aug. 1962), app. A.
41. Isom G. Rigell, memo, "LC-39 Networks," 4 Sept. 1962; Norman Gerstenzang, memo for record, 5 Sept. 1962; unsigned memo, "Information Required by LO-FEE for LC-39, VAB Criteria," 6 Sept. 1962.
42. Joe J. Koperski, Chief, Engineering Div., Corps of Engineers, to R. P. Dodd, "Back-to-Back vs. In-Line Configuration, Comparisons and Conclusions—Launch Complex 39–Vertical Assembly Building," 21 Sept. 1962.
43. Deese interview, 4 Oct. 1973.
44. NASA, "A Report on Launch Facility Concepts for Advanced Saturn Launch Facilities," by Marvin Redfield, John Hammersmith, and Jay A. Salmonson, 13 Feb. 1962.
45. House, *Hearings: 1963 NASA Authorization*, p. 941.
46. LOC, "Summary of Conference with Members of Manned Space Flight," 23 Mar. 1962, pp. 13–34.
47. Ibid., pp. 21, 30. In an interview with James Frangie on 13 Aug. 1969, Col. Bidgood pointed out that LC-39 provided launch rate flexibility but had limitations in its ability to accommodate different vehicles.
48. NASA, "Minutes of the Management Council Office of Manned Space Flight," 29 May 1962.
49. Ibid., 22 June 1962; DDJ, 15 June 1962.
50. Logsdon, "NASA's Implementation," pp. 56–60.
51. The mode selection story continued several more months as NASA had to defend the choice against strong criticism from the President's Science Advisory Committee. For a lengthier treatment of one of Apollo's most interesting episodes, see Logsdon, "NASA's Implementation."
52. "LOD Weekly Notes," Sendler, 5, 19 July 1962, Bidgood, 5 July, 2, 23 Aug. 1962; DDJ, 15, 21 Aug. 1962.
53. Poppel to Bidgood, "Preliminary Design for a Mobile Arming Tower for Launch Complex 39," 10 Aug. 1962.
54. LOC, "Minutes of the Saturn C-5 Launch Operations Working Group Meeting, 18–19 July 1962," 8 Aug. 1962, pp. 1–5 and app. 9.
55. Redfield interview.

Chapter 7: The Launch Directorate Becomes an Operating Center

1. Gen. Ostrander, "Proposed Organization for Launch Activities at AMR and PMR," 6 July 1961; Debus, "Proposed Organization for Launch Activities at Atlantic Missile Range and Pacific Missile Range Due to Proposed Realignment of NASA Programs," 12 June 1961.
2. Debus to Rees, Dep. Dir. for R&D, MSFC, "Operating Procedures and Responsibilities of MSFC Divisions and Others at Launch Site," 4 Aug. 1961.
3. Ibid.; see also DDJ, 27 July 1961.
4. D. M. Morris, Dep. Dir. of Admin., MSFC, to Albert Siepert, Dir., Office of Business Admin., NASA Hq., 6 June 1961; Harry H. Gorman, Assoc. Dep. Dir. for Admin., MSFC, to Seamans, 26 Sept. 1961.
5. Debus, "A Paper on Launch and Spaceflight Operations," 27 Sept. 1961.
6. Debus, "Analysis of Major Elements Regarding the Functions and Organization of Launch and Spaceflight Operations," 10 Oct. 1961.
7. Concurrence by Wernher von Braun, appended to n. 6 reference.
8. Seamans to Young and Siepert, 13 Oct. 1961; Debus interview, 16 May 1972.
9. Rees to von Braun, "New Organization Proposals for LOD," 17 Oct. 1961; Debus interview, 16 May 1972.
10. Debus interview, 22 Aug. 1969.
11. NASA release 62-53, "Establishment of the Launch Operations Center at AMR and the Pacific Launch Operations Office at PMR," 7 Mar. 1962.
12. Young to Seamans, "Internal Organization of the Launch Operations Center," 29 June 1962.
13. Debus interview, 22 Aug. 1969.
14. "MSFC-LOC Separation Agreement," 8 June 1962, printed in Francis E. Jarrett, Jr., and Robert A. Lindemann, "Historical Origins of NASA's Launch Operations Center to 1 July 1962" (KSC, 1964), app.; NASA release LOC-63-64, 24 Apr. 1963. The transfer of LVOD personnel from MSFC to LOC was completed by 6 May 1963. See James M. Ragusa, "John F. Kennedy Space Center (KSC) NASA Reorganization Policy and Methods," (M.S. thesis, Florida State Univ., Apr. 1968), p. 25.
15. Melton interview.
16. DDJ, 1 Sept. 1961; Clark interview.
17. *Missiles and Rockets,* 6 Nov. 1961, p. 18.
18. Ernest W. Brackett, Dir., Procurement and Supply, to Dir., Off. of Admin., with enclosure, "Establishment of Launch Operations Center," 15 June 1962.
19. Clarence Bidgood, "Facilities Office Memo No. 2," 1 Feb. 1962; *Spaceport News,* 22 Feb. 1963.
20. Debus interview, 22 Aug. 1969.
21. Parker interview, 14 Feb. 1969.
22. By early June 1962, 930 sq m of off-site space had been leased and plans were made to lease 1400 more by 30 July 1962. Robert Heiser to Rachel Pratt, "Notes from von Braun," 11 June 1962; Gordon Harris, Chief of Public Affairs, to Bagnulo, 1 Oct. 1964.
23. Hall to Facilities Program Off., "Justification for Leasing Additional Space in CAC Building," 13 Mar. 1963. During the period of limited office space, thought was given toward acquiring a barge from the Navy to be moored at the Saturn Barge Terminal and used as an office. Hall to Bidgood, "Barge Anchorage," 18 Sept. 1962; Hawkins to Hall, "Trailer Request for LC-34," 1 Mar. 1963; *Spaceport News,* 13 Oct. 1966, p. 6.
24. NASA General Management Instruction (hereafter GMI) 2-2-9.1, "Basic Operating Concepts for the Launch Operations Center at Merritt Island and the Atlantic Missile Range," 10 Jan. 1963.
25. House Committee on Science and Astronautics, Subcommittee on Manned Space Flight, *Hearings: 1964 NASA Authorization,* 88th Cong., 1st sess., pt. 2a, pp. 127, 129.

26. Bidgood interview, 13 Aug. 1969.
27. Seamans to Dir., LOC, "General Responsibilities and Functions of the NASA Center Director," 10 Jan. 1963, with enclosure, "General Responsibilities and Functions of a NASA Center Director," 10 Jan. 1963. These documents were circulated in LOC on 18 Jan. 1963; see Office of the Dir., "General Responsibilities and Functions of a NASA Center Director," 18 Jan. 1963. They were subsequently revised and published as attachments to NASA GMI 2-0-3, "Informational Material on Assignment of Responsibilities in the NASA Organization Structure," 3 June 1963.
28. NASA GMI 2-2-9.1, "Basic Operating Concepts for the Launch Operations Center at Merritt Island and the Atlantic Missile Range," 10 Jan. 1963. A 4 Mar. 1963 plan provided for an "Assistant Director for Program Management" (a title actually adopted in the reorganization of 28 Jan. 1964); a 28 Mar. 1963 plan provided for an "Assistant Director for LOC Programs," as did also a 2 Apr. 1963 plan.
29. The NASA Daytona Beach Operations was established and designated an integral part of LOC in NASA circular 2-2-9, 23 June 1963, which stated that the Manager would "report to the Director, Launch Operations Center, Cocoa Beach, Florida." See also Debus to George Mueller, "Change in Organizational Structure," 13 Dec. 1963; Debus to staff, "LOC Organization Structure," 6 Aug. 1963, and accompanying manual, "LOC Organization Structure," 2 Aug. 1963; LOC Organization Chart, approved by Hugh L. Dryden, 24 Apr. 1963
30. *Spaceport News,* 1 May 1963, p. 6.
31. Bidgood, who had joined LOC on 1 Nov. 1962, organized a Facilities Office by late 1962 and published his first organization chart on 13 Feb. 1963. Bidgood interview, 14 Nov. 1968.
32. "LOC Organization Structure," 2 Aug. 1963.
33. *Spaceport News,* 1 May 1963, p. 6. For Debus's views on the "development operational loop," see his "Analysis of Major Elements Regarding the Functions and Organization of Launch and Spaceflight Operations," 10 Oct. 1961; also "LOC Organizational Structure," 2 Aug. 1963, p. 9.
34. C. C. Parker, "Boards, Committees, Panels, Teams and Working Groups," 25 Sept. 1963.
35. *Spaceport News,* 30 June 1963.
36. Hugo Young, Bryan Silcock, and Peter Dunn, *Journey to Tranquility* (Garden City, NY: Doubleday & Co., 1970), p. 158.
37. Rosholt, *An Administrative History of NASA,* pp. 288–89.
38. *The Washington Post,* 21 Sept. 1963, p. A–10.
39. Thomas's letter and the President's reply appear in Senate Committee of Appropriations, *Hearings: Independent Offices Appropriations, 1964,* 88th Cong., 1st sess., pt. 2, pp. 1616–18.
40. *Fortune,* Nov. 1963, pp. 125–29, 270, 274, 280.
41. *Spaceport News,* 21, 27 Nov. 1963.
42. Executive Order 11129, *Designating Certain Facilities of the National Aeronautics and Space Administration and of the Department of Defense in the State of Florida, as the John F. Kennedy Space Center,* 29 Nov. 1963; NASA announcement 63-283, "Designation of the John F. Kennedy Space Center, NASA," 20 Dec. 1963; message SAF 82841, Sec. of the Air Force to Cmdr., AFSC, Andrews AFB, 7 Jan. 1964; "Decisions on Geographic Names in the United States, Dec. 1962 through December 1963," decision list 6303, U.S. Board on Geographic Names (Washington: Dept. of the Interior, 1964), p. 20.
43. Debus to Mrs. W. L. Stewart, 26 Dec. 1963.
44. Ernest G. Schwiebert, *A History of the U.S. Air Forces Ballistic Missiles*, pp. 130, 201–203, 247; Akens, *Saturn Illustrated Chronology,* pp. 67–68.
45. OMSF, "Management Council Minutes, 29 Oct. 1963," 31 Oct. 1963.

46. Rosholt, *Administrative History,* pp. 289-97. Newell headed thc Office of Space Sciences and Applications, Bisplinghoff, the Office of Advanced Research and Technology.
47. NASA release KSC-10-64, 6 Feb. 1964; KSC, "NASA Organization Chart," 28 Jan. 1964; *Spaceport News,* 13 Feb. 1964.
48. "KSC Notes," Petrone to Debus, 29 Oct. 1964.
49. "Kennedy Space Center Apollo Document Tree," approved by Rocco Petrone, 3 Nov. 1965, Joel Kent's private papers.
50. Childers interview, 7 Nov. 1972; Gramer interview, 21 Sept. 1972.

Chapter 8: Funding the Project

1. Lyndon B. Johnson, *The Vantage Point: Perspectives of the Presidency, 1963-69* (New York: Popular Library, 1971), p. 283.
2. Unless otherwise indicated the information on the early budgetary process was gathered from interviews with Robert G. Long, Resources Management Br., KSC; William E. Pearson, Chief, Management Information Control Br., KSC; Alton D. Fryer, Resources Management Div., KSC; and Elizabeth A. Johnson, Financial Management Div., KSC. Much of this information was later formalized in OMSF, *Apollo Program Development Plan,* M-D MA 500, 1 Jan. 1966.
3. The real estate paragraph was usually omitted since land acquisition for the MLLP was provided for under separate CoF documents.
4. Robert C. Seamans, Jr., "Guidelines for Preparation of Detailed Fiscal Year 1964 Budget Estimates—Section I," 30 Aug. 1962, attach. 5, p. 6.
5. "It should be borne in mind, however, that, as a matter of policy, NASA may not choose to exercise its authority to the full extent permitted by law," General Counsel (John A. Johnson) to Paul C. Dembling, "Request from Congressman Karth regarding the 'extent of NASA's authority to reprogram or transfer appropriate funds within the agency,' " 3 Apr. 1963, with encl. NASA was authorized to transfer sums from one budget line item to another to the extent of 5% of the item to which the transfer was to be made, to meet unusual cost variations, provided the total amount authorized was not exceeded. It was also authorized to transfer up to 5% of the CoF appropriation to the RD&O appropriation, and vice versa. Additionally, the NASA Administrator was authorized under certain circumstances to use CoF funds for such things as emergency repairs if of greater urgency than the construction of new facilities.
6. C. C. Parker, Chief, Operations Off., LOD, to J. Martin, "FY-63 C&E Requirement," 19 Dec. 1960.
7. Don R. Ostrander, NASA Hq., to MSFC, "Preliminary Fiscal Year 1963 Budget," 27 Feb. 1961.
8. Akens, *Saturn Illustrated Chronology,* pp. 19, 22-23.
9. *Spaceport News,* 19 Jan. 1967, p. 7.
10. Burke interview.
11. Parker/Greenglass to D'Onofrio, "FY-63 CoF Budget Requirements," 13 Dec. 1961.
12. LOD, "Atlantic Missile Range Fiscal Year 1963 Estimates, Construction of Facilities," undated. See also House, *1963 NASA Authorization,* p. 879; Senate, *NASA Authorization for Fiscal Year 1963*, p. 126; and Senate Committee on Appropriations, 87th Cong., 2d sess., *Hearings: Second Supplemental Appropriation Bill for 1962,* 4 Apr. 1962, p. 149.
13. LOD, "Atlantic Missile Range, Fiscal Year 1963 Estimates, Construction of Facilities," undated (ca. Jan. 1962).

14. House, Manned Space Flight Subcommittee, "Summary of Conference with Members of Manned Space Flight Subcommittee of House Committee on Science and Astronautics at the NASA Launch Operations Center, Cocoa Beach, Fla., 23 Mar. 1962."
15. Ibid., pp. 19–21. The conference record frequently does not identify the speaker.
16. Ibid., p. 21.
17. Ibid.
18. Ibid., pp. 21, 24–27.
19. Ibid., pp. 29–30.
20. Ibid., pp. 55–57.
21. Ibid., p. 58.
22. William E. Lilly, Dir., Program Review and Resources Management Off., OMSF, to LOC, "Advance Approval of FY-63 CoF," 14 Mar. 1962; TWX, C. C. Parker to NASA Hq., "Advance Approval of FY-63 CoF," 21 May 1962.
23. Rosholt, *Administrative History,* pp. 233, 284; *Congress and the Nation, 1945–1964* (Washington: Congressional Quarterly Service, 1965), 1:320.
24. OMSF, "Management Council Minutes," 21 Sept. 1962, item 8; LOC, "FY 1963 CoF Resubmission and Supplemental Program," 8 Sept. 1962.
25. C. C. Parker, Dep. Assoc. Dir. for Admin. Services, LOC, to William E. Lilly, Dir. Program Review and Resources Management, OMSF, 8 Sept. 1962.
26. LOC to NASA Hq., "LOC Fiscal Year 1963 Estimates, Advanced Saturn Complex No. 39," 8 Sept. 1962, with encl.
27. C. C. Parker, LOC, to William E. Lilly, 18 Sept. 1962.
28. *Congress and the Nation, 1945–1964,* pp. 320, 390.
29. Debus to Holmes, 26 Oct. 1962.
30. Project Approval Document, "Construction of Facilities, Advanced Saturn Launch Complex No. 39," code 46-46-990-933-3450, 6 Nov. 1962, p. AMR–63–10; NASA Form 504, "Allotment/Sub-Allotment Authorization," amendment 05, 16 Nov. 1962.
31. Frederick L. Dunlap, Chief, Budget Br., NASA Hq., to Ed Melton, Financial Management Off., LOC, 27 Dec. 1962.
32. NASA Procurement Div., KSC, "Status Summary of Active Contracts as of 31 Mar. 1964," sec. III, "Active Intergovernmental Purchase Orders," p. 3.
33. OMSF directive M-D 9330.01, "Manned Space Flight Program Launch Schedule for Apollo and Saturn Class Vehicles," 15 Oct. 1962. See also MSFC, "Reference Director, MSFC/MSC/OMSF, Flight and Mission Schedule History," 21 Feb. 1963; and OMSF, "Management Council Minutes," 31 July 1962, item 5F.
34. Debus, "Fiscal Year 1964 Preliminary Budget," 8 Mar. 1962; Seamans to Dir., OMSF, "Guidelines for Preparation of Detailed Fiscal Year 1964 Budget Estimates," sec. I, 30 Aug. 1962; LOC, "FY 1964 CoF Program," 1 Nov. 1962.
35. LOC, "FY 1964 CoF Program," 1 Nov. 1962, pp. DF-B 16 through DF-B 18.
36. Ibid.
37. W. F. Barney, Chief, Control Off., MSFC, to Mr. Lada, "Fiscal Year 1963 Spacecraft CoF Projects," 19 Feb. 1962; OMSF, "Management Council Minutes," 28 Aug. 1962.
38. G. Merritt Preston, MSC, AMR Ops., to NASA Hq., Attn: G. M. Low, "Chronology and Background Information on Gemini and Apollo Facilities at AMR," 18 Jan. 1963. This letter was prepared in response to an 11 Jan. 1963 request from Congressman George P. Miller of California, Chairman of the House Committee on Science and Astronautics, for an explanation of the decision to combine some Gemini and Apollo spacecraft facilities on Merritt Island. Combining facilities also made it difficult to extract costs directly chargeable to the Apollo program.
39. Manned Spacecraft Center/Atlantic Missile Range, "FY 1963 Construction of Facilities, Project Documentation," 15 Oct. 1962; Debus to William E. Lilly, 15 Oct. 1962. The use of the project title "Apollo Mission Support Facilities" persisted for several months.
40. LOC, "FY 1964 CoF Program," 1 Nov. 1962.

41. House Subcommittee on Manned Space Flight, 88th Cong., 1st sess., *Hearings: 1964 NASA Authorization,* pt. 2(b), p. 986. In fact, the $432 million figure had been given to the subcommittee at Cape Canaveral on 23 Mar. 1962.
42. Ibid., pp. 987–89.
43. Ibid., pp. 989–90.
44. Ibid., p. 1276.
45. Ibid., pp. 991–94, 1275–83.
46. *Congress and the Nation, 1945–1964,* p. 326.
47. Ibid., pp. 326, 329.

Chapter 9: Apollo Integration

1. Senate Committee on Aeronautical and Space Sciences, *Hearings: NASA Authorization for Fiscal Year 1963*, 87th Cong., 2d sess., 14 June 1962, p. 486; see also House Committee on Science and Astronautics, Subcommittee on Manned Space Flight, *1963 NASA Authorization*, 87th Cong., 2d sess., 26 Mar. 1962, pt. 2, pp. 543–44.
2. "Weekly Notes," Petrone to Debus, 12 Apr., 17 May 1962; "Minutes of the Sixth Meeting of the Management Council of the Office of Manned Space Flight, 29 May 1962." OMSF realized that G.E.'s favored status would offend stage contractors and stipulated in the contract (NASw-410) certain provisions that restricted G.E.'s use of sensitive information.
3. "Agreements Reached at the August Meeting at the Cape Concerning the G.E. Contract," unsigned and undated (Debus and Petrone represented LOC).
4. DDJ, 7 Nov. 1962; OMSF, "Management Council Minutes," 21 Sept., 27 Nov. 1962.
5. House Committee on Science and Astronautics, Subcommittee on Manned Space Flight, *Hearings: 1964 NASA Authorization*, 88th Cong., 1st sess., Mar.–June 1963. The G.E. contract and its ramifications crop up throughout these hearings, particularly in vol. 3, pt. 2(b). The subcommittee hearings at Daytona Beach are contained in app. C to pt. 2(b), pp. 1285–1352.
6. DDJ, 3, 9 July 1963; OMSF, "Management Council Minutes," 27 Aug. 1963.
7. Manned Spacecraft Center, Langley AFB, VA, "Minutes of MSFC-MSC Space Vehicle Board No. 1, 3 Oct. 1961," 7 Nov. 1961.
8. Von Braun to Mueller, "Flight Missions Planning Panel," 30 Dec. 1963; Wagner interview.
9. MSC, "Minutes of MSFC-MSC Space Vehicle Board No. 1, 3 Oct. 1961," 7 Nov. 1961.
10. LOD Weekly Notes, Petrone, 8 Feb. 1962.
11. LOD Weekly Notes, Petrone, 3 May, 21 June 1962, and Poppel, 18 Apr. 1962; "Summary of Launch Operations Panel Activities," prepared by Emil Bertram for Joseph Shea, 18 July 1963.
12. Bertram, "Summary of Launch Operations Panel Activities," 18 July 1963; LOC, "Minutes of Meeting, Apollo-Saturn Launch Operations Panel, 6 Aug. 1963."
13. "Minutes of Systems Review Meeting, Houston, Texas, 10 Jan. 1963," JSC Archives.
14. "Panel Review Board Minutes, 9–10 Aug. 1963"; OMSF, "Management Council Minutes," 28 May 1963.
15. Gruene interview, 19 Nov. 1970. The U.S. Comptroller General ruled these "body-shop" contracts illegal in Mar. 1964. House Committee on Post Office and Civil Service, *Decision of Comptroller General of the United States Regarding Contractor Personnel in Department of Defense*, 89th Cong., 1st sess., report 188, 8 Mar. 1965.
16. Orvil Sparkman, "S-IV Propellant Loading Sequence," 26 Sept. 1961.
17. LOC Weekly Notes, Gruene, 29 Aug., 5 Sept. 1963; Petrone, 10 Oct. 1963.

18. Debus to Shea, Dep. Dir. for Systems, NASA Hq., "Apollo Interface Control Procedures," 24 June 1963; minutes of meeting, "Delineation of Interface Responsibility between Astrionics Division and LOC," 29 May 1962, signed by Poppel and H. J. Fichtner, Chief, Electrical Systems Integration.
19. Debus and von Braun, "Memo of Agreement: MSFC/KSC Relations," 11 Aug. 1964.
20. Debus and von Braun, "Clarification and Implementation Instruction, MSFC/KSC Relations Agreement dated 11 August 1964," 9 Mar. 1965.
21. Bertram, "LIEF Implementation," 3 Apr. 1964; Bertram interview, 15 Nov. 1973.
22. Bertram, "LIEF Implementation," 3 Apr. 1964.
23. Mary Louise Morse and Jean Kernahan Bays, *The Apollo Spacecraft: A Chronology*, vol. 2, *November 8, 1962—September 30, 1964*, NASA SP-4009 (Washington, 1973), pp. iii–vi; Jay Holmes, "Minutes of Special Staff Meeting, Office of Associate Administrator for Manned Space Flight," 31 Jan. 1964; Shea to Phillips, 27 Mar. 1964, Phillips File, NASA Hq. History Off.; Joachim P. Kuettner, Mgr., Saturn Apollo Systems Integrations, MSFC, "Trip Report," 22 Oct. 1962, in Petrone's notes.
24. Memo attached to DDJ, 6 Aug. 1962; Poppel interview, 24 Jan. 1973.
25. Petrone to M. Dell, Apollo Support Off., MSC, 5 Nov. 1962.
26. Debus to Holmes, 14 Nov. 1962 (letter summarizing discussions between the two men on 19 Oct.), Debus papers.
27. Petrone to J. T. Doke, Apollo Project Off., 22 Oct. 1962; Petrone to B. Porter Brown, 13 Nov. 1962.
28. B. Porter Brown, Prelaunch Ops. Div., Ops. Support Off., to Walter Wagner, KSC, "Mission Operations Control Room Information," 3 Feb. 1964; Petrone to Brown, 7 Feb. 1963.
29. KSC Weekly Notes, Petrone, 12 Dec. 1963.
30. KSC Weekly Notes, Petrone, Jan.–May 1965.
31. LOC Weekly Notes, Petrone, 20 Sept. 1962; Petrone to Kuettner, "Weekly Report to MSC," 19 Oct., 20 Nov., 18 Dec. 1962, in Petrone's notes (1962–1964).
32. Debus notes of 20 June 1963, in Petrone's notes; Bertram interview, 28 Sept. 1973; Horn interview; Moore interview; Hand interview.
33. MSFC, *Saturn V Flight Manual, SA 506*, 25 Feb. 1969, sec. 3, 9.
34. KSC, *Apollo/Saturn V Flight Safety Plan, Vehicle AS-501* (1967), pp. 1-1, 2-1, 3-1.
35. Taylor, *Liftoff!* p. 83; Adolf H. Knothe, "Range Safety—Do We Need It?" paper 70-249, American Institute of Aeronautics and Astronautics, Launch Operations Meeting, Cocoa Beach, FL, 2–4 Feb. 1970, p. 2.
36. R. M. Montgomery, "Range Safety of the Eastern Test Range," paper 70-246, American Institute of Aeronautics and Astronautics, Launch Operations Meeting, Cocoa Beach, FL, 2–4 Feb. 1970, p. 2; Arthur Moore to Benson, "Comments on Launch Operations History," 4 Oct. 1974.
37. Debus to Davis, "Range Safety Policies and Procedures," 11 June 1962, with attached letter from Davis to Debus, 10 May 1962, Debus papers.
38. Emil Bertram, memo for record, "Range Safety Information Channels," 30 Mar. 1962, KSC Range Safety Off. Notes; LOC Weekly Notes, Knothe, 3 May 1962; Bertram to Petrone, "Apollo Saturn Range Safety," 7 May 1962, KSC Range Safety Off. Notes; Bertram to Petrone, 9 May 1963, Petrone's notes.
39. AFETR Manual 127-1, *Range Safety Manual*, 1 Sept. 1972, 1:4–6; according to KSC officials the wording on this matter in the current manual is practically unchanged from the manual in force ten years earlier, no copy of which was available.
40. LOC Weekly Notes, Knothe, 9 May, 3 July 1963.
41. Knothe, "Minutes of Meeting on the Use of Liquid Explosives for a Fuel Dispersion System," 12 July 1963, in KSC Range Safety Off. Notes; LOC Weekly Notes, Knothe, 25 July 1963; Christopher C. Kraft, "Range Safety Aspects of the Apollo Program," 5 Aug. 1963.

42. Knothe to attendees, "Minutes of Meeting: Range Safety Aspects of Apollo Program, Held at NASA/LOC on 29 Aug. 1963," 5 Sept. 1963, KSC Range Safety Off. Notes.
43. Kraft, "Aspects of Apollo Range Safety," 1 Nov. 1963.
44. Kraft, "Apollo Range Safety," 11 Dec. 1963, in KSC Range Safety Off. Notes; Hans Gruene, "Apollo Service Module Propellant Dispersion System Interface Disagreement between MSC and KSC/MSFC," 25 Mar. 1964; LOC Weekly Notes, Knothe, 16 Apr. 1964 (marginal note by Debus); George E. Mueller, Assoc. Admin. for Manned Space Flight, to Cmdr., National Range Div., USAF, 18 Sept. 1964.

Chapter 10: Saturn I Launches (1962–1965)

1. LOD, "SA-2 Daily Status Reports," Robert Moser papers, Federal Archives and Records Center, East Point, GA, accession 68A1230, boxes 436257, 436259.
2. MSFC, *Results of the Second Saturn Launch Vehicle Test Flight SA-2*, report MPR-SAT-63-13 (Huntsville, AL, 16 Oct. 1963), pp. 1–5, 24, 49; Speer, *Saturn I Flight Test Evaluation*, pp. 1–6.
3. ARINC Research Corp., *Reliability Study of Saturn SA-3 Pre-Launch Operations*, by Arthur W. Green et al. (Washington, 3 Jan. 1963), pp. 4-7 through 4-11, 7-1.
4. MSFC, *Results of the Third Saturn I Launch Vehicle Test Flight, SA-3*, report MPR-SAT-64-13 (Huntsville, AL, 26 Feb. 1964), pp. 1–8; Speer, "Saturn I Flight Test Evaluation," p. 2; DDJ, 1 Nov. 1962.
5. MSFC, *Results of the Fourth Saturn I Launch Vehicle Test Flight, SA-4*, report MPR-SAT-63-6 (Huntsville, AL, 10 May 1963), pp. 5–7; *Spaceport News*, 9 Apr. 1964, p. 3; Chambers interview.
6. MSFC, *Results of SA-4*, pp. 1–7, 16–17; Speer, "Saturn I Flight Test Evaluation," p. 2.
7. MSFC, *Results of the Saturn I Launch Vehicle Test Flights*, report MPR-SAT-FE-66-9 (Huntsville, AL, 9 Dec. 1966), pp. 26–27.
8. MSFC, *Results of SA-3*, pp. 7–8; MSFC, *Results of SA-4*, p. 7; House Committee on Science and Astronautics, Subcommittee on Manned Space Flight, *Hearings: 1964 NASA Authorization*, 88th Cong., 1st sess., 6 Mar. 1963, pt. 2(a), p. 198; MSFC, *Saturn Monthly Progress Reports* (Jan.–Aug. 1962).
9. Debus to Rees, "Additional Saturn Launch Complex," 29 Jan. 1960.
10. Capt. Arthur G. Porcher, Chief, Facilities Br., Army Test Off., AFMTC, "Additional Launch Facilities for Saturn Type Vehicle," 5 Feb. 1960; Col. Donald Heaton to Gen. Ostrander, "Price Increase in Second Saturn Launch Complex," 12 Feb. 1960.
11. Philip Claybourne, Saturn Project Off., to MFL Br. Chiefs, "Back-Up of Saturn Launch Facilities," 10 May 1960.
12. Harvey F. Pierce to Debus, 26 Feb. 1960.
13. Debus to Zeiler, "Formation of Committee to Review Service Structure Design," 9 Mar. 1960; DDJ, 11, 13 Apr. 1960.
14. Debus, memo for record, "Drift of the Saturn C-2 Vehicle at Launching," 12 July 1960; MSFC, "Summary Report and Recommendations of Saturn Service Structure No. II Design Committee," by Harvey F. Pierce, 12 July 1960, pp. 4–18.
15. MSFC, "Summary Report of Saturn Service Structure Committee," pp. 4–18.
16. Ibid.; LOC, "Concept Development of Saturn Service Structure, No. II," by James Deese, Apr. 1963, pp. 26–27, James Deese papers.
17. DDJ, 29 Aug. 1960.
18. Debus to von Braun, "Hazard Study of Liquid Hydrogen, LO_2 and RP-1," 10 Jan. 1961; DDJ, 9, 11, 13 Jan. 1961.
19. DDJ, 13 Jan. 1961.

20. LFSEO Monthly Progress Report, 12 June 1961, p. 1; Poppel to Parker, "Criteria for VLF 37," 22 Dec. 1960; J. W. Ault, memo for record, "Contract for LC-37 Design," 23 Feb. 1961; Dodd to Corps of Engineers, "Vibroflotation for Complex 37A," 18 Oct. 1961, Debus papers.
21. MSFC, *Saturn Quarterly Progress Report* (July–Sept. 1961), report MPR-SAT-61-11, 1 Dec. 1961, p. 94; MSFC, *Saturn Quarterly Progress Report* (Jan.–Mar. 1962), MPR-SAT-62-3, p. 36; Michael Getler, "Complex 37 Will Dwarf Predecessors," *Missiles and Rockets*, 18 Dec. 1961, pp. 24–25, 47.
22. Getler, "Complex 37," pp. 24–25, 47; "The Biggest Thing on Wheels in the World," prepared by Batten, Barton, Durstine, and Osborne, Inc., Pittsburgh, for U.S. Steel, Jan. 1963.
23. D. E. Eppert, Chief, Construction Div., Canaveral Dist., Corps of Engineers, to James J. Frangie, "List of Saturn Construction Contracts," 12 Sept. 1968, p. 16; *NASA Fifth Semi-Annual Report to Congress*, 1 Oct. 1960 through 30 June 1961, p. 145; Emil Bertram, memo for record, "Apollo-Saturn Subpanel Activities," 15 July 1963, p. 3.
24. MSFC, *Results of Saturn I Launch Vehicle Tests*, pp. 3–5.
25. "Daily Status Reports, LC-37B Wet Test Vehicle," Robert Moser papers; Moser interview, 18 July 1973; Akens, *Saturn Illustrated Chronology*, pp. 58–61.
26. Gruene to Debus, 12 Sept. 1963.
27. D. L. Childs to LVO, S-IV-5 Status Reports #23, 29 Aug., and #33, 11 Sept. 1963; S-IV-5 Daily Log, 21–22 Sept. 1963, Rober Moser papers; LVO, "SA-5 Daily Status Report," 23, 24, 25 Sept. 1953.
28. "SA-5 Daily Status Reports," 11, 14, 17 Oct. 1963; Gruene to Debus, 17 Oct. 1963.
29. LVO, "SA-5 Daily Status Reports," 11 Oct. 1963.
30. Ibid., 22 Oct., 7 Nov. 1963; Gruene to Debus, 31 Oct. 1963; Fannin interview.
31. LVO, "SA-5 Daily Status Reports," 27 Nov. 1963; Corn interview, 23 July 1973; Zeiler interview, 23 July 1973; Pickett interview.
32. LVO, "SA-5 Daily Status Report," 27 Nov., 6, 8, 10, 13 Dec. 1963.
33. Ibid., 23, 27 Dec. 1963, 14, 17, 19 Jan. 1964.
34. Akens, *Saturn Illustrated Chronology*, pp. 72–73; *Cocoa Tribune*, 29 Jan. 1964.
35. MSFC, *Results of the Fifth Saturn I Launch Vehicle Test Flight, SA-5*, report MPR-SAT-FE-64-17 (Huntsville, AL, 22 Sept. 1964), pp. 5–7; *Cocoa Tribune*, 28, 29 Jan. 1964.
36. KSC, "Presentation to the Subcommittee on Manned Space Flight of the House Committee on Science and Astronautics at KSC," 27 Jan. 1964; Sherrer interview.
37. *Cocoa Tribune*, 28 Jan. 1964, p. 2.
38. KSC, "Presentation to the Subcommittee," 27 Jan. 1964.
39. R. P. Eichelberger, "The Saturn Telemetry System," pp. 1–3; KSC, "Technical Progress Report," 24 Jan. 1964; *Spaceport News*, 23 Jan. 1964, p. 2; "Consolidated Instrumentation Plan," pt. IIA of *Firing Test Report, Saturn I SA-5*, 22 Jan. 1964 (TR-4-36), pp. 6, 19, 31.
40. *Spaceport News*, 4 June 1964, p. 1; NASA release 63-268, 23 Jan. 1964; *New York Times*, 27 Jan. 1964; Speer, "Saturn I Flight Test Evaluation," pp. 1–8.
41. *Orlando Sentinel*, 30 Jan. 1964, pp.1, 42.
42. James Grimwood, JSC Historian, supplied information for this section.
43. *Spaceport News*, 4 June 1964, p. 2; NASA, *Astronautics and Aeronautics, 1964*, pp. 70, 126; Sasseen interview, 26 July 1973.
44. *Orlando Sentinel*, 21 May 1964; *Melbourne Daily Times*, 26, 27 May 1964.
45. *Spaceport News*, 4 June 1964, p. 5; MSFC, *Results of the Saturn I Launch Vehicle Test Flights*, p. 23.
46. *Spaceport News*, 13 Aug. 1964, p. 3; Davidson interview.
47. *Cocoa Tribune*, 20 July 1964; *Spaceport News*, 23 July 1964, p. 2; Newall interview.
48. *Cocoa Tribune*, 28 Aug. 1964; *Orlando Sentinel Star*, 8, 9 Sept. 1964; *Miami Herald*, 16 Sept. 1964; *Spaceport News*, 27 Aug., 3, 10, 17 Sept. 1964; *Aviation Week and Space Technology*, 28 Sept. 1964, p. 27.

49. Gen. Samuel Phillips to George Mueller, 14 Jan. 1965; Mueller to Debus, 10 Feb. 1965; weekly notes from Petrone to Debus, 4 Feb. 1965.
50. MSFC, *Results of the Eighth Saturn I Launch Vehicle Test Flight, SA-9*, report MPR-SAT-FE-65-6 (Huntsville, AL, 30 Apr. 1965), p. 14; Akens, *Saturn Illustrated Chronology*, p. 104.
51. Ibid., pp. 9–14; "Pegasus Returning Meteoroid Flux Data," *Aviation Week and Space Technology*, 22 Feb. 1965, p. 28.
52. MSFC, *Results of the Ninth Saturn I Launch Vehicle Test Flight, SA-8*, report MPR-SAT-FE-11 (Huntsville, AL, 27 July 1965), pp. 7–15; "First Industry-Built Saturn I Puts Pegasus-2 in Precise Orbit," *Aviation Week and Space Technology*, 31 May 1965, p. 21.
53. MSFC, *Results of the Tenth Saturn Launch Vehicle Test Flight, SA-10*, report MPR-SAT-FE-65-14 (Huntsville, AL, 24 Sept. 1965), p. 8.

Chapter 11: Ground Plans for Outer-Space Ventures

1. In this section the authors relied extensively on research by William Lockyer, Jr., and James Covington.
2. Col. J. V. Sollohub to Debus, 15 Oct. 1962.
3. "To Design for the Moon Age, Four Firms Work as One Team," *Engineering News-Record* 172 (6 Feb. 1964): 46–48.
4. Alexander interview; Anton Tedesko to Urbahn, Knecht, and Rutledge, 10 Aug. 1962.
5. URSAM, "VAB-LC39: Report of a Meeting with Representatives of LOC, Corps of Engineers and Component Contractors," Cape Canaveral, FL, 28 Aug. 1962.
6. Ibid., pp. 2–6.
7. Wesley Allen, Brown Engineering Co., memo for record, "Meeting with Facilities and MSC," 17 Sept. 1962.
8. Bidgood to Poppel, 26 Sept. 1962.
9. Col. Wm. Alexander, "Report on VAB," undated, p. 5; J. Bing to R. P. Dodd, 7 Nov. 1962; Theodor A. Poppel to Bidgood, 21 Nov. 1962; and Gerstenzang and Carraway to Dodd, 23 Nov. 1962.
10. William D. Alexander, "Vertical Assembly Building—Project Description, Organization, and Procedures," *Civil Engineering* 35 (Jan. 1965): 42–44.
11. Ibid., p. 44.
12. Gerald C. Frewer, "Kennedy Space Center—Assembly Line on a Gigantic Scale," *The Engineering Designer,* May 1967, p. 7.
13. Anton Tedesko, "Base for USA Manned Space Rockets (Structures for Assembly and Launching)," *International Association for Bridge and Structural Engineering Publications* 26 (May 1971): 535; Tedesko, "Design of the Vertical Assembly Building," *Civil Engineering* 35 (Jan. 1965): 45–49.
14. Anton Tedesko, "Space Truss Braces Huge Building for Moon Rocket," *Engineering News-Record* 172 (6 Feb. 1964): 24–27.
15. James H. Deese, "The Problem of Low Level Wind Distribution," paper presented at the Structural Engineers Councils of Florida, First Annual Conference, Tampa, 9 Nov. 1964.
16. Kurt Debus, "Some Design Problems Encountered in Construction of Launch Complex 39," paper given in Darmstadt, Germany, 25 June 1964; R. P. Dodd, "HVAC Temperature Control System for VAB and LCC," with attachment, "VAB HVAC Temperature Control System," 14 July 1963; G. J. Burrus, LCC and Sup. Fac. Sec., memo for record, "LCC Air Conditioning Unit Reliability," 28 July 1965.
17. Debus, "Some Design Problems Encountered," p. 35. *Apollo Launch Complex 39 Facilities Handbook,* issued by the U.S. Army Corps of Engineers, South Atlantic Div.,

p. 14, gives different numbers: height of each door opening, 140 meters; lower door opening 46.32 meters wide and 34.74 meters high; upper door opening 23.16 meters wide and 104.24 meters high.

18. Alexander, "Report on VAB," p. 13; Tedesko, "Design of the Vertical Assembly Building," pp. 48–49; Dodd interview.
19. Philip C. Rutledge, "Vertical Assembly Building—Design of Foundations," *Civil Engineering* 35 (Jan. 1965): 50–52.
20. Alexander, "Vertical Assembly Building—Project Description," pp. 43–44.
21. Stein interview; Bidgood to Clearman, "Design of the Vertical Assembly Building, Advanced Saturn Launch Complex 39," 6 Mar. 1963.
22. Bidgood to Mr. Lenezewski, CE Canaveral Dist., "LC-39 Transfer Aisles," 7 July 1963; Andrew Pickett to Dodd, "Platform Access to S-IC Inter-Tank Area VAB," 27 June 1963; Dodd to Bertram, "LC-39 VAB and LCC," 3 July 1963.
23. "Launch Complex 39," brochure issued by Corps of Engineers for contractors' conference, Oct. 1963, p. 4.
24. R. P. Young, NASA Exec. Off., to Webb, 13 June 1963.
25. M. Menghini, Field Rep., URSAM, memo for record, "Telephone calls to and from Col. Alexander," 31 Oct. 1962; Anton Tedesko, "Assembly and Launch Facilities for the Apollo Program, Merritt Island, Florida: Design of the Structure of the Vertical Assembly Building," paper presented at the ADCE Structural Engineering Conference and Annual Meeting, 19–21 Oct. 1964, p. 10.
26. Stein interview.
27. Brown interview. The NASA–Corps of Engineers movie *The Big Challenge* confirms Brown's testimony.
28. D. T. Brewster to W. W. Kavanaugh, "Minutes of Meeting between M-ASTRA and M-LVOD, 13 Sept. 1962," 31 Oct. 1962; MSFC, "Saturn V Electrical Ground Support Equipment for Launch Complex 39," pp. 1-11.
29. C. Q. Stewart, Mechanical Engineering, memo for record, 1 Aug. 1962.
30. "Minutes of Crawlerway Design Conference," NASA-LOC, Cape Canaveral, 21 Feb. 1963.
31. "Minutes of Crawler Transporter Crawlerway Meeting," LOC E & L Building, 27 Mar. 1963.
32. J. B. Bing, memo for record, 9 July 1963; Bing, memo for record, "Trip Report," 16 Aug. 1963. In line with the complaint of Mr. Bing, the authors found no reference on the part of URSAM people to Giffels and Rossetti in any of the many articles that appeared on URSAM's work on the VAB. In an article in *Engineering News–Record* for 6 Feb. 1964, p. 28, for instance, one of the URSAM principals mentions the companies that constructed the first launch pad and the crawlerway, but not the firm that designed both of them.
33. A. H. Bagnulo to U.S. Army Engineer Dist., Canaveral, 7 Oct. 1963; Lt. Col. Leo J. Miller, Corps of Engineers, Asst. Dist. Engineer, "Construction Coordination Group for NASA-LOC Complex 39," 14 Oct. 1963; Ernst interview.
34. Launch Support Equipment Engineering Div. Monthly Progress Reports, 10 Oct. 1962, 13 Mar. 1964.
35. J. H. Deese, Chief, Facilities Engineering Sec., memo for record, "Engineering Analysis of Launch Pad Diaphragm Construction, Launch Pad 39B," 11 Mar. 1963.
36. "Theoretical Analysis of Surface Temperatures, Flame Trench, Complex 39 A/B, KSC," technical memo 2-62, Mar. 1967, U.S. Army Engineer, Ohio River Div. Laboratories, Cincinnati.
37. Giffels and Rossetti, "Structural Design of Pad Terminal Connection Room and Environmental Control System Buildings," 11 Apr. 1963, sheet 85; Launch Support Equipment Engineering Div., "Preliminary Release Levels for Ground Support Equipment, Launch Complex 39," 19 Dec. 1963.

38. Poppel to Petrone, "Policy Statement and Design Concept for C-5 Propellant Loading Systems," 1 June 1962. A Complex 39 Foundation Prestudy Conference was held on 29 May 1962 at Jacksonvile: C. Q. Stewart, memo for record, "Foundation Design Prestudy Conference, Jacksonville DE, 29 May 1962," 31 May 1962.
39. *NASA Merritt Island Launch Area Master Plan,* vol. 3, pt. 1, *Industrial Area,* sec. 1, "General Site Plan," 22 Mar. 1963. Cf. *John F. Kennedy Space Center, NASA, Master Plan,* pt. 2, *Industrial Area Plans,* sec. 1, "General Site Plan," 25 Oct. 1965.
40. House Committee on Science and Astronautics, *Master Planning of NASA Installations,* House report 167, 89th Cong., 1st sess., 15 Mar. 1965, p. 24.
41. MSC Florida Ops., *Merritt Island Facilities,* undated pamphlet describing facilities funded through FY 1963, 1964, and 1965.
42. Dir., Information Systems, KSC, "Project Development Plan for Launch Instrumentation," 6 June 1966, p. 2-1.
43. Bruns interview, 22 Aug. 1969.
44. Bidgood to CE Jacksonville, "Central Instrumentation Facility, MILA," 10 May 1963; B. Baker, memo for record, "Siting of the CIF," 22 Aug. 1963.
45. LOC Staff Study, *Concepts for Support Service at the Merritt Island Launch Area,* 6 May 1963.
46. Ibid.; Albert F. Siepert, memo for record, 20 Oct. 1966.
47. "Interim Agreement Implementing the 17 Jan. 1963 Agreement between the Department of Defense and NASA Regarding Management of the Atlantic Missile Range of DOD and the Merritt Island Launch Area of NASA, Part III, Logistic and Administrative Functions," signed 10 May 1963 by Maj. Gen. L. I. Davis, USAF, and Debus; NASA release 63-111, 23 May 1963.
48. C. C. Parker, LOC Asst. Dir. for Admin., to Debus, 23 May, 5, 19 June, 13 Aug., 9 Oct., 14 Nov. 1963, 8 Jan. 1964; LOC release 74-63, 4 Oct. 1963.
49. Parker to Debus, 18 Sept., 17 Oct., 7, 27 Nov. 1963, 2, 16 Jan. 1964; KSC release 56-64, 24 Apr. 1964.

Chapter 12: From Designs to Structures

1. LOC, "Construction Progress Reports," 6 Nov. 1962, p. 5; 27 Nov. 1962, p. 5; 10 July 1963, p. 2.
2. *Spaceport News,* 31 Oct. 1963, p. 4.
3. "Construction Progress Reports," 15 June 1963, p. 7; 13 Sept. 1963, p. 1; D. E. Eppert, Chief, Construction Div., Canaveral Dist., Corps of Engineers, to J. J. Frangie, with attachment: "Tabulation of Contracts Supervised by the Corps of Engineers for Construction of Complexes 34 and 37 as Well as Work on Merritt Island for NASA Facilities" (hereafter cited as Tabulation of Corps of Engineers Contracts).
4. *Spaceport News,* 8 Aug. 1963, pp. 4–5.
5. Ibid., 28 Feb., 8, 15 Aug. 1963, 29 July 1965.
6. Ibid., 15 Aug. 1963; "Launch Operations Progress Report," 26 Aug. 1963.
7. John F. Kennedy, address at Rice University 24 Sept. 1962, *Public Papers of the Presidents,* Washington, 1963, p. 329.
8. NASA Hq., OMSF Instruction MD-M9330.001, 15 Oct. 1962, with enclosure.
9. Facilities Programing Off., Facilities Engineering and Construction Div., LOC, "Summary Project Status Report," 29 Nov. 1963, p. III-1; Tabulation of Corps of Engineers Contracts, Sept. 1968.
10. *Spaceport News,* 16 Jan. 1964.
11. Corps of Engineers, South Atlantic Div., Canaveral Dist., *Apollo Launch Complex 39 Facilities Handbook* (hereafter cited as *LC-39 Facilities Handbook*), pp. 4–5.

12. *Spaceport News,* 16 Jan. 1964, p. 7.
13. "Construction Progress Reports," 16 Aug. 1963, p. 9; 20 Jan. 1964, p. 11; *Spaceport News,* 16 Jan. 1964, 26 Sept. 1963; *LC-39 Facilities Handbook,* p. 15.
14. *Spaceport News,* 16 Jan. 1964.
15. "Summary Project Status Report," 29 Nov. 1963, p. III-2.
16. "LOC Monthly Status Report to the Management Council, Office of Manned Space Flight," presented by Kurt H. Debus, 24 Sept. 1963, p. 15; Summary Project Status Report, 29 Nov. 1963, p. III-2.
17. "Construction Progress Report," 22 Nov. 1963, p. 15.
18. "Summary Project Status Reports," 29 Nov. 1963, p. III-3; 17 Apr. 1964, p. III-2; *Spaceport News,* 19 Sept. 1963, p. 1.
19. "Construction Progress Report," 29 Jan. 1964, p. 20.
20. "Narrative Project Status Report, 1–30 Apr. 1964," p. I-9.
21. "Construction Progress Report," 20 Jan. 1964, pp. 10–11; *LC-39 Facilities Handbook,* pp. 3, 5.
22. "Narrative Project Status Report, 1–28 Feb. 1964," p. I-9; "Summary Project Status Report," 2 Oct. 1964, pp. IV-1, IV-2.
23. *Detailed Construction Schedule, VAB Area Facilities,* 1 May 1964, rev. 30 June 1964, with cover letter from William E. Pearson, Chief, Schedules Off., 27 July 1964 (this series of schedules, revised and issued periodically, will be cited as *Detailed Construction Schedule,* facility, date).
24. "Narrative Project Status Report, 1–30 Apr. 1964," p. I-9.
25. *Apollo/Saturn V MILA Facilities Description,* report K-V-011, p. 1-13.
26. *Spaceport News,* 3, 10, 17 Sept. 1964.
27. Ibid., 3 Sept. 1964; Jones interview.
28. *Spaceport News,* 8 Oct. 1964.
29. "Narrative Project Status Report, 1–31 Dec. 1964," pp. I-4, I-5.
30. Ibid; *LC-39 Facilities Handbook,* pp. 9–10.
31. A. H. Bagnulo, Dir., Facilities Engineering and Construction Div., KSC, "Revised Designations for NASA Facilities," 3 Feb. 1965; George E. Mueller, Assoc. Admin. for Manned Space Flight, to Dir., KSC, "Facility Titles, KSC," with attachment, 8 Sept. 1965.
32. "Narrative Project Status Report," 1–31 Jan. 1965, pp. I-3 through I-5.
33. Ibid., 1–31 Mar. 1965, p. I-2; *Spaceport News,* 1 Apr. 1965.
34. "Construction Progress Report," 27 Nov. 1962.
35. Ibid., 22 Jan., 25 Feb. 1963.
36. MSC Florida Ops., "Description and Justification for Spacecraft Operations and Checkout Building," included in *John F. Kennedy Space Center, NASA Fiscal Year 1963 Estimates, Apollo Mission Support Facilities, Project 7623;* "Manned Spacecraft Center Consolidated Activity Report for 16 Feb.–21 Mar. 1964," p. 78.
37. KSC, "Project Status Report," 1–31 Dec. 1964, p. I-39; Tabulation of Corps of Engineers Contracts, Sept. 1968.
38. Reyes interview, 24 June 1974; Chauvin interview, 24 June 1974.
39. KSC, *Apollo/Saturn V MILA Facilities Description*, K-V-011, p. 3-1.
40. KSC, "Technical Progress Report," 19 Feb. 1964, p. 19.
41. Ling-Temco-Vought, Inc., "Historical Events—Calendar year 1964, Gemini and Apollo Programs and Facilities, Manned Spacecraft Center Florida Operations, Cape Kennedy and Merritt Island," 22 Dec. 1964; Morris interview.
42. *Spaceport News,* 15 Apr. 1965.
43. Ibid.
44. Ibid., 27 May 1965.
45. Ibid., 13 Sept., 26 Dec. 1965.

46. KSC Weekly Notes, Miraglia, 6 July 1964; Parker, 7 June 1964; KSC, *C. O. E., Report of Fatal Accident at LC-39,* signed by Col. W. L. Starnes.

Chapter 13: New Devices for New Deeds

1. D. D. Buchanan, memo for the record, 16 April 1962.
2. LOC, *Procurement Plan,* signed by Kurt Debus, Director, 11 Sept. 1962, p. 2.
3. DDJ, 1, 7 Nov. 1962; Debus to Brackett, 29 Nov. 1962; LOD, "Crawler/Transporter Proposal Conference Attendees, 17 December 1962" (in Buchanan file).
4. R. P. Young, Exec. Off., memo for record, 13 Mar. 1963, in NASA History Office; James E. Webb and Robert C. Seamans, Jr., "Statement of the NASA Administrator on Selection of a Contractor for the Crawler-Transporter," 13 May 1963, ibid.; *Congress and the Nation, 1945–1964* (Washington: Congressional Quarterly Service, 1965), p. 320.
5. "Briefing, Crawler-Transporter Procurement," 5 Feb. 1963, copy in Fred Renaud's private papers.
6. "Decision to Negotiate an Individual Contract under 10 USC 2394 (a) (11)," 5 Dec. 1963; "Determination and Findings for Method of Contracting, Cost-Plus-Incentive Fee," 7 Dec. 1962, both in NASA History Off.
7. W. Kraft, Admin. Asst., Marion Power Shovel Co., to Theodor A. Poppel, 11 Dec. 1963, pp. 1–4, in Fred Renaud's private papers.
8. Ibid., pp. 5–6.
9. Renaud interview, 4 Apr. 1973.
10. *Aviation Week* 84 (20 June 1966): 78.
11. Gramer interview, 19 July 1973.
12. "Fire Alarm System for Crawler-Transporter," 29 Jan. 1965, in Fred Renaud's private papers.
13. Gorman to Petrone, 22 Mar. 1965, including a memo of 29 Feb. (*sic*) 1965, in Fred Renaud's private papers.
14. "Fire Protection Survey and Recommendations," attachment to letter of C. W. Conway to Ronald Worchester, 16 June 1965; copies in Fred Renaud's private papers.
15. *Spaceport News*, 15 Feb. 1968, p. 6.
16. Unless otherwise cited, the descriptive information in this and the following paragraphs concerning the crawlerway and launch pad A facilities is based on the *LC-39 Facilities Handbook,* pp. 35–51.
17. *Spaceport News,* 19 Sept. 1963, p. 1; 17 Oct. 1963, p. 1; Tabulation of Corps of Engineers Contracts, Sept. 1968; "Summary Project Status Report," 29 Nov. 1963, pp. III-2, IV-3; *Technical Progress Report, Second Quarter CY 1965* (TR-194), 30 July 1965, p. 38.
18. *Apollo/Saturn V MILA Facilities Descriptions,* report K-V-011, p. 1-26.
19. Memo, Col. Bagnulo, 3 Feb. 1965.
20. Hahn interview.
21. Boylston interview.
22. Wm. Clearman, "Prototype of Service Arm 6," 30 July 1963, on microfilm in Vehicle Servicing and Accessories Sec. of Design Engineering Off., KSC.
23. "Qualification Test for Cable Retract Sled for Saturn V and Pneumatic Console No. 2," prepared by C. Dyer, Brown Engineering, in Design Engineering Files; photographs in Brad Downs's Design Engineering Files, KSC. The authors are indebted to Mr. Downs for his help in this section.
24. "Weekly Notes," Haworth, 5 Aug.; Clark, 6 Aug. 1964; Gramer interview, 19 Sept. 1972.
25. "Weekly Notes," 19 Aug. 1964.

26. *Technical Progress Report Third and Fourth Quarter CY 1964* (TR-159), 5 Mar. 1965, p. 61; "Saturn V Swing Arm Program Problem," an analytical statement, unsigned and undated. This contract was NAS 10-1751.
27. Rowland interview; R. D. Rowland, Hayes International Corp., to Benson, 25 July 1972.
28. James W. Dalton and Willard Halcomb, Apollo–Saturn V Test and Systems Engineering Off., to Petrone, 28 Oct. 1964; "Saturn V Swing Arm Program Problem," p. 1.
29. William L. Clearman, Jr., Chief, Apollo-Saturn V Test and Systems Engineering Off., to Chief, Launch Equipment Support Sec., Procurement Div., "Contract NAS10-1751, Proposed Changes to Incorporate Revised Drawing Lists," 23 Nov. 1966.
30. James W. Dalton, Apollo-Saturn V Test and Systems Engineering Off., "Minutes of Meeting—Change Review Board—Service Army Contract with International—17 Sept. 1964," 22 Sept. 1964; *Method of Handling Engineering Changes, Contracts NAS10-1751—NAS10-1847.*
31. James W. Dalton to William T. Clearman, "Status of Hayes Service Arm Contract as Result of Sole Source Vendor Items," 2 Dec. 1964.
32. "Saturn V Swing Arm Program Problem," p. 2.
33. Kurt H. Debus to L. F. Jeffers, Hayes International Corp., Birmingham, AL, 5 Nov. 1965.
34. "Saturn V Swing Arm Program Problem," p. 1; "Management Inquiry into the Procurement of Service Arms for Launch Complex 39," pp. 42, 43, 62.
35. "Saturn V Swing Arm Program Problem," p. 1; Gramer interview, 21 Sept. 1972.
36. *Spaceport News,* 28 Oct. 1965, p. 3.
37. *Technical Progress Report Third and Fourth Quarter CY 1964* (TR-159), 5 Mar. 1965, p. 61; Procurement Div., "Status Summary of Active Contracts as of 31 Mar. 1964," sec. II, p. 25.
38. *Technical Progress Report Third Quarter CY 1965* (TR-250), 30 Sept. 1965, pp. 3–18.
39. *Saturn V Launch Support Equipment General Criteria and Description* (SP-4-37-D), rev. 15 Sept. 1964, Launch Support Equipment Engineering Div., pp. 2–62; *Technical Progress Report First Quarter CY 1965,* 26 Apr. 1965, p. 46; KSC Procurement Div., "Status Summary of Active Contracts as of 30 Sept. 1966," sec. II, p. 6.
40. "Apollo/Saturn V MILA Facilities Descriptions," pp. 2-81, 2-82; "Construction Progress Reports," 1 July 1965, p. 4.
41. R. T. Cruden and J. R. Ellis, memo for record, "Ordnance Meeting, LC-39 Arming Tower," 25 Mar. 1963; J. R. Ellis, memo for record, "Ordnance Requirements, Arming Tower LC-39," 26 Mar. 1963; W. T. Clearman, Jr., and James H. Deese, "Meeting at Complex 34 Operations Support Building to Discuss Saturn-V Ordnance Installation Problem," 27 Mar. 1963. The authors are indebted to Francis Jarrett for research on this subject, which is covered more fully in Jarrett and Lindemann, "History of the John F. Kennedy Space Center, NASA, to 1965," typescript.
42. J. R. Ellis, memo for record, "Meeting and Discussions Concerning Arming Tower, LC-39," 9 Apr. 1963; minutes of meeting, "Rust Contract BE-9002, LC-39 Arming Tower, Contract DA-08-123-ENG-(NASA-1752)," 15 Apr. 1963.
43. Off. of the Canaveral Dist. Engineer, "Report on Restudy of Arming Tower to Resolve Dead Load and Wind Load Problems," 20 Dec. 1963, pp. 1–5.
44. Ibid.
45. Vehicle Design Integration Working Group, "Minutes of the Saturn V Common Ordnance Meeting," Huntsville, AL, 10–11 Dec. 1963, pp. 1–4.
46. "Summary Project Status Report," 17 Apr. 1964, p. IV-3; "Development Summary Schedule, Complex 39, 1963"; *Technical Progress Report, Third and Fourth Quarter CY 1964,* 5 Mar. 1965, pp. 45, 58; Tabulation of Corps of Engineers Contracts, Sept. 1968;

Technical Progress Report, Second Quarter CY 1965, 30 July 1965, p. 38; *Technical Progress Report, Third Quarter CY 1965* (TR-250), 30 Sept. 1965, pp. 3–18; *LC-39 Facilities Handbook*, p. 54.

47. *Technical Progress Report, Third and Fourth Quarter CY 1964* (TR-159), 5 Mar. 1965, p. 45.
48. *Technical Progress Report, First Quarter CY 1965* (TR-168), 26 Apr. 1965, p. 33; *Technical Progress Report, Second Quarter CY 1965* (TR-194), 30 July 1965, p. 34.
49. H. D. Brewster and E. G. Hughes, *Lightning Protection for Saturn Launch Complex 39,* report TR-4-28-2-D, 18 Oct. 1963.
50. Ibid., app. A.
51. Ibid., pp. 3-6 through 3-16 and app. A.
52. Ibid., pp. 2-3, 3-3, 3-4, A-1; A. R. Raffaelli, "Introduction to Lightning," report LOC LT1R-2-DE-62-6, 14 Dec. 1962.
53. H. D. Brewster to Lightning Protection Committee, "Minutes of the Third Lightning Protection Committee Meeting, 29 Sept. 1965, at KSC," 20 Oct. 1965, KSC Technical Documents Library.
54. KSC, "Weather Effects on Apollo/Saturn V Operations, Apollo 4 through Apollo 13," report 630-44-0001, 27 July 1970.
55. KSC release 11-66, 21 Jan. 1966.
56. *Building Construction Magazine,* Feb. 1966, p. 29.

Chapter 14: Socio-Economic Problems on the Space Coast

1. Petrone interview, 25 May 1972, pp. 62–68. Petrone discussed the difference between the industrial and construction workers in a sympathetic and understanding way.
2. Senate Committee on Government Operations, Subcommittee on Investigations, *Hearings on Work Stoppages at Missile Bases*, 87th Cong., 2d sess., 25 Apr.–9 June 1961.
3. Ibid., pt. 1, pp. 11–15, 36–46.
4. The John F. Kennedy Space Center Missile Site Labor Relations Committee, "Function Responsibilities and Procedure," p. 1.
5. Glenn M. Parker, "The Missile Site Labor Commission," *ILR Research* 8 (1962): 11.
6. John Miraglia, "Project Stabilization Agreement," pp. 1–2.
7. Senate Committee on Government Operations, Subcommittee on Investigations, *Hearings on Work Stoppages at Missile Bases*, 87th Cong., 1st sess., pt. 2, 4 May 1961, pp. 520 ff.
8. Yates interview.
9. Edward Kiffmeyer, Labor Relations Off., AFETR, "Strike Summary Reports," Patrick Air Force Base, FL, monthly reports from Jan. 1962–July 1965; *History of Air Force Missile Test Center*, vol. 1, 1964, p. 166.
10. *Spaceport News*, 6 Feb. 1964.
11. *Melbourne Daily Times*, 18 Feb. 1964.
12. *Orlando Sentinel*, 5 Feb. 1964.
13. Charles L. Buckley, Jr., Chief, Security Off., memo for record, "FEC Incident, MILA," 1 Feb. 1964.
14. *Cocoa Tribune*, 10 Feb. 1964.
15. *Orlando Sentinel*, 12 Feb. 1964.
16. *Melbourne Daily Times*, 18 Feb. 1964.
17. KSC Weekly Notes, Miraglia, 14 Feb. 1964.
18. *Orlando Sentinel*, 19, 28 Feb. 1964.
19. KSC Weekly Notes, Miraglia, 22 Apr. 1964.
20. Ibid.

21. KSC Weekly Notes, Miraglia, 3 June 1964.
22. KSC Weekly Notes, Miraglia, 11 June 1964; Titusville *Star-Advocate*, 10 June 1964.
23. Paul Styles, Dir., Off. of Labor Relations, memo, 9 June, 1964, copy in files of KSC Security Office; *Miami Herald*, 19 June 1964, p. 2; Gooch interview; Horner interview.
24. KSC Weekly Notes, Miraglia (signed by Oliver E. Kearns), 25 June 1964.
25. Ibid.; *History of the Air Force Missile Test Center*, vol. 1, 1964, pp. 159, 165.
26. KSC Weekly Notes, Miraglia, 16 Sept. 1964.
27. KSC Weekly Notes, Miraglia (signed by Oliver E. Kearns), 23 July, 20 Aug., 2 Sept. 1964.
28. Ibid., 10 Sept. 1964. Several contractor representatives who dealt with labor matters shared Kearns' view of Baxley.
29. Ibid., 16 Sept. 1964.
30. *Cocoa Tribune*, 14 Sept. 1965.
31. KSC Weekly Notes, Kearns, 22 Sept. 1965; *Orlando Sentinel*, 17 Sept. 1965.
32. *Cocoa Tribune*, 4 Oct. 1965.
33. *Time*, 4 July 1969, p. 38.
34. George L. Simpson, Jr., to Webb and Dryden, 25 June 1965.
35. Annie May Hartsfield, Mary Alice Griffin, and Charles M. Grigg, eds., *Summary Report NASA Impact on Brevard County* (Tallahassee: Institute of Social Research, Florida State Univ., 1966), pp. 10–11, table 2, p. 21, citing U.S. census reports.
36. Ibid., pp. 13, 52.
37. Ibid., pp. 104, 106, 107.
38. Ibid., pp. 17, 18, 26, 96.
39. Charles Grigg and Wallace A. Dynes, *Selected Factors in the Deceleration of Social Change in a Rapidly Growing Area* (Tallahassee, 1966), table 3, p. 144.
40. *Spaceport News*, 16 May 1963, p. 6.
41. Ibid., 13 June 1963, p. 3.
42. Ibid., 14 May, 9 July 1964. Sixteen thousand individuals, 56% of the total work force, responded to the questionnaires.
43. Siebeneichen interview.
44. Peter Dodd, *Social Change in Space-Impacted Communities* (Cambridge, MA: The Committee on Space of the American Academy of Arts and Sciences, Aug. 1964), pp. 20–21.
45. KSC Weekly Notes, Miraglia, 22 Apr., 27 May 1964.
46. KSC Weekly Notes, Van Staden, 7 Apr. 1965.
47. *Spaceport News*, 20 Aug. 1964.
48. Ibid., 1 July 1965.
49. *Florida Statistical Abstract, 1969* (Gainesville: Univ. of Florida, 1969), pp. 21, 28.
50. *Spaceport News*, 29 Feb. 1968.
51. *Time*, 4 July 1969, p. 38.
52. Quoted in John G. Rogers, "What Life at Cape Kennedy Does to Marriage," *Parade*, 9 July 1969.
53. Dr. Ronald C. Erbs, M.D., to Faherty, 17 July 1974, in author's personal files.
54. Nazaro interview.

Chapter 15: Putting It All Together: LC-39 Site Activation

1. KSC, "Apollo/Saturn V Facility Activation Plan," 3d Coordination Draft, 30 Dec. 1965; Petrone interview, 17 Sept. 1970; "Presentation of the NASA Oversight Subcommittee, Committee on Science and Astronautics, House of Representatives," 29 Oct. 1968, pp. 38–57.

2. William T. Clearman, Jr., to T. A. Strong, "Lt. Col. Donald R. Scheller, USAF (NASA)," 12 Oct. 1964; Clearman interview, 13 Sept. 1973.
3. KSC, "Minutes of Apollo/Saturn V Site Activation Board Meeting #1," 19 Mar. 1965.
4. KSC, *LC-39 Site Activation Master Schedule (Preliminary), Level A,* rev. 17 Feb. 1967.
5. KSC, "Minutes of Site Activation Board Meeting #1"; "Presentation to the NASA Oversight Subcommittee," 29 Oct. 1968, pp. 43–44.
6. Gruene to Apollo/Saturn V Test Off., "Comments on Site Activation Board Charter," 1 Apr. 1965.
7. Bagnulo to Scheller, "Site Activation Board Charter," 30 Mar. 1965.
8. Clark to Scheller, "Comments on Apollo/Saturn V Site Activation Board," 31 Mar. 1965.
9. KSC, "Minutes of Apollo/Saturn V Site Activation Board Meeting #3," 5 Aug. 1965.
10. Donald R. Scheller, "Management by Exception, Activation of Apollo/Saturn V Launch Complex 39," 15 May 1967; "Presentation to the NASA Oversight Subcommittee," 29 Oct. 1968, pp. 41–42.
11. L. S. Harris, Chief, Site Operation Gp., Engineering and Development Dir., KSC, to Bagnulo, "Activation Projects, LC-39," 23 Sept. 1965, in KSC Engineering and Development Dir. Reading Files, 1965–66; "Minutes of KSC Site Activation Working Group Meeting #1," 3 Dec. 1965; Scheller to SAB, "Apollo/Saturn V SAB Management Meeting Membership," 19 Jan. 1966.
12. Scheller, "Management by Exception"; Clearman interview, 26 Oct. 1973; Fulton interview; Chandler interview.
13. Boeing Atlantic Test Center Management Systems Staff, "ERS Recovery Plan," by A. J. Culver and K. G. Baird, Apr. 1966.
14. Scheller, "Management by Exception"; Murphy interview.
15. Petrone's notes, 14 Feb. 1963; Wagner interview, 21 Sept. 1973; Gassman interview; *NASA Apollo Inter-Center ICD Management Procedure*, report CM-001-001-1B, Jan. 1969, pp. 3–4.
16. Petrone, "KSC Apollo/Saturn Configuration Management Program Directive," 29 Sept. 1965, Management Configuration Off. files.
17. KSC Apollo Program Directive No. 2, 9 Dec. 1965.
18. Gassman interview; Leet interview, 8 Nov. 1973.
19. KSC/MSC, *ICD-IRN Processing,* 6 May 1968; Wagner interview, 21 Sept. 1973.
20. Petrone to Poppel, "VLF-39 Facility Checkout Vehicle Minimum Requirements and Operational Characteristic," 1 Mar. 1962; Poppel (signed by Owens) to Petrone, "Minimum Requirements for Facility Checkout Vehicle for Complex 39," 2 Apr. 1962.
21. Phillips to NASA Manned Space Flight Centers, "Apollo Delivery and Launch Schedules," 16 Feb. 1965; Phillips to NASA Manned Space Flight Centers, "Apollo Schedules and Mission Assignments," 12 Jan. 1965, in Phillips file, NASA Hq. History Off.; KSC, Plans, Program, and Resources Dir., "Verification of '500-F' Schedule Dates Based upon OMSF Approved Apollo Schedule," 10 Mar. 1966.
22. Petrone to Arthur Rudolph, Saturn V Program Manager, MSFC, 11 Jan. 1966.
23. Haggard interview; "Marion Power and Shovel Company, PMSLC Hearing, Miami, Florida, May 1965," contract NASA 10-477, KSC Labor Relations Off.
24. Michael E. Haworth, Jr., NASA Contracting Off., to Marion Power Shovel, "Contract NAS 10-477," 25 Jan. 1965.
25. Clearman to Petrone, Weekly Notes of 25 June 1965.
26. George W. Walter, *Modifications to Bearings for Traction Support Rollers on Crawler-Transporters*, report KSC TR-260-D, 15 Dec. 1965, pp. 1–2; KSC Weekly Notes, Poppel, 29 July 1965; F. Jones, Technical Supervisor, to Richard McCoy, "Contract NAS 10-477, Salvage of Bearings," 26 Oct. 1965.
27. Walter, *Modifications to Bearings*, p. 2; "KSC Press Briefing and Crawler Demonstration," 25 Jan. 1966, pp. 3–4, 12.

28. M. E. Haworth to Patrick Kraft, Treasurer, Marion Power Shovel Co., 15 June 1965.
29. M. E. Haworth to F. Boyle, Pres., Marion Power Shovel Co., "NAS 10-477," 14 Oct. 1965; Gordon Harris, Chief of Public Affairs, to Debus, 30 Sept. 1965.
30. Morgan F. Jones to Poppel, 15 Nov. 1965; Poppel, "Weekly Notes to Debus," 8 Oct. 1965; "Technical Progress Reports," Third Quarter 1965, pp. 3–16; Buchanan interview, 4 Oct. 1974; Walter, *Modifications to Bearings*, p. 12; "KSC Press Briefing and Crawler Demonstration," 25 Jan. 1966, pp. 16–17.
31. KSC Weekly Notes, Poppel, 3, 10 Dec. 1965; Bagnulo, 17 Dec. 1965; *Spaceport News*, 3 Feb. 1966.
32. NASA, *PERT, Program Evaluation and Review Technique, Handbook*, NPC-101, 1 Sept. 1961; "KSC Presentation to the NASA Oversight Subcommittee, Committee on Science and Astronautics," 29 Oct. 1968, pp. 38–58; Potate interview, 6 June 1972.
33. "Minutes of Apollo/Saturn V Site Activation Board Meeting #3," 5 Aug. 1965; KSC Weekly Notes, Petrone, 30 July 1965.
34. L. Steven Harris to Bagnulo, "Site Activation Board Meeting," 28 Oct. 1965, in Engineering and Development Dir. reading files, 1965–1966; Wiley interview, 31 Oct. 1973.
35. Petrone to Phillips, 4 Nov. 1965.
36. Petrone to Phillips, "Proposed SA500F-1/501 Work Around Schedule," 7 Dec. 1965.
37. KSC Weekly Notes, Bagnulo, 17 Nov. 1965; Bagnulo to Debus and others, "Pad A Settlement," 9 Dec. 1965; Roberts interview.
38. "Minutes of Apollo/Saturn Site Activation Board Meeting #14," 6 Jan. 1966; "LC-39 Site Activation Master Schedule Meeting," 17 Jan. 1966.
39. Potter interview; Steven Harris interview; Tom Wills interview; "Reading File of Engineering Division's Site Activation Group," Sept. 1965, Steven Harris's files, KSC.
40. "Minutes of Apollo/Saturn V Site Activation Board Meeting #16," 3 Feb. 1966; KSC, "PERT Analysis Report," 20 Jan. 1966.
41. Clearman, Weekly Notes to Petrone, 7 Jan., 4, 18, 25 Feb., 3 Mar. 1966.
42. Clearman, Weekly Notes to Petrone, 3, 25 Mar., 1, 15 Apr. 1966.
43. "LC-39 Site Activation Status Reports," weekly for March and April 1966; Brewster interview; Hahn interview.
44. "LC-39 Site Activation Status Report," 27 Apr. 1966; Hahn interview.
45. "LC-39 Site Activation Status Reports," weekly reports for Mar. 1966; Weekly Notes, Bagnulo, 25 Mar. 1966; *Spaceport News*, 10, 31 Mar. 1966.
46. Rigell interview; "LC-39 Site Activation Weekly Reports," Mar., Apr. 1966.
47. "LC-39 Site Activation Reports," 13 Apr., 5 May, 3 Aug. 1966.
48. "Minutes of Apollo/Saturn V Site Activation Board Meeting #24," 26 May 1966; *Spaceport News*, 26 May 1966.
49. "Minutes of Apollo/Saturn V Site Activation Board Meeting #25," 16 June 1966; *Spaceport News*, 16 June 1966.
50. "Minutes of Apollo/Saturn V Site Activation Board Meetings" 25, 26, and 27, dated 16, 23 June, 7 July 1966.
51. Barfus interview; Enlow interview; Sparkman interview, 6 Dec. 1973.
52. "Minutes of Apollo/Saturn V Site Activation Board Meeting #27," 7 July 1966.
53. "Minutes of Apollo/Saturn V Site Activation Board Meetings" 27 and 28, dated 7 and 21 July 1966; *Spaceport News*, 4 Aug. 1966.
54. "Minutes of Apollo/Saturn V Site Activation Board Meetings" 28 and 29, 21 July, 4 Aug. 1966.
55. William I. Moore and Raymond J. Arnold, "Failure of Apollo/Saturn V Liquid Oxygen Loading System," 1967 Cryogenic Engineering Conference, 21–23 Aug. 1967, Stanford Univ., CA, paper K-1, in *Advances in Cryogenic Engineering* 13 (1967): 534–44; Boeing Atlantic Test Center, "Technical Report of Complex 39A LOX System Failure, 10 Sept. 1966"; "Fund Board of Inquiry Findings on Failure of LOX Distribution System—19

Aug. 1966," J. G. Shinkle, Chairman; "Presentation to the Congressional Subcommittee on Manned Space Flight," pp. 114–19.

56. Moore and Arnold, "Failure of LOX Loading System," pp. 534–44; "Presentation to the Subcommittee on Manned Space Flight," pp. 114–19.
57. "LC-39 Site Activation Status Reports," weekly for Sept., Oct. 1966.
58. Robert Hotz, *Aviation Week and Space Technology*, 22 Mar. 1965, p. 11.
59. "KSC Presentation to the NASA Oversight Subcommittee," 29 Oct. 1968; Petrone interview, 17 Sept. 1970.

Chapter 16: Automating Launch Operations

1. Sidney Sternberg, "Automated Checkout Equipment—The Apollo Hippocrates," in *Man on the Moon,* ed. Eugene Rabinowitch and Richard Lewis (New York: Basic Books, 1969), pp. 196–97.
2. W. Haeussermann, Dir., Guidance and Control Div., MSFC, memo for record, "Meeting on Saturn Checkout Equipment," 22 July 1960; Paul interview.
3. Debus to Dieter Grau, "Automatic Checkout Committee," 2 Sept. 1960; Richard interview, 12 Dec. 1973. See B. J. Funderburk, *Automation in Saturn I First Stage Checkout* (NASA TN D-4328, Jan. 1968), for story of the Packard Bell 250 and MSFC's early automation efforts.
4. Ludie Richard and Charles O. Brooks, *The Saturn Systems Automation Plan,* MSFC, 15 Sept. 1961, sec. II.
5. Ibid., sec. VII.
6. "Brief Chronological History of the Saturn V Breadboard," attached to *MSFC Automation Plan,* 8 May 1962; Burns interview; Greenfield interview.
7. Jafferis interviews, 19 Dec. 1973, 22 Jan. 1974; Greenfield interview; Whiteside interview, 4 Jan. 1974.
8. Jafferis interview, 19 Dec. 1973; "Description for Use of Saturn Ground Computer on SA-5," draft copy in Jafferis's private papers; *Spaceport News,* 21 May 1964, p. 2.
9. B. E. Duran, "Saturn I/IB Launch Vehicle Operational Status and Experience," read at Aeronautic and Space Engineering and Manufacturing Meeting, New York, 7-11 Oct. 1968, Society of Automotive Engineers reprint 680739; KSC, "Utilization of Saturn/Apollo Control and Checkout System for Prelaunch Checkout and Launch Operations," GP-663, 25 Mar. 1969.
10. W. O. Frost and D. E. Norvell, "Telemetry System Design for Saturn Vehicles," *Proceedings, 1966 International Telemetering Conference,* Los Angeles, 18–20 Oct. 1966, p. 70. See also E. A. Robin, "Development and Utilization of Computer Test Programs for Checkout of Space Vehicles," p. 297; Canaveral Council of Technical Societies, *Proceedings of the Second Space Congress,* Cocoa Beach, FL, 5–7 Apr. 1965, pp. 617, 634; D. M. Schmidt, "Automatic Checkout Systems for Stages of the Saturn V Manned Space Vehicle," *International Convention Record of Electrical and Electronics Engineers* 13 (pt. 4, Mar. 1965), p. 87.
11. Canaveral Council of Technical Societies, *Proceedings of the Second Space Congress,* p. 656; William G. Bodie, "Techniques of Implementing Launch Automation Programs, Saturn IB Space Vehicle System," *Practical Techniques and Applications,* 4: 740. See also *Apollo/Saturn IB Launch Operations Plan AS-203,* KSC document K-IB-021.3, p. 6-8.
12. Duran, "Saturn I/IB Launch Vehicle Operational Status"; KSC, "Utilization of Saturn/Apollo Control and Checkout."
13. Richard Dutton and William Jafferis, "Utilization of Saturn/Apollo Control and Checkout System for Prelaunch Checkout and Launch Operations," paper read at New York

Univ., Project SETE, 24–28 July 1967, pp. 3-34 through 3-43; Medlock interview; Thompson interview.

14. F. Brooks Moore and William Jafferis, "Apollo/Saturn Prelaunch Checkout Display Systems," read at IEEE Conference on Displays, Univ. of Loughborough, England, 7–10 Sept. 1971, pp. 7–9.
15. Ibid., pp. 9, 15–16.
16. Ibid., p. 14.
17. Richard Jenke to Benson, 17 Jan. 1975; Richard Smith interview; Medlock interview; Thompson interview.
18. Dutton and Jafferis, "Utilization of Saturn/Apollo Control and Checkout System," pp. 3-44 through 3-48; Jenke to Benson, 17 Jan. 1975.
19. Jenke to Benson, 17 Jan. 1975; Medlock and Thompson interviews.
20. Jenke to Benson, 17 Jan. 1975; Medlock and Thompson interviews.
21. Fridtjof Speer, Chairman, Saturn System Evaluation Working Gp., MSFC, to LOD Dir., "Justification for Early Delivery of the Saturn Blockhouse Records and Sequence Records," 19 Sept. 1961, Debus reading file.
22. R. W. Bivans, G. D. Matthews, and F. T. Innes, "A Scanning and Digitizing System for Multiple Asynchronous Telemetry Data Sources," read at National Telemetry Conference, Los Angeles, June 1964, p. 1, G. D. Matthews's private papers.
23. Bruns interview, 3 Jan. 1974; Bobby Griffin and G. D. Matthews, *The Real-Time Telemetry Data Processing Effort at the Launch Operations Center,* MTP-LVO-63-2, MSFC, p. 1.
24. Griffin and Matthews, *Real-Time Telemetry Data,* pp. 9–11; Bruns interview, 3 Jan. 1974; Corbett, Hughes, and Jelen interviews.
25. Griffin and Matthews, *Real-Time Telemetry Data,* pp. 3–6; George Matthews interview.
26. Griffin and Matthews, *Real-Time Telemetry Data,* pp. 14–17; Bivans, Matthews, and Innes, "Scanning and Digitizing"; George Matthews interview; LOC Weekly Notes, Sendler to Debus, 23 Aug., 15 Nov. 1962.
27. KSC Computation Br., "Scientific Computation Support of Saturn/Apollo Vehicle, SA-7," TR-103-2, 3 Dec. 1964.
28. Joralan interview, 3 Jan. 1974; LOC Weekly Notes, Sendler to Debus, 22 Mar. 1962.
29. LOC Weekly Notes, Gruene to Debus, 21 Feb. 1962.
30. Raymond Clark, Asst. LOC Dir., to Col. Max Carey, "Request for Additional Data on NASA Telemetry Requirements," 6 Oct. 1962; Debus to Davis, "The AFMTC Launch Area Telemetry System Plan, 28 September 1962," 18 Oct. 1962.
31. Telephone directory, Project Mercury Field Ops., STG, Cape Canaveral, FL, Sept. 1961; "Patent Application on ACE, NASA Case No. 8012," encl. to letter, James O. Harrel to Harold G. Johnson, 20 Jan. 1967, Johnson's private papers; Walton interview, 17 Dec. 1970. The Cape launch team first appeared as Preflight Operations Division on a Sept. 1962 MSC organization chart. Earlier it was called Mercury Field Operations or MSC's Atlantic Missile Range Operations.
32. Parsons interview; Preston interview, 22 Jan. 1974.
33. Parsons interview; Walton interview, 17 Dec. 1970; Preston interview, 22 Jan. 1974; W. E. Parsons, Head, Flight Instrumentation Sec., to C. W. Frick, Head, Apollo Project Off., MSC, "Implementation Plan for Apollo SPACE System," 26 June 1962, Johnson's private papers; "PACE-S/C History," compiled by Harold Johnson ca. 1963, Johnson's private papers.
34. Parsons interview; Harold Johnson interview; "PACE-S/C History"; Walton interview, 17 Dec. 1970.
35. Tom S. Walton, MSC Florida Ops., *Experimental Station Implementation and Planning,* 18 Dec. 1964; Walton interview, 17 Dec. 1970.
36. "PACE-S/C History"; Parsons interview; Walton interview, 23 Jan. 1974.
37. Parsons interview; Norwalk interview; Walton interview, 23 Jan. 1974.

38. Parsons interview.
39. Page interview; *Spaceport News,* 6 Jan. 1966; Apollo Support Dept., General Electric Co., *ACE-S/C, Acceptance Checkout Equipment, Spacecraft,* Daytona Beach, FL, undated.
40. Apollo Support Dept., General Electric Co., *ACE-S/C;* James O. Hassell to Harold G. Johnson, "Patent Application on ACE," 20 Jan. 1967, with encl. 1, "Patent Application on ACE, NASA Case No. 8012," Johnson's private papers; Harold Johnson interview.
41. Moore and Jafferis, "Apollo/Saturn Prelaunch Checkout," pp. 4–7.

Chapter 17: Launching the Saturn IB

1. Debus to Dep. Assoc. Admin. for Manned Space Flight, "Saturn I/IB Pad Utilization," 13 Nov. 1963; T. F. Goldcamp, memo for record, "Modification of LC-34 for Saturn IB," 12 Dec. 1963.
2. OMSF, *Mission Operation Report, Apollo/Saturn Flight Mission AS-201,* NASA report M-932-66-01, pp. 14–17; KSC Weekly Notes, Poppel, 1 July 1965; NASA release 66-32, *Apollo/Saturn 201 Press Kit,* 17 Feb. 1966, pp. 41–43.
3. OMSF, *Mission Operation Report, AS-201,* pp. 14–17; NASA, *AS-201 Press Kit,* pp. 41–43.
4. KSC Weekly Notes, Bagnulo, 5 Aug. 1965; *Spaceport News,* 26 Aug. 1965; KSC, "Daily Status Report, AS-201," 27 Dec. 1965–Jan. 1966.
5. KSC Weekly Notes, Hans Gruene, 26 Aug. 1965.
6. Akens, *Saturn Illustrated Chronology,* p. 117; KSC Weekly Notes, Gruene, 26 Aug. 1965.
7. KSC Weekly Notes, Gruene, 17 Sept., 1 Oct. 1965.
8. KSC Weekly Notes, Gruene, 17, 24 Sept., 1 Oct. 1965; Petrone, 7 Oct. 1965.
9. *Brevard Sentinel,* 20 Feb. 1966; KSC release 17-66, 16 Feb. 1966; *Spaceport News,* 18 Feb., 3 Mar. 1966.
10. MSFC, *Saturn IB Vehicle Handbook,* vol. 1, *Vehicle Description* (prepared by Chrysler Corp. Space Div.), 25 July 1966, p. II-7 (S-IB stage data summary); MSFC, *Saturn-Apollo Space Vehicle Summary, AS-201,* p. 21; NASA, *AS-201 Press Kit,* pp. 37–38; KSC Weekly Notes, Von Staden, 19 Aug. 1965.
11. MSFC, *Saturn-Apollo Space Vehicle Summary, AS-201;* Akens, *Saturn Illustrated Chronology,* p. 121; KSC Weekly Notes, John J. Williams, 28 Oct. 1965; NASA, *AS-201 Press Kit,* pp. 22, 39–40.
12. KSC Weekly Notes, Preston, 30 Sept. 1965; Petrone, 30 Sept. 1965.
13. KSC Weekly Notes, Petrone, 7, 28 Oct. 1965; Gruene, 8, 15, 22, 29 Oct. 1965.
14. KSC Weekly Notes, John J. Williams, 28 Oct., 10, 18 Nov. 1965; Preston, 10, 18 Nov. 1965.
15. KSC Weekly Notes, Gruene, 10 Nov. 1965.
16. KSC Weekly Notes, Gruene, 26 Nov., 3 Dec. 1965.
17. KSC, "Daily Status Report, AS-201," 8–23 Dec. 1965; KSC Weekly Report, John J. Williams to Debus, 6 Jan. 1966.
18. KSC, "Daily Status Report, AS-201," 8–23 Dec. 1965; KSC Weekly Report, Gruene to Debus, 10 Dec. 1965.
19. KSC, "Daily Status Report, AS-201," 27 Dec. 1965–7 Jan. 1966; KSC Weekly Reports, Gruene to Debus, 7 Jan. 1966; Williams to Debus, 6 Jan. 1966.
20. *Miami Herald,* 13 Jan. 1966.
21. Carlson interview, 16 Dec. 1970.
22. Bryan interview.
23. KSC, "Daily Status Report, AS-201," 24 Jan.–18 Feb. 1966; KSC Weekly Notes, Gruene to Debus, 15 Oct. 1965; KSC, *Apollo/Saturn IB Launch Plan, AS-201,* 27 Oct. 1965.

24. Phillips to Petrone, TWX, 17 Feb. 1966, Phillips chronological files.
25. KSC, Launch Vehicle Operations, "Problems in AS-201 Checkout," 11 Mar. 1966.
26. Donnelly interview, 17 Nov. 1970.
27. Gruene interview, 19 Nov. 1970.
28. KSC, *Apollo/Saturn IB Ground Systems Evaluation Report, AS-201,* Apr. 1966.
29. Melvyn Savage, Apollo Test Dir., to Phillips, Apollo Program Dir., "A/S 201 Hold," 3 Mar. 1966.
30. Gruene interview, 19 Nov. 1970.
31. Savage to Phillips, "A/S 201 Hold," 3 Mar. 1966.
32. *Brevard Sentinel,* 20 Feb. 1966; NASA, *Apollo/Saturn 201 Press Kit,* pp. 6–8; KSC, *AS-201 Ground Systems Evaluation Report,* p. iii.
33. Carlson interview, 16 Dec. 1970.
34. KSC, *AS-201 Ground Systems Evaluation Report,* p. iii; NASA, *Sixteenth Semi-Annual Report to Congress,* 1 July–31 Dec. 1966, p. 58; Weekly Notes, E. P. Bertram to Petrone, 3 Mar. 1966.
35. Debus to KSC Management Board, 17 Jan. 1966; Siepert to Debus, "Approach and Status of KSC Task Force on Management Appraisal," 1 Mar. 1966. The research for this portion on KSC's 1966 reorganization was done by Robert Lindemann and Frank Jarrett.
36. KSC, draft briefing memo, "Proposed KSC Reorganization," n.d., p. 3; NASA announcement, "Approval of Revised KSC Organizational Structure," 29 Apr. 1966.
37. KSC, draft briefing memo, "Proposed KSC Reorganization"; KSC, "Approval of Revised KSC Organizational Structure," KSC release 123-66, 29 Apr. 1966.
38. Akens, *Saturn Illustrated Chronology,* p. 138; *Spaceport News,* 30 June, 7 July 1966; NASA, *Apollo/Saturn 203 Press Kit*, 21 June 1966, pp. 2–3, 18–19; KSC, "Daily Status Report, AS-203," 6–15 Apr. 1966.
39. KSC, "Daily Status Report, AS-203," 19 Apr.–31 May 1966.
40. Guy Thomas to Chief, NASA Requirements Br., 1 June 1966, in Rocco Petrone's notes, 1966.
41. NASA, *Apollo/Saturn, AS-203, Post-Launch Report No. 1,* 22 July 1966; Akens, *Saturn Illustrated Chronology,* p. 144; *Spaceport News,* 7 July 1966.
42. KSC, "Daily Status Reports, AS-202," 28 Feb.–22 Aug. 1966, in particular see 30 Mar., 14, 27 Apr., 22 June, 5, 15, 29 July, 8, 15 Aug.; Sasseen interview, 4 Feb. 1974.
43. *Spaceport News,* 18 Aug. 1966.
44. NASA, *Sixteenth Semiannual Report to Congress, 1 July–31 Dec. 1966,* pp. 47–48; NASA release 66-213, 25 Aug. 1966.

Chapter 18: The Fire That Seared the Spaceport

1. Senate Committee on Aeronautical and Space Sciences, *Report on Apollo 204 Accident,* report 956, 90th Cong., 2d sess., 30 Jan. 1968, pp. 3–7.
2. Idem, *Apollo Accident: Hearings,* 90th Cong., 1st sess., pt. 1, pp. 13–54. Dr. Charles A. Berry, chief of medical programs at MSC, introduced and discussed Dr. E. Roth's four-part report, "The Selection of Space-Cabin Atmosphere."
3. Frank J. Handel, "Gaseous Environments during Space Missions," *Journal of Space Craft and Rockets* 1 (July–Aug. 1964): 361.
4. *Report of Apollo 204 Review Board to the Administrator, NASA,* 5 Apr. 1967, app. D, panel 2, pp. D-2-25, D-2-26.
5. *Science Journal* 2 (Feb. 1966): 83.
6. *Space/Aeronautics* 45 (Feb. 1966): 26, 28, 32.
7. Gen. Samuel Phillips, Apollo Program Dir., to John Leland Atwood, Pres., North American Aviation, "NASA Review Team Report," 19 Dec. 1965.

8. Ibid., p. 1.
9. Ibid., p. 66.
10. Senate Committee on Aeronautical and Space Sciences, *Report on Apollo 204 Accident,* pt. 4, p. 318.
11. House Subcommittee on NASA Oversight of the Committee on Science and Astronautics, *Investigation into Apollo 204 Accident: Hearings,* 90th Cong., 1st sess., 1: 404.
12. Ibid., p. 450.
13. *Report of Apollo 204 Review Board,* p. 4-1.
14. "Daily Status Report, AS-204," 29 Aug. 1966; unless otherwise noted, the material in this section is based on these reports between 29 Aug. 1966 and 26 Jan. 1967.
15. *Report of Apollo 204 Review Board,* p. 4-1.
16. Ibid., pp. 4-1, 4-2.
17. Chauvin interview, 6 June 1974.
18. *Report of Apollo 204 Review Board,* p. 4-2.
19. Chauvin and Reyes interviews, 6–7 June 1974.
20. Ibid.
21. *Report of Apollo 204 Review Board,* p. 4-2.
22. Ibid.
23. Notes by M. Mogilevsky, signed, undated, relative to his conversation with Thomas R. Baron, 12–13 Dec. 1966, in files of Frank Childers, KSC.
24. Statement of Frank Childers, 9 Feb. 1967, submitted at the request of the KSC Director, copy in files of Childers.
25. John H. Brooks, Chief, NASA Regional Inspections Off., to Kurt Debus, "Thomas Ronald Baron, North American Aviation Employee," 3 Feb. 1967.
26. Ibid.
27. Hansel interview.
28. Brooks to Debus, 3 Feb. 1967.
29. *Orlando Sentinel*, 6 Feb. 1967. John Hansel said later than North American had ample reason for firing Baron, because he had violated procedural requirements that brought automatic dismissal. Hansel interview.
30. Brooks to Debus, 3 Feb. 1967.
31. Ibid.
32. Titusville *Star-Advocate,* 7 Feb. 1967.
33. Childers interview.
34. Reyes interview, 19 Jan. 1973.
35. House Subcommittee on NASA Oversight of the Committee on Science and Astronautics, *Investigation into Apollo 204 Accident: Hearings,* 90th Cong., 1st sess., 1: 498 ff.
36. Erlend A. Kennan and Edmund H. Harvey, Jr., *Mission to the Moon* (New York: William Morrow and Co., Inc., 1969), pp. 115–16, 147n. This book is highly critical of NASA and the space program, with special emphasis on the 204 fire.
37. Chauvin and Reyes interviews, 6–7 Jun. 1974.
38. *Report of Apollo 204 Review Board,* app. D, panel 7, p. D-7-12.
39. Ibid., app. B, p. B-142, testimony of Clarence Chauvin.
40. Ibid., p. B-145, testimony of William Schick.
41. *Report of Apollo 204 Review Board,* app. D, panel 7, p. D-7-13.
42. Ibid., pp. D-7-4, D-7-5.
43. Ibid., app. B, pp. B-153, B-154, testimony of Gary W. Propst; p. B-159, testimony of A. R. Caswell.
44. Ibid., p. B-91, testimony of Bruce W. Davis.
45. Ibid., p. B-39, testimony of D. O. Babbitt.
46. *Report of Apollo 204 Review Board,* app. D, panel 11, p. D-11-36. At least one member of the Pan American Fire Department, James A. Burch, testified that he had arrived in

time to help open the hatch—even though he admitted the trip to the gantry took from five to six minutes and ascent on the slow elevator consumed two minutes more. Ibid., app. B, p. B-177.

47. *Time,* 10 Feb. 1967, p. 19.
48. *Newsweek,* 13 Feb. 1967, pp. 96–97.
49. *The Sunday Star,* Washington, 21 May 1967.
50. Quoted in *Today,* 14 Apr. 1967; 14 May 1967.
51. *New York Times,* 4 Apr. 1967.
52. H. Bliss, "NASA's in the Cold, Cold Ground," *ATCHE Journal* 13 (May 1967): 419.
53. Lyndon B. Johnson, *The Vantage Point: Perspectives of the Presidency, 1963–1969* (New York: Popular Library, 1971), p. 284.
54. House Subcommittee on NASA Oversight of the Committee on Science and Astronautics, *Investigation into Apollo 204 Accident: Hearings,* 90th Cong., 1st sess., 1: 207.
55. Announcement of Dr. Kurt H. Debus, 3 Feb. 1967, "KSC Cooperation with the Apollo 204 Investigation."
56. *Time,* 10 Feb. 1967, reported rumors of lengthy suffering that preceded the astronauts' deaths. The autopsy disproved these charges.
57. *Aviation Week and Space Technology,* 13 Feb. 1967, p. 33.
58. *Time,* 14 Apr. 1967.
59. *Report of Apollo 204 Review Board,* 6 Apr. 1967, pp. 5-1, 5-2.
60. Ibid., p. 5-9.
61. Ibid., pp. 6-1, 6-2, 6-3.
62. Atkins interview, 29 May 1974.
63. *Report of Apollo 204 Review Board,* pp. 6-2, 6-3.
64. Ibid., p. 6-3.
65. House Subcommittee on NASA Oversight of the Committee on Science and Astronautics, *Investigation into Apollo 204 Accident: Hearings,* 90th Cong., 1st sess., 1: 81.
66. Atkins interview, 5 Sept. 1973.
67. *Report of Apollo 204 Review Board,* app. B, pp. B-39 through B-146.
68. Senate Committee on Aeronautical and Space Sciences, *Apollo Accident: Hearings,* pts. 1, 2.
69. Ibid., pt. 4, p. 365; House Subcommittee on NASA Oversight of the Committee on Science and Astronautics, *Investigation into Apollo 204 Accident: Hearings,* 90th Cong., 1st sess., 1: 13.
70. Senate Commitee on Aeronautical and Space Sciences, *Apollo Accident: Hearings,* 90th Cong., 1st sess., pt. 6, p. 541.
71. Ibid., pt. 2, p. 127; pt. 5, pp. 416–17; House Subcommittee on NASA Oversight of the Committee on Science and Astronautics, *Investigation into Apollo 204 Accident: Hearings,* 90th Cong., 1st sess., 1: 265.
72. House Subcommittee on NASA Oversight of the Committee on Science and Astronautics, *Investigation into Apollo 204 Accident: Hearings,* 90th Cong., 1st sess., 1: 386–87.
73. Ibid., pp. 390–91.
74. Ibid., p. 391.
75. Ibid., 1: 460–80, 501.
76. Senate Committee on Aeronautical and Space Sciences, *Report on Apollo 204 Accident,* report 956, 90th Cong., 2d sess., p. 7; Senate Committee on Aeronautical and Space Sciences, *Apollo Accident: Hearings,* 90th Cong., 1st sess., pt. 4, p. 319. "Some early tendency to shift blame for the fire upon North American Aviation," Tom Alexander wrote in *Fortune,* July 1969, p. 117, "was gradually supplanted by NASA's admission that the fire was largely its own management's failure. NASA had overlooked and thereby in effect approved an inherent fault in design, namely the locking up of men in a capsule full of inflammable materials in an atmosphere of pure oxygen at sixteen pounds per

square inch of pressure. NASA, after all, had more experience in the design and operation of space hardware than any other organization and was, therefore, more to blame than North American if the hardware worked badly."

In 1972, however, North American Rockwell Corp., North American Aviation, Inc., Rockwell Standard Corp., and Rockwell Standard Co. settled out of court with the widows of the three astronauts who charged the spacecraft builders with negligence. The widows of White and Chaffee each received $150000, the widow of Grissom $300000. *Washington Post,* 11 Nov. 1972.

77. Senate Committee on Aeronautical and Space Sciences, *Apollo Accident: Hearings,* 90th Cong., 1st sess., pt. 5, pp. 397, 428.
78. Senate Committee on Aeronautical and Space Sciences, *Report on Apollo 204 Accident,* report 956, 90th Cong., 1st sess., pp. 11, 20.
79. "New Hatch Slashes Apollo Egress Time," *Aviation Week and Space Technology,* 15 May 1967, p. 26.
80. William J. Normyle, "NASA Details Sweeping Apollo Revisions," *Aviation Week and Space Technology,* 15 May 1967, p. 24.
81. George E. Mueller, "Apollo Actions in Preparation for the Next Manned Flight," *Astronautics and Aeronautics* 5 (Aug. 1967): 28–33; "Records of Spacecraft Testing, July 1968," in files of R. E. Reyes, Preflight Operations Br., KSC.
82. Normyle, "NASA Details," p. 25; Reyes interview, 30 Oct. 1973; Atkins interview, 5 Nov. 1973. Actually the official reports to Debus during 1966 show no written reports from the Safety Office. Atkins must have reported orally at irregular intervals.
83. Mueller, "Apollo Actions," p. 33.
84. House Special Studies Subcommittee of the Committee on Government Operations, *Investigation of the Boeing-TIE Contract: Hearings,* 90th Cong., 2d sess., pp. 3–9.
85. Ibid., pp. 10, 13–14, 24.
86. "Technical Integration and Evaluation Contract," NASw 1650, Statement of Work, 15 June 1967.
87. Wagner interview; "Boeing-TIE Goals and Accomplishments," copy in file of Walter Wagner, KSC.

Chapter 19: Apollo 4: The Trial Run

1. OMSF, *Apollo Program Flight Summary Report, Apollo Missions AS-201 through Apollo 8,* pp. 13–17; MSFC, *Technical Information Summary, AS-501, Apollo Saturn V Flight Vehicle,* R-ASTR-S-67-65, 15 Sept. 1967.
2. "NASA Announces Changes in Saturn Missions," NASA release 63-246, 30 Oct. 1963.
3. Dir., Apollo Program, "Clarification of Apollo Saturn IB and V Flight Mission Designations," 12 Apr. 1965.
4. MSFC, *Technical Information Summary, AS-501,* pp. 24–75.
5. OMSF, *Apollo Program Directive No. 4D,* 1 July 1966; *No. 4E,* 22 Sept. 1966; *No. 4F,* 30 Nov. 1966; Proffitt interview; NASA, *Sixteenth Semiannual Report to Congress, 1 July–31 Dec. 1966,* pp. 49, 51–52. See also NASA, *Seventeenth Semiannual Report to Congress, 1 Jan.–30 June 1967,* p. 11, for information on the spacer.
6. KSC, "LC-39 Site Activation Status Report," 14 Sept. 1966; *Spaceport News,* 15 Sept. 1966; KSC, "Apollo 4 (AS-501) Daily Status Reports," Sept.–6 Oct. 1966.
7. KSC, "Apollo 4 Daily Status Reports," 29 Nov., 1, 2, 6, 7, 13 Dec. 1966.
8. Ibid., Dec. 1966 and Jan. 1967; KSC, "Program Milestone Data—Apollo," 15 July 1971.
9. KSC, "Apollo 4 Daily Status Reports," Feb.–Mar. 1967 (see 16 Mar. for number of wiring discrepancies in spacecraft); NASA, *Seventeenth Semiannual Report to Congress, 1 Jan.–30 June 1967,* pp. 11–12.

10. *Spaceport News,* 3 Mar. 1966; Fowler interview.
11. KSC, "Apollo 4 Daily Status Reports," Feb. 1967; *Spaceport News,* 15 Feb. 1968.
12. KSC, "LC-39 Site Activation Status Report," 19, 26 Apr. 1967.
13. NASA release 67-132, summarized in *Astronautics and Aeronautics, 1967,* p. 164; KSC, "Apollo 4 Daily Status Reports," May–Aug. 1967; *Spaceport News,* 31 Aug. 1967.
14. KSC, *Catalog of Launch Vehicle Tests, Saturn V, Apollo/Saturn V, Revision 1,* 15 June 1966, GP-244.
15. Ibid., p. 1-27; Carlson interview, 5 Sept. 1974.
16. KSC, *Catalog of Launch Vehicle Tests, Saturn V,* p. 2-18.
17. Ibid., pp. 1-18, 1-28.
18. KSC, "Apollo 11 (AS-506) Daily Status Report," 25 Mar. 1969.
19. *Catalog of Launch Vehicle Tests, Saturn V,* pp. 7-1 through 7-13.
20. Ibid., pp. 9-1 through 9-43; KSC, *Apollo/Saturn Program Development/Operations Plan,* 2: 3-90 and 3-93 provide a comparison of the two tests and their objectives.
21. KSC, *Catalog of Launch Vehicle Tests, Saturn V,* pp. 9-1 through 9-43.
22. Ibid., pp. 9-21 and 9-25.
23. Ibid., p. 9-3.
24. Ibid.; Donnelly interview, 19 June 1974.
25. Harris interview; Carlson interview, 5 Sept. 1974.
26. KSC, "Program Milestone Data, Apollo," 6 June 1973.
27. KSC, "Apollo 4 Daily Status Reports," Sept. 1967; Donnelly interview, 19 June 1974.
28. KSC, "Apollo 4 Daily Status Reports," 27 Sept.–13 Oct. 1967; Richard S. Lewis, *Appointment on the Moon: The Inside Story of America's Space Venture* (New York: Viking Press, 1968), p. 406.
29. Donnelly interview, 19 Nov. 1970.
30. NASA release 67-274 and *Baltimore Sun,* 26 Oct. 1967, p. A6, summarized in *Astronautics and Aeronautics, 1967,* p. 319.
31. Phillips to Mueller, memo for record, 5 Sept. 1968; KSC, "Apollo 4 Daily Status Reports," 19–20 Oct. 1967.
32. *Spaceport News,* 23 Nov. 1967.
33. Ibid.
34. Ibid.
35. *KSC Information and Protocol Operations Plan—Apollo 4 Mission,* pp. 1–4.
36. Ibid.
37. J. E. Ballou, GAO Area Audit Mgr., Atlanta, to L. Melton, KSC, 30 Mar. 1965; *Aviation Week* 89 (23 Sept. 1968): 74; H. L. DeLung, Acting Regional Mgr., GAO, Atlanta, to Debus, 20 Aug. 1965. The authors recognize their reliance on thorough researches of Maj. James J. Frangie in this section.
38. Debus to DeLung, 15 Oct. 1965.
39. R. J. Madison, Mgr., GAO Regional Off., Atlanta, to Debus, 6 July 1966. See also p. 16 of draft report, Comptroller General of the U.S., "Review of Launch Complex 39 Facilities for the Saturn V Vehicle, John F. Kennedy Space Center, Florida, NASA," undated, accompanying letter of Clerio P. Pin, Assoc. Dir., GAO, to Webb, 8 June 1967 (hereafter cited as "GAO Draft Report, KSC, 1967"); Mueller to William Parker, Asst. Dir., Civil Accounting and Auditing Div., GAO, Washington, 16 Aug. 1966.
40. Melton, memo for record, 29 Aug. 1966; Malcolm S. Stringer, memo for record, "GAO Review of Justification for Redundant and Duplicate Launch Complex 39 Facilities, 31 Oct. 1967."
41. Raymond Einhorn, Dir. of Audits, to Debus, "GAO Draft Report on Review of Launch Complex 39 Facilities for the Saturn V Vehicle, KSC, NASA," 13 June 1967; "GAO Draft Report, KSC, 1967," p. 24.
42. Debus to Mueller, 19 July 1967, in NASA Hq. History Office; Einhorn to Debus, 15 Sept. 1967.

43. Raymond Middleton to KSC Liaison Representative with GAO, 30 Oct. 1967; Melton, memo for record, 19 Oct. 1967; Debus to Mueller, 2 Nov. 1967, in NASA Hq. History Office; Harold B. Finger, Assoc. Admin. for Organization and Management, to Clerio P. Pin, Assoc. Dir., GAO, Washington, 24 Jan. 1968, with enclosures, pp. 10–12, NASA Hq. History Office.
44. Pin to Finger, 27 Mar. 1968, in NASA Hq. History Office.
45. Stringer interview; Finger to Pin, 24 Jan. 1968, NASA Hq. History Office.
46. House Committee on Science and Astronautics, *1964 NASA Authorization,* 88th Cong., 1st sess., 1963, pt. 1, p. 5; pt. 2, pp. 126–27.

Chapter 20: Man on Apollo

1. NASA, OMSF, *Apollo Program Directive No. 4H,* 3 Nov. 1967.
2. KSC, "Apollo 5 Daily Status Reports," 3 Mar.–14 Apr. 1967.
3. KSC release 1-68, 3 Jan. 1968; Widick interview.
4. "Apollo 5 Daily Status Reports," 23 June, 14–21 Aug., 19 Nov., 22 Dec. 1967, 19 Jan. 1968.
5. Statement of Rocco Petrone and Gen. S. Phillips, "Apollo 5 Post-Launch Press Conference," 22 Jan. 1968; NASA, OMSF, *Apollo Program Flight Summary Report, Apollo Missions AS-201 through Apollo 8,* Jan. 1969, pp. 20–22.
6. "LC-39 Site Activation Status Report," 5, 26 Apr. 1967; KSC, "Apollo 6 (AS-502) Daily Status Report," 22 Mar. 1967.
7. "Apollo 6 Daily Status Report," Mar. 1967.
8. Ibid., May–June 1967.
9. Ibid., July–Aug. 1967.
10. Ibid., 12 Sept.–Oct. 1967.
11. Ibid., Dec. 1967.
12. Ibid., 28 Dec. 1967–11 Jan. 1968.
13. Ibid., 15–16, 30 Jan., 6–9 Feb. 1968; *Spaceport News,* 15 Feb. 1968.
14. Phillips to Mueller, 5 Sept. 1968; "Apollo 6 Daily Status Report," 8–21 Mar. 1968.
15. "Apollo 6 Daily Status Report," 25 Mar. 1968; Phillips to Mueller, 5 Sept. 1968; KSC, *Apollo/Saturn V Ground Systems Evaluation Report, AS-502,* KSC document 140-44-0010, pp. 2-2, 4-1; NASA, "Apollo 6 Pre-Launch Press Conference," Cocoa Beach, 3 Apr. 1968, pp. 3–4, 7–10.
16. House Committee on Government Operations, *Hearing: Investigation of the Boeing-TIE Contract,* 90th Congress, 2d sess., 15 July 1968, p. 10.
17. NASA Hq., "Apollo 6 SLA Problem and Resolution," 17 Dec. 1968; NASA, *Nineteenth Semiannual Report to Congress, 1 Jan.–30 June 1968,* p. 19.
18. NASA, *Nineteenth Semiannual Report,* pp. 8–18.
19. NASA, "Apollo 6 Post-Launch Press Conference," LC-39 Press Site, 4 Apr. 1968, pp. 3–5.
20. *Spaceport News,* 11 Apr. 1968.
21. Erlend A. Kennan and Edmund H. Harvey, Jr., *Mission to the Moon,* pp. 284–85.
22. NASA, *Astronautics and Aeronautics, 1968,* pp. 83, 119–20; NASA, *Nineteenth Semiannual Report,* p. 19.
23. Phillips to Mueller, 5 Sept. 1968; Erich E. Goerner, "LOX Prevalve to Prevent POGO Effect on Saturn V," *Space/Aeronautics,* Dec. 1968, p. 72; House Committee on Science and Astronautics, *Hearings on 1970 NASA Authorization,* 91st Cong., 1st sess., pt. 2, pp. 27–29.
24. MSC, *Apollo Spacecraft Program Quarterly Status Report,* no. 25, 30 Sept. 1968, pp. 1–9.

25. *Spaceport News,* 29 Feb. 1968.
26. Robert Sherrod, "The Selling of the Astronauts," *Columbia Journalism Review,* May–June 1973, pp. 17–25.
27. Ibid., pp. 16–17. When former astronaut John Glenn entered the Ohio senatorial primary in the spring of 1964, news broke that he and the other original astronauts had financial interests in Cape Colony Inn in Cocoa Beach. Profitability of the Inn was obviously related to the space program.
28. One of the astronauts was so fearful of heights that he hesitated to cross the catwalk at the 31st floor of the VAB, so the ground crew covered the grating on the swing arm with boards whenever he crossed to the spacecraft.
29. *Spaceport News,* 4 Oct. 1968.
30. Wendt interview.
31. Neil Armstrong et al., *First on the Moon* (Boston: Little, Brown and Co., 1970), nondocumented, interesting account of Apollo 11, previous Apollos, and the astronauts and their families, based on interviews.
32. *Spaceport News,* 28 Mar. 1968.
33. Ibid.
34. *Kennedy Space Center Story* (1971 ed.), pp. 227–28.
35. NASA, OMSF, "Apollo Program Directive 4H," 3 Nov. 1967; KSC, "Apollo 7 (AS-205) Daily Status Reports," 11 May–3 June 1968.
36. "Apollo 7 Daily Status Reports," June–July 1968; KSC, "Minutes of Apollo Launch Operations Committee (ALOC) Meetings," 13 June, 11 July, 1 Aug. 1968; *Spaceport News,* 1 Aug. 1968.
37. "Minutes of ALOC Meetings," 1, 15, 29 Aug. 1968; *Spaceport News,* 29 Aug., 12 Sept. 1968; Ragusa interview.
38. NASA, OMSF, *Mission Operation Report: Apollo 7 (AS-205) Mission,* 30 Sept. 1968.
39. Ibid.
40. Wernher von Braun and Frederick L. Ordway III, *History of Rocketry and Space Travel* (New York: Thomas V. Crowell and Co., 1969), pp. 226–27.
41. *Spaceport News,* 24 Oct. 1968.
42. Ibid.
43. KSC release 22-68, 29 Jan. 1968; NASA, *Nineteenth Semiannual Report to Congress,* 19 Jan.–30 June 1968, p. 13; Roderick O. Middleton, KSC Apollo Program Mgr., to Samuel C. Phillips, TWX, 7 Mar. 1968; MSFC, *Saturn V Launch Vehicle Flight Evaluation Report—AS-503*, p. 3-1.
44. Harold B. Finger to Mueller, 1 May 1968.
45. Ibid.
46. MSFC, *Saturn V Launch Vehicle Flight Evaluation Report—AS-503,* pp. 3-1, 3-2; KSC, "Apollo 8 Daily Status Reports," 29 Apr.–6 May 1968; unsigned document on working schedule for manned (CSM-103) and unmanned (BP-30) AS-503 missions, 21 Apr. 1968; Phillips to Debus, TWX, 29 Apr. 1968.
47. KSC, "Apollo 8 Daily Status Reports," 8, 10, 17, 31 May 1968.
48. Ibid., 14, 17 June 1968.
49. Proffitt interview, 1 Dec. 1970.
50. NASA, *Astronautics and Aeronautics, 1968,* p. 191.
51. George M. Low to C. H. Bolender and K. S. Kleinknecht, "Chuck Mathews Review of KSC Activities," 14 Sept. 1968, Apollo discussion papers, JSC Historical Archives.
52. Debus to Mueller, 16 July 1968.
53. Phillips to Debus, 24 Aug. 1968.
54. KSC, "Apollo 8 Daily Status Reports," 10 June–22 July 1968.
55. Ibid., 22 July–4 Aug. 1968; George M. Low, memo for record, 19 Aug. 1968, in Apollo discussion papers, JSC Historical Archives.

56. KSC, "Apollo 8 Daily Status Reports," 28 June–25 July 1968.
57. Ibid., 12–16 Aug. 1968; Low, memo for record, 19 Aug. 1968; "Transcript of News Conference on Apollo Program Changes," 19 Aug. 1968.
58. Phillips to Debus, 10, 20 Aug. 1968.
59. KSC, *Apollo/Saturn V Launch Operations Test and Checkout Requirements, AS-503,* document K-V-051-01/3, p. 6-1.
60. KSC, "Apollo 8 Daily Status Report," 16 Sept. 1968.
61. NASA, "Transcript of Saturn AS-503 Delta Design Certification Review," 19 Sept. 1968; KSC, "Apollo 8 Daily Status Reports," 8–10 Oct. 1968.
62. NASA, *Astronautics and Aeronautics, 1968,* p. 266.
63. Ibid., p. 278.
64. KSC, "Apollo 8 Daily Status Reports," 15, 19 Nov. 1968; KSC, *Apollo/Saturn V Launch Mission Rules, Apollo 8 (AS-503/CSM 103),* document K-V-05.10/3, pp. 1-2 through 1-29.
65. KSC, "Apollo 8 Daily Status Reports," 5–11 Dec. 1968.
66. NASA, *Astronautics and Aeronautics, 1968,* p. 313.
67. John N. Wilford, "Final Countdown On for Moon Shot Tomorrow," *New York Times,* 20 Dec. 1968.
68. "Apollo 8 Onboard Voice Transcription, As Recorded on the Spacecraft Onboard Recorder (Data Storage Equipment)," MSC, Jan. 1969, tape 58-4.
69. *Kennedy Space Center Story* (1971 ed.), pp. 101–102.
70. *Boeing Atlantic Test Center News* 6 (13 Jan. 1969): 1.
71. MSC, *Apollo 8 Technical Debriefing,* 2 Jan. 1969, p. 139.
72. *Spaceport News,* 16 Jan. 1969; Reyes interview and Chauvin interview, June 1973.
73. "Briefing by KSC-NASA for the Congressional Subcommittee on Manned Space Flight," 28 Feb. 1969, pp. 4–5.

Chapter 21: Success

1. MSC, *Apollo Spacecraft Program Quarterly Status Report No. 25*, 30 Sept. 1968, p. 28.
2. NASA, *Current News*, 20 December 1968, p. 1; NASA, *Astronautics and Aeronautics, 1969*, p. 16.
3. NASA, *Astronautics and Aeronautics, 1967*, pp. 330–31; Phillips to Debus, 19 Aug. 1968.
4. KSC, "Apollo 9 (AS-504) Daily Status Reports," May–Oct. 1968.
5. Ibid., 8–26 Nov. 1968.
6. Ibid., 2–31 Dec. 1968.
7. Titusville *Star-Advocate*, 16 Dec. 1968.
8. Renaud interview, 16 May 1973.
9. Titusville *Star-Advocate*, 16 Dec. 1968.
10. Ibid.
11. Ibid.
12. Ibid.
13. KSC, "Apollo 9 (AS-504) Daily Status Reports," 3–9 Jan. 1969.
14. Ibid., 10–11 Feb. 1969; KSC release 33–69, 4 Feb. 1969.
15. Ibid., 12, 20 Feb. 1969; KSC, *Apollo/Saturn V Test and Checkout Plan, AS-504 and All Subsequent Missions*, pp. 4-9 through 4-12; Widick interview, 15 Dec. 1970; KSC, *Apollo/Saturn V Space Vehicle Countdown Demonstration Test (Apollo 9)*, p. vi.
16. KSC, "Apollo 9 Daily Status Report," 27–28 Feb. 1969; NASA, "Apollo 9 Postponement News Conference," 27 Feb. 1969, CST 12:05, pp. 9A/1 through 9C/2; KSC, *Apollo/Saturn V Space Vehicle Countdown (Apollo 9)*, pp. v–vi.

17. KSC, *Apollo/Saturn V Ground Systems Evaluation Report, Apollo 9*, p. 2-1.
18. KSC, "Briefing for the Subcommittee on Manned Space Flight, Committee on Science and Astronautics, House of Representatives," 28 Feb. 1969, pp. 44–70.
19. Youmans interview, 5 Feb. 1971; Proffitt interview, 1 Dec. 1970; George Low to C. H. Bolender and K. S. Kleinknecht, "Chuck Mathews Review of KSC Activities," 14 Sept. 1968, JSC Archives, Apollo activity file.
20. KSC, *Apollo/Saturn V Launch Operations Test and Checkout Requirements, AS-504 and All Subsequent Missions*, document K-V-051-01, p. 1-1; Proffitt interview, 1 Dec. 1970.
21. Chart included in folder with John M. Marshall's interview with Henry C. Paul at KSC, 9 Dec. 1970, in KSC Historian's Office, illustrates graphically the growth of automation overall and of Atoll in particular for the period both before and subsequent to AS-504.
22. R. B. Johansen, "Developments in On-Board and Ground Checkout Systems," American Institute of Aeronautics and Astronautics, Cocoa Beach, FL, 2–4 Feb. 1970, AIAA paper 70-245, pp. 3–5; James E. Rorex and Robert P. Eichelberger, "Digital Data Acquisition System in Saturn V," in *Proceedings of the Second Space Congress*, 5–7 Apr. 1965, Cocoa Beach, FL, sponsored by Canaveral Council of Technical Societies, pp. 632–49; Debus, "Launching the Moon Rocket," p. 25.
23. W. V. George and C. A. Stinson, "An Automated Telemetry Checkout Station for the Saturn V Systems," *NTC/66: Proceedings, National Telemetering Conference*, Boston, 10–12 May 1966, p. 117.
24. NASA *Proceedings of the Apollo Unified S-Band Technical Conference*, p. 248; Edmund F. O'Conner, "Launch Vehicles for the Apollo Program," pp. 165–66; Walyer O. Frost, "SS-FM: A Telemetry Technique for Wide-Band Data," Institute of Radio Engineers, *Transactions on Space Electronics and Telemetry*, SET-2 (Dec. 1962), p. 289, notes the first use of SS-FM telemetry on the SA-2 flight.
25. NASA, *Proceedings of the Apollo Unified S-Band Technical Conference*, p. 248.
26. Samuel C. Phillips to MSFC, 28 June 1965; Adolf H. Knothe, "Range Safety—Do We Need It?" American Institute of Aeronautics and Astronautics, Launch Operations Meeting, Cocoa Beach, FL, Feb. 1970, p. 3.
27. "Tracking and Data Acquisition," *Spaceflight* 11 (June 1969): 190; NASA, *Proceedings of the Apollo Unified S-Band Technical Conference*, p. 3.
28. "How We Will Communicate with Astronauts on the Moon," *Space World*, Jan. 1969, pp. 33, 35.
29. "Tracking and Data Acquisition," p. 190; NASA, *Sixteenth Semiannual Report to Congress*, 1 July–31 Dec. 1966, pp. 167–68.
30. Frank Leary, "Support Net for Manned Space Flight," *Space/Aeronautics*, Dec. 1966, pp. 71–72. KSC, "Apollo/Saturn V Launch Operations Plan," AS-501/502, pp. 7-18 through 7-20 contains a general description of the LIEF and ALDS systems and their relationship to each other and to operations at KSC.
31. Edmond C. Buckley to Mueller, 26 Aug. 1964.
32. *Space Daily*, 28 Jan. 1966, p. 177.
33. John F. Mason, "Modernizing the Missile Range: Part 1," *Electronics*, Feb. 1965, pp. 94–95.
34. This conclusion is derived from the chart shown in NASA, *Proceedings of the Apollo Unified S-Band Technical Conference*, p. 296; NASA, *Sixteenth Semiannual Report to Congress, 1 July–31 Dec. 1966*, p. 165.
35. Twigg interview.
36. KSC, "Apollo 11 (AS-506) Daily Status Reports," Jan.–Apr. 1969; KSC, *Kennedy Space Center Story*, 1971 ed., p. 119.
37. NASA, *Apollo Flight Summary Report*, pp. 82–83.
38. KSC, "Apollo Program Milestone Data," 15 July 1973; KSC, "Apollo 11 Daily Status Reports," 20 May–4 July 1969.

39. KSC, *Kennedy Space Center Story*, 1971, pp. 121–25; *Spaceport News*, 23 July 1969.
40. *Kennedy Space Center Story*, 1971, pp. 222–23.
41. Ibid., p. 124; KSC, "Apollo Countdown Document, C-07."
42. *Spaceport News*, 23 July 1969.
43. Ibid.
44. Ibid., 30 July 1969.
45. Anne M. Lindbergh, *Earthshine* (Harcourt Brace Jovanovich: New York, 1969), pp. 42–43.

Chapter 22: A Slower Pace: Apollo 12–14

1. KSC, "AS-507 Daily Status Reports."
2. NASA, *Apollo 12 Press Kit*, pp. 1–2; Lt. Gen. Samuel Phillips, transcript of news conference at MSC, 24 July 1969, summarized in *Astronautics and Aeronautics, 1969*, p. 243; KSC, "AS-507 Daily Status Reports," June–July 1969.
3. *Armed Forces Journal*, 27 Sept. 1969, p. 8; NASA releases 69-151, 10 Nov. 1969; 70-4, 8 Jan. 1970; *Spaceport News*, 28 Aug., 11 Sept., 4 Dec. 1969.
4. *Spaceport News*, 28 Aug., 11 Sept. 1969, 18 June 1970.
5. KSC, "AS-507 Daily Status Reports"; Chauvin interview, 2 Apr. 1974; *Washington Post*, 30 Oct. 1969, p. A8.
6. KSC, "AS-507 Daily Status Report," 13 Nov. 1969; *Washington Post*, 13 Nov. 1969, p. A1; *Spaceport News*, 20 Nov. 1969; Sieck interview; KSC, "Apollo 12 (AS-507) Quick Look Assessment Report," 26 Nov. 1969; NASA, "Pre-Launch Press Conference," KSC and MSC, 13 Nov. 1969, pp. 6B-3, 6B-4.
7. NASA, *Apollo 12 Press Kit*, pp. 43–45; Widick interview, 23 May 1974.
8. "Launch VIP List Headed by Nixon," *Orlando Sentinel*, 13 Nov. 1969; NASA, *Analysis of Apollo 12 Lightning Incident*, MSC-01540, Feb. 1970, p. 12; Manned Spacecraft Center, "Apollo 12 Technical Crew Debriefing," 1 Dec. 1969, p. 2-1.
9. MSC, "Apollo 12 Technical Crew Debriefing," 1 Dec. 1969, p. 2-1.
10. NASA, "Apollo 12 Post Launch Briefing," 14 Nov. 1969, at KSC, p. 8A/2.
11. Ibid.; Kapryan interview.
12. MSC, "Apollo 12 Technical Crew Debriefing," p. 3-2.
13. NASA, "Apollo 12 Mission Commentary," p. 15/1.
14. NASA, "Apollo 12 Post Launch Briefing," pp. 8A-3, 8D-2.
15. Richard M. Nixon, "Remarks to NASA Personnel at the Kennedy Space Center," 14 Nov. 1969, *Public Papers of the Presidents, 1969* (Washington, 1971), p. 936.
16. NASA, "Apollo 12 Mission Commentary," summarized in *Astronautics and Aeronautics, 1969*, pp. 372–78.
17. *Spaceport News*, 1 Jan. 1970.
18. Ibid.
19. NASA, *Analysis of Apollo 12 Lightning Incident*; for one contribution from the scientific community see app. B, M. Brook, C. R. Holmes, and C. B. Moore, "Exploration of Some Hazards to Naval Equipment and Operations beneath Electrified Clouds."
20. KSC, *Launch Mission Rules Apollo 13 (SA-508/CSM 109.LM-7)*, 17 Feb. 1970, p. 1-17.
21. NASA, *Analysis of Apollo 12 Lightning Incident*, pp. 29, 36.
22. KSC, "Apollo 13 (AS-508) Daily Status Report," July 1969–Mar. 1970; KSC, "Proceedings of Manned Space Flight Subcommittee Hearings at Kennedy Space Center, 10 Apr. 1970," pp. 20–26; Moser interview, 17 Apr. 1974.
23. KSC Board of Investigation, "Investigation of Circumstances Surrounding Incident Resulting in Destruction by Fire of Three Motor Vehicles in Vicinity of Perimeter Fence on Pad A of LC-39 on 3/25/70," see part II, 2A and 2B, for narrative of events and committee recommendations; "Transcript of Proceedings of Manned Space Flight Subcommittee at KSC, 10 Apr. 1970," pp. 31–32; Corn interview, 22 Apr. 1974.

24. KSC, "Apollo 13 (AS-508) Daily Status Report," 27, 30, 31 Mar. 1970; NASA, *Report of the Apollo 13 Review Board*, 15 June 1970, pp. 4-21 through 4-23; Lamberth interview.
25. NASA, "Apollo 13 Status Report," A13-1, 9:30 a.m., 5 Apr. 1970; NASA, "Apollo 13 Change of Shift Briefing," 13 Apr. 1970, 2:30 p.m., p. 20A/1; KSC, "Apollo 13 (AS-508) Daily Status Report," 6 Apr. 1970; Lamberth interview; KSC Weekly Report, Kapryan, 2 Apr. 1970.
26. KSC releases 41-70, 6 Mar.; 43-70, 11 Mar.; 153-70, 23 Mar. 1970; *Spaceport News*, 12, 26 Mar., 23 Apr. 1970; KSC Weekly Report, Kapryan, 9 Apr. 1970.
27. KSC, Apollo News Center, "Apollo 13 Status Reports," 1–4, 6–7, 9–10 Apr. 1970; NASA, "Apollo 13 Medical Status Briefing #1," 6 Apr. 1970, 6:50 p.m., p. 8c/1; NASA, "Apollo 13 Medical Status Briefing #2," 8 Apr. 1970, 6:46 p.m.
28. *Baltimore Sun*, 10 Apr. 1970. The $800 000 represented overtime pay for workers at KSC and the cost of the recovery force for the Pacific Ocean splashdown.
29. MSC, "Apollo 13 Prelaunch Press Conference," at KSC, 10 Apr. 1970, 2:10 p.m., pp. 12B/1–12B/4.
30. McCafferty interview, 28 Jan. 1971.
31. MSC, "Apollo 13 Prelaunch Press Conference," pp. 12A/2 and 12B/2.
32. KSC, "Apollo 13 (AS-508) Post Launch Report," 24 Apr. 1970; KSC, "Apollo 13 (AS-508) Flight Summary."
33. MSC, "Apollo 13 Mission Commentary," 13 Apr. 1970, CST 8:34 p.m. GET 55:11, pp. 165/1 through 168/1.
34. Ibid., p. 196/1.
35. NASA, *Report of Apollo 13 Review Board,* pp. 4-25 through 4-46.
36. Ibid., pp. 4-46 through 4-48; NASA, *Current News*, 11, 17 Apr. 1970; McCafferty interview, 28 Jan. 1971; KSC, *Kennedy Space Center Story*, 1971, pp. 152–54.
37. *Washington Post*, 17 Apr. 1970.
38. NASA, *Report of Apollo 13 Review Board*, p. 5-1.
39. Ibid., pp. 4-17 through 4-23, 5-1 through 5-9.
40. Ibid., preface.
41. NASA, "Transcript of Press Conference at KSC on 9 Nov. 1970," quoted in *Astronautics and Aeronautics, 1970*, p. 364; *New York Times*, 10 Nov. 1970, p. 33; for changes in Apollo 14 launch dates see Apollo Program Directives 4K, 4M, 4N; also *Astronautics and Aeronautics, 1970*, pp. 7, 205, 218.
42. KSC, "Apollo 14 (AS-509) Daily Status Reports."
43. NASA, *Apollo 14 Press Kit*, 8 Jan. 1971, p. 93; Humphrey interview.
44. KSC, "Apollo 14 (AS-509) Daily Status Reports"; OMSF, "Apollo Program Weekly Status Reports," June–Aug. 1970.
45. KSC, "Apollo 14 (AS-509) Daily Status Reports"; KSC, "Apollo 14 Post Launch Report," 16 Feb. 1971.
46. *Spaceport News*, 3 Dec. 1970; *Washington Post*, 12 Jan. 1971; *Houston Chronicle*, 18 Jan. 1971.
47. McCafferty interview, 28 Jan. 1971.
48. KSC, "Apollo 14 Post Launch Report"; *Spaceport News*, 11 Feb. 1971.
49. *New York Times*, 15 Nov. 1969, quoted in *Astronautics and Aeronautics, 1969*, p. 380.
50. NASA, *Astronautics and Aeronautics, 1967*, pp. 17, 337; *1968*, pp. 19, 241; *NASA Historical Pocket Statistics*, Jan. 1974, pp. D-2 through D-7.
51. OMSF, "Apollo Program Directive 4-K, Subject: Apollo Program Schedule and Hardware Planning Guidelines and Requirements," 10 July 1969, pp. 5-7 and 10-11.
52. Lee A. DuBridge, testimony on NASA FY 70 Authorization before the Senate Committee on Aeronautical and Space Sciences, 9 May 1969, quoted in *Astronautics and Aeronautics, 1969*, p. 134.
53. *New York Times*, 10 Aug. 1969, p. 44; see *Lunar Exploration: Strategy for Research 1969-1975*, published by the National Academy of Sciences, National Research Council,

Space Science Board, for further evidence of attitudes in the scientific community.

54. *Washington Post*, 5 Oct. 1969.
55. NASA, *America's Next Decades in Space, A Report for the Space Task Group*, Sept. 1969.
56. NASA, "Apollo Program Directive 4-M," 16 Mar. 1970, with cover sheet from Mathews to Debus, 6 Apr. 1970; George Low, Dep. Admin., NASA, "Fiscal Year 1971 Budget Briefing for Community Leaders," KSC, 2 Feb. 1970; transcript of NASA news conference, Washington, D.C., 2 Sept. 1970, summarized in *Astronautics and Aeronautics, 1970*, pp. 284–85.
57. Debus, "Briefing for Community Leaders on FY-1970 Budget at the Kennedy Space Center," 30 Apr. 1960; Miles Ross, Dep. Dir., KSC, "Briefing at Breakfast Meeting of Brevard County Chamber of Commerce," 25 Sept. 1973; Kaufman interview.
58. *New York Times*, 26 Oct. 1969, p. F15; *Orlando Sentinel*, 13 Nov. 1969, p. 2A; *Miami Herald*, 11 Oct. 1970, p. H14; Charles Johnson interview.
59. *Philadelphia Evening Bulletin*, 17 Nov. 1970.
60. Kapryan discussed the question of morale at the Apollo 12 prelaunch briefing, 13 Nov. 1969; see the minutes, p. PC/6E/2.

Chapter 23: Extended Lunar Exploration: Apollo 15–17

1. NASA, *Apollo 15 Press Kit*, 15 July 1971, pp. 1-8; *Astronautics and Aeronautics, 1970*, pp. 284–85; *Time*, 9 Aug. 1971, pp. 10–15.
2. NASA, *Apollo 15 Press Kit*, pp. 5, 60–69.
3. Ibid., pp. 94–100, 134–44; Petrone, Apollo Program Dir., to Manned Space Flight Centers, "6/12/70 Weight and Performance Review Agreements and Actions," 7 July 1970.
4. NASA, *Apollo 15 Press Kit*, p. 133.
5. KSC, "Apollo 15 Post Launch Report," 12 Aug. 1971, pp. 1-1 and 1-2; OMSF, "Apollo Program Weekly Status Reports," June–Sept. 1970; KSC, "Apollo 15 (AS-510) Daily Status Reports," May 1970–Jan. 1971.
6. KSC, "Apollo 15 Daily Status Reports," Jan.–May 1971; Jackie Smith interview.
7. Chauvin interview, 23 May 1974.
8. NASA, *Apollo 15 Press Kit*, pp. 61–69; KSC, "Apollo 15 Daily Status Reports," Jan.–May 1971; KSC, "Apollo 15 Post Launch Report," pp. 1–2; Edwin Johnson interview.
9. NASA, *Apollo 15 Press Kit*, pp. 78–82.
10. KSC release 41-71, "LRV Flight Model Delivery," 10 Mar. 1971; NASA, *Apollo 15 Press Kit*, 15 July 1971, pp. 77–97; *Time*, 9 Aug. 1971; Arthur Scholz to Benson, 18 Oct. 1974. The authors found several different costs cited for the rover. The $12.9 million price is *Time*'s figure and reflects the total project cost of $38+ million divided by three flight vehicle rovers. The cost per vehicle drops if the training rover is included, or if the $13 million R&D costs are excluded. Goldsmith interview.
11. NASA, *Apollo 15 Press Kit*, pp. 77–97; Widick interview, 23 May 1974; Reyes interview, 6 June 1974; Carothers interview; Scholz interview; "Apollo 15 Daily Status Reports," 15 Mar.–25 Apr. 1971.
12. *Washington Post*, 5 May 1971; *New York Times*, 5 May 1971; *Spaceport News*, 6 May 1971.
13. KSC, "Apollo 15 Post Launch Report," p. 1-2; KSC, "Apollo 15 Daily Status Reports," Feb.–Mar. 1971; Collner interview.
14. KSC, "Apollo 15 Daily Status Reports," 29 Mar. 1971; Cochran interview.
15. KSC, "Apollo 15 Daily Status Reports," 29 Mar.–8 Apr. 1971; Lang interview.
16. *Houston Post*, 3 Apr. 1971; *Los Angeles Times*, 3 Apr. 1971.

17. KSC, "Apollo 15 Post Launch Report," pp. 1-2, 1-3; KSC, "Apollo 15 Daily Status Reports," June–July 1971.
18. NASA, "Apollo 15 Mission Commentary," 26 July 1971, 4:11 GET, 12:43 CDT, p. 41/2.
19. KSC, "Apollo 16 (AS-511) Daily Status Reports," Aug.–Nov. 1971; KSC, "Apollo 16 (AS-511) Post Launch Report," 2 May 1972.
20. KSC, "Apollo 16 Post Launch Report"; Crawford interview; Hangartner interview.
21. Ely interview; KSC, S-IC Flight Control, "Test Problem Report," 3 Dec. 1971; KSC, "Apollo 16 Post Launch Report."
22. *New York Times*, 8, 10 Jan. 1972; KSC, "Apollo 16 Post Launch Report."
23. Moxley interview.
24. *New York Times*, 28, 31 Jan. 1972; *Spaceport News*, 27 Jan. 1972; KSC, "Apollo 16 Daily Status Reports," Jan.–Feb. 1972; KSC, "Apollo 16 Post Launch Report."
25. Moxley interview; KSC, "Apollo 16 Daily Status Reports," Feb. 1972.
26. NASA, *Apollo 16 Press Kit*, 22 Mar. 1972.
27. Montgomery interview; McKnight interview.
28. KSC, "Apollo 16 Post Launch Report"; *Spaceport News*, 9 Mar. 1974.
29. *Spaceport News*, 23 Mar. 1972; KSC release 62-72, "Apollo 16 Crewmen Outline Lunar Mission for Spaceport Launch Team," 16 Mar. 1972.
30. KSC, "Apollo 16 Daily Status Reports"; KSC, "Apollo 16 Post Launch Report."
31. *Miami Herald*, 3 Oct. 1972, p. B1; *Baltimore Sun*, 21 Nov. 1972; *Spaceport News*, 21 Sept.–14 Dec. 1972; *Time*, 18 Dec. 1972.
32. KSC, "Apollo 17 Post Launch Report," 19 Dec. 1972, pp. 5-1, 5-2.
33. NASA, *Apollo 17 Press Kit*, 14 Nov. 1972, pp. 56–61.
34. Jackie Smith interview.
35. KSC, "Apollo 17 Post Launch Report," pp. 5-2 to 5-4; *Miami Herald*, 29 Aug. 1972, p. 1.
36. *Lawrence* (Kansas) *Daily Journal-World*, 30 Sept. 1972; *Huntsville Times*, 1 Oct. 1972; NASA, *Astronautics and Aeronautics, 1972*, p. 330; KSC, "Apollo 17 Daily Status Reports," 29 Sept. 1972.
37. *Wall Street Journal*, 13 Nov. 1972, p. 12.
38. *Los Angeles Times*, 27 Nov. 1972; *Spaceport News*, 16 Nov. 1972.
39. NASA, "Apollo 17 Commentary"; KSC, "Apollo 17 Post Launch Report," 19 Dec. 1972.
40. NASA, "Apollo 17 Commentary."
41. *Spaceport News*, 14 Dec. 1972.
42. NASA, "Apollo 17 Commentary"; NASA, "Apollo 17 Post Launch Press Conference"; NASA Apollo Program Dir., "Apollo 17 Mission (AS-512) Post Mission Operation Report No. 1," 19 Dec. 1972, pp. 3–4; *Spaceport News*, 14 Dec. 1972.
43. NASA, "Apollo 17 Flight Summary."

Bibliography

While a great number of books have been written about the Apollo program, there has been no previous history of the launch facilities and operations. *Liftoff*, by L. B. Taylor, Jr. (see *Books*, below) provides a lively journalistic account of the spaceport in the 1960s. Unfortunately, the book ends before Apollo reached its goal. William R. Shelton's *Countdown: The Story of Cape Canaveral* is an entertaining eyewitness account of launch operations at the Cape during the 1950s. Gordon Harris's *Selling Uncle Sam* recalls Apollo events as seen from the Office of Public Affairs at KSC. Michael Collins's *Carrying the Fire* contains an astronaut's views of the Gemini and Apollo programs. The Apollo 11 astronaut set himself a high goal—writing a book without a dull or confusing passage—and then accomplished it. His treatment of technical problems is to be envied. While there is no balanced account of the AS-204 fire, the near tragedy of Apollo 13 is well covered in Henry Cooper's *Thirteen: The Flight That Failed*. A good general account of the Apollo program is John Noble Wilford's *We Reach the Moon*.

The reader will find a wealth of detailed information about Apollo launch facilities and operations in journal articles and conference papers. The popular aerospace magazines (*Missiles and Rockets, Aviation Week and Space Technology*, *Space/Aeronautics*, and *Astronautics and Aeronautics*) trace the progress of the Apollo program. Numerous scientific and engineering journals contain articles by members of the launch team. An even better source for technical exposition is the papers prepared for conferences such as the annual Space Congress held in Cocoa Beach, Florida. The proceedings for most of these conferences were printed.

NASA's dealings with Congress are revealed in thousands of pages of briefings, testimony, and hearings. The agency's *Semiannual Report to Congress* (1958–1969) provides a detailed account of the progress toward a manned lunar landing. At the annual budget hearings, top NASA officials made similar statements and answered numerous questions about specific activities. Special committee hearings at KSC regarding launch operations appear as appendixes in the annual hearings or as special congressional reports.

The authors relied on *Astronautics and Aeronautics* as a basic guide to aerospace events of the 1960s. NASA's History Office has compiled these annual chronologies since 1961; the first two years the work appeared as a report to the House Committee on Science and Astronautics and subsequently as a NASA special publication. Although there are several thousand entries in each volume, the series is well indexed. Another helpful source is *Current News*, a compilation of newspaper articles about NASA activities prepared by the agency's Office of Public Affairs. The authors obtained information on specific missions from surprisingly detailed NASA press kits (e.g., the Apollo 8 press kit is 105 pages), mission summaries, and the transcripts of press conferences. The publications are available in the KSC archives and other NASA installations.

KSC's public affairs publications proved very helpful. *The Kennedy Space Center Story* (1969, 1972, 1974) is a well-written informative account of events at the space center since the early 1960s. The first edition attempted no historical evaluation and ignored unpleasant events, such as the AS-204 fire. Hundreds of KSC news releases about the Apollo program provided interesting sidelights for the history. The Center's newspaper, *Spaceport News*, prepared under the direction of the Public Affairs Office, served a similar function. Distinctly a house organ, the paper avoided controversy, but was, nevertheless, useful for background and specific facts.

Three unpublished works, prepared at KSC, blazed a research path for the authors. Frank Jarrett and Robert Lindemann's "History of the John F. Kennedy Space Center, NASA (Origins through December 1965)" provides a detailed, carefully researched account of early center history. Even more helpful are the unpublished manuscripts of James Covington, James Frangie, and William Lockyer (Apollo Launch Facilities) and George Bittle and John Marshall (Apollo Launch Operations). Both manuscripts are in the KSC historical archives.

Concerning primary sources, the General Accounting Office's criticisms notwithstanding, the authors found an overabundance of source material at the Kennedy Space Center. Documents on the AS-204 fire, alone, occupy more than 60 large cartons in the KSC records-holding area. The card catalog in the center's documents department references several thousand studies and procedures for the Apollo/Saturn. Fortunately, KSC's records retrieval and library systems provide quick access to documents.

Dr. Kurt Debus's "Daily Journal" (1959–1963) and the weekly reports rendered to him by the KSC staff (1962–1972) were key sources of information. These documents are located in the center director's office at KSC. The authors found other valuable data in Debus's correspondence files, in storage at the Federal Records Center in Atlanta. While the originals can be retrieved through KSC's records management office, the letters used for this history have been reproduced for the KSC archives. Rocco Petrone's program office was another rich source of reports, memoranda, and letters. Some carbon copies are on file in the KSC archives, but the bulk of this material has been retired to Atlanta. Similar documents from other KSC sources, numbering in the thousands, have been collected by the KSC staff during the past ten years.

The progress of design and construction of the three Saturn launch complexes is reflected in a series of reports: Saturn Monthly and Quarterly Progress Reports (published at Huntsville with a section on the Cape), the Monthly Progress Reports of the Launch Facilities Support Equipment Office (mainly about ground support equipment), and Construction Progress Reports and Project Status Reports on LC-39. These documents are available in the KSC historical archives along with the minutes of the Site Activation Board meetings and the Site Activation Status Reports. Important documents for the launch operations include the minutes of the Apollo Launch Operations Committee, the daily status reports for Apollo missions, Apollo/Saturn V test procedures, and the postlaunch reports. The daily status reports and the test procedures for the Saturn I launches were secured from Robert Moser's papers in the Federal Records Center at Atlanta.

A number of documents in the KSC archives concern the center's relations with other members of the Apollo team. The minutes of the Management Council Meetings relate important discussions while Brainerd Holmes was head of the Office of Manned Space Flight. Other sets of minutes from 1961–1963 cover the activities of the Launch Operations Panel and the Panel Review Board. Management instructions from the Headquarters and KSC program offices are contained in the Apollo Program Directives. The offices established the crucial scheduling dates in the series of directives referred to as "dash 4"; the frequent revisions chart the vicissitudes of the Apollo program from 1965 through 1972. Most of these directives are available in the historical archives or the documents department. The researcher may wish to make use of other documents in the archives including mission flight manuals and safety plans, interface control documents, and the Apollo Program Development Plans prepared by the Office of Manned Space Flight.

Interviews with participants were among the most valuable sources of information. Whenever possible, the authors evaluated the objectiveness and accuracy of an interview against other accounts of the same events. A list of the interviews is included in the bibliography. The transcripts are available in the KSC archives.

Books

Akens, David S. *Historical Origins of the George C. Marshall Space Flight Center*. Huntsville, AL: Marshall Space Flight Center, 1960.

———. *Saturn Illustrated Chronology*, 5th ed. Huntsville, AL: Marshall Space Flight Center, 1971.

Aldrin, Edwin E., with Wayne Wargo. *Return to Earth*. New York: Random House, 1973.

Alexander, Tom. *Project Apollo: Man to the Moon*. New York: Harper and Row, 1964.

Armstrong, Neil, Michael Collins, and Edwin E. Aldrin, Jr., written with Gene Farmer and Dora Jane Hambilin. *First on the Moon*. Epilogue by Arthur C. Clarke. Boston: Little, Brown and Co., 1970.

Behrendt, Lloyd L., comp. *Development and Operation of the Atlantic Missile Range*, History of the Air Force Missile Test Center, vol. 2. Patrick Air Force Base, FL: Air Force Missile Test Center, 1963.

Bergaust, Erik. *Murder on Pad 34*. New York: G. P. Putnam's Sons, 1968.

Booker, P. J., G. C. Frewer, and G. K. C. Pardoe. *Project Apollo: Way to the Moon*. American Elsevier Publishing Co., 1969.

Bramlitt, E. R. *History of Canaveral District*, South Atlantic Division, U.S. Corps of Engineers, 1971.

Cantafio, Leopold J., ed. *Range Instrumentation*. Englewood Cliffs, NJ: Prentice-Hall, 1967.

Collins, Michael. *Carrying the Fire: An Astronaut's Journeys*, with a foreword by Charles A. Lindbergh. New York: Farrar, Straus, and Giroux, 1974.

Cooper, Henry S. F., Jr. *Apollo on the Moon*. New York: Dial Press, 1969.

———. *Thirteen: The Flight That Failed*. New York: Dial Press, 1973.

Emme, Eugene, ed. *The History of Rocket Technology*. Detroit: Wayne State Press, 1964.

———. *A History of Space Flight*. New York: Holt, Rinehart and Winston, 1965.

Ertel, Ivan D., and Mary Louise Morse. *The Apollo Spacecraft: A Chronology*, vol. 1, *Through November 7, 1962*. NASA SP-4009. Washington, 1969.

Etzioni, Amitoi. *The Moon-Doggle: Domestic and International Implications of the Space Race*. Garden City, NY: Doubleday & Co., 1964.

Green, Constance McLaughlin, and Milton Lomask. *Vanguard: A History*. NASA SP-4202. Washington, 1970.

Grey, Jerry, and Vivian Grey, eds. *Space Flight Report to the Nation*. New York: Basic Books, 1962.

Grimwood, James M., and Barton C. Hacker, with Peter Vorzimmer. *Project Gemini Technology and Operations: A Chronology*. NASA SP-4002. Washington, 1969.

Harris, Gordon L. *Selling Uncle Sam*. Hicksville, NY: Exposition Press, 1976.

Holmes, Jay. *America on the Moon: The Enterprise of the Sixties*. Philadelphia: J. B. Lippincott Co., 1962.

Huzel, Dieter K. *Peenemuende to Canaveral*. Englewood Cliffs, NJ: Prentice-Hall 1962.

Johnson, Lyndon B. *The Vantage Point: Perspectives of the Presidency, 1963–1969*. New York: Popular Library, 1971.

Kennan, Erlend A., and Edmund H. Harvey, Jr. *Mission to the Moon: A Critical Examination of NASA and the Space Program*. New York: William Morrow and Co., 1969.

Klee, Ernest, and Otto Mark. *Birth of the Missile: The Secrets of Peenemuende*. New York: E. P. Dutton and Co., 1963.

Koelle, Heinz H., ed. *Handbook of Astronautical Engineering*. New York: McGraw-Hill Book Co., 1961.

Lay, Beirne, Jr. *Earthbound Astronauts: The Builders of Apollo-Saturn*. Englewood Cliffs, NJ: Prentice-Hall, 1971.

Lewis, Richard S. *Appointment on the Moon: The Inside Story of America's Space Venture*. New York: Viking Press, 1968.

Ley, Willy. *Rockets, Missiles, and Men in Space.* New York: Viking Press, 1968.
Logsdon, John M. *The Decision to Go to the Moon: Project Apollo and the National Interest.* Cambridge: MIT Press, 1970.
McGovern, James. *Crossbow and Overcast.* New York: William Morrow & Co., 1964.
Medaris, John B. *Countdown for Decision.* New York: G. P. Putnam's Sons, 1960.
Morse, Mary Louise, and Jean Kernahan Bays. *The Apollo Spacecraft: A Chronology*, vol. 2, *November 8, 1962–September 30, 1964*. NASA SP-4009. Washington, 1973.
Nieburg, H. L. *In the Name of Science*. Chicago: Quadrangle Books, 1966.
Ordway, Frederick L., III, ed. *Advances in Space Science and Technology*, vol. 6. New York: Academic Press, 1964.
Rabinowitch, Eugene, and Richard Lewis, eds. *Man on the Moon: The Impact on Science, Technology and International Cooperation*. New York: Basic Books, 1969.
Rosholt, Robert L. *An Administrative History of NASA, 1958–1963*. NASA SP-4101. Washington, 1966.
Sänger, Eugene. *Space Flight: Countdown for the Future*, trans. Karl Frucht. New York: McGraw-Hill Book Co., 1965.
Schwiebert, Ernest G. *A History of the U.S. Air Force Ballistic Missiles*. Washington: Frederick A. Prager, Publishers, 1965.
Shelton, William R. *Countdown: The Story of Cape Canaveral*. Boston: Little, Brown and Co., 1960.
———. *American Space Exploration: The First Decade*. Boston: Little, Brown and Co., 1967.
Swenson, Loyd S., Jr., James M. Grimwood, and Charles C. Alexander. *This New Ocean: A History of Project Mercury*. NASA SP-4201. Washington, 1966.
Sorenson, Theodore C. *Kennedy*. New York: Harper & Row, 1965.
Taylor, L. B., Jr. *Liftoff! The Story of America's Spaceport*. New York: E. P. Dutton & Co., 1968.
Von Braun, Wernher, and Frederick L. Ordway III. *History of Rocketry and Space Travel.* New York: Thomas V. Crowell and Co., 1966, 1969, 1975.
———. *Space Frontier.* New York: Holt, Rinehart, and Winston, 1967.
Wilford, John N. *We Reach the Moon.* New York: Bantam Books, 1964.
Webb, James E. *Space Age Management: The Large-Scale Approach*. New York: McGraw-Hill Book Co., 1969.
Young, Hugo, Bryan Silcock, and Peter Dunn. *Journey to Tranquility.* Garden City, NY: Doubleday & Co., 1970.

Journal Articles

Alelyunas, Paul. "Checkout: Man's Changing Role." *Space/Aeronautics* 44 (Dec. 1965): 66–73.
Alexander, George. "Cape Canaveral to Expand for Lunar Task." *Aviation Week*, 31 July 1961, p. 28.
———. "Telemetry Data Confirms Saturn Success." *Aviation Week and Space Technology*, 6 Nov. 1961, pp. 30–32.
———. "Inquiry Focuses on Electrical Systems." *Aviation Week and Space Technology*, 6 Feb. 1967, pp. 30–34.
Alexander, Tom. "The Unexpected Payoff of Project Apollo." *Fortune* 80 (July 1969): 114–117.
Alexander, William D. "Vertical Assembly Building—Project Description, Organization, and Procedures." *Civil Engineering* 35 (Jan. 1965): 42–44.

"Apollo 15: The Most Perilous Journey." *Time*, 9 Aug. 1971, pp. 10–15.
"Apollo: Giant Equipment Problems," *Missiles and Rockets*, 18 Sept. 1961, p. 19.
Bleymaier, Joseph S. "ITL and Titan III." *Astronautics and Aerospace Engineering* 1 (Mar. 1963): 33–36.
Bliss, H. "NASA's in the Cold, Cold Ground." *ATCHE Journal* 13 (May 1967): 419.
Campbell, John B. "What Happened to Apollo." *Space/Aeronautics* 48 (Aug. 1967): 54–70.
Cerquettini, C. "Sprayable Polyurethane Foam Insulation, Saturn II Booster." *SAMPE Journal* 5 (June-July 1969): 28–29.
"Death on Old Shaky." *Time*, 27 Jan. 1961, pp. 15–16.
Debus, Kurt H. "Launching the Moon Rocket." *Astronautics and Aerospace Engineering* 1 (Mar. 1963): 20–32.
———. "Saturn Launch Complex." *Ordnance* 46 (Jan.-Feb. 1962): 522–23.
"First Industry-Built Saturn I Puts Pegasus-2 in Precise Orbit." *Aviation Week and Space Technology*, 31 May 1965, p. 21.
Fisher, Allen C., Jr. "Cape Canaveral's 6000-Mile Shooting Gallery." *National Geographic* 116 (Oct. 1959).
Fleming, William A. "Launch Operations Challenge." *Astronautics* 6 (June 1961): 20–23.
Frewer, Gerald C. "Kennedy Space Center—Assembly Line on a Gigantic Scale." *The Engineering Designer*, May 1967, p. 7.
———. "The Crawler Transporter for Project Apollo." *The Designing Engineer*, July 1967, p. 15.
Getler, Michael. "Apollo: Was It Worth It?" *Space/Aeronautics* 3 (Sept. 1969): 48.
———. "Complex 37 Will Dwarf Predecessors." *Missiles and Rockets*, 18 Dec. 1961, p. 24.
Goerner, Erich E. "LOX Prevalve to Prevent POGO Effect on Saturn V." *Space/Aeronautics* 50 (Dec. 1968): 72–74.
Heaton, Donald H. "Approaches to Rendezvous." *Astronautics* 7 (Apr. 1962): 24.
Hendel, Frank J. "Gaseous Environments during Space Missions." *Journal of Spacecraft and Rockets* 1 (July-Aug. 1964): 353–64.
Holmes, D. Brainerd. "Man in Space—A Challenge to Engineers." *Challenge* 1 (Spring 1963): 28.
Houbolt, John C. "Lunar-Orbit Rendezvous and Manned Lunar Landing." *Astronautics* 7 (Apr. 1962): 26.
"How Soon the Moon?" *Time*, 14 Apr. 1967, pp. 86–87.
"Inquest on Apollo." *Time*, 10 Feb. 1967, pp. 18–19.
Knothe, Adolf H. "Range Safety—A Necessary Evil." *Aerospace Engineering* 20 (June 1961): 20.
Kolcum, Edward H. "S-1 Award Puts Chrysler in Space Field." *Aviation Week and Space Technology* 75 (27 Nov. 1961): 22.
Kovit, Bernard. "The Saturns." *Space/Aeronautics* 42 (Aug. 1964): 40–52.
"Launch Complex 39 Built Specifically for Saturn V." *Space Age News* 12 (Aug. 1969): 92.
"Launch Vehicles." *Spaceflight* 11 (Mar. 1969): 74.
Leary, Frank. "Support Net for Manned Space Flight." *Space/Aeronautics* 46 (Dec. 1966): 68–80.
Lewis, Richard S. "The Kennedy Effect." *Bulletin of the Atomic Scientists* 24 (Mar. 1968): 2.
"Life in the Space Age." *Time*, 4 July 1969, pp. 38–39.
Mason, John F. "Modernizing the Missile Range: Part I." *Electronics* 22 (Feb. 1965): 94–105.
Mast, Larry T. "Automatic Test and Checkout in Missile and Space Systems." *Astronautics and Aerospace Engineering* 1 (Mar. 1963): 41–44.
McGuire, Frank G. "Kapustin Yar Serves as Russia's Cape Canaveral." *Missiles and Rockets* 3 (Feb. 1958): 61–62.
McMillan, Brockway. "The Military Role in Space." *Astronautics* 7 (Oct. 1962): 18–21.
Means, Paul. "Group Taking on Another Vital Role." *Missiles and Rockets*, 14 Mar. 1960, pp. 22–24.

"Men of the Year." *Time*, 3 Jan. 1969, pp. 9-16.

Mendelbaum, Leonard. "Apollo: How the United States Decided to Go to the Moon." *Science*, 14 Feb. 1969, pp. 649–54.

Moore, W. I., and R. J. Arnold. "Failures of Apollo/Saturn V Liquid Oxygen Loading System." *Advances in Cryogenic Engineering* 13 (1966): 534–44.

Norcross, J. S., and Berl W. Martin. "Air Force Eastern Test Range UHF Telemetry Status Report." *Telemetry Journal* 4 (Apr./May 1969): 25–31.

Normyle, William J. "NASA Details Sweeping Apollo Revisions." *Aviation Week and Space Technology*, 15 May 1967, pp. 24–26.

Parker, Glenn M. "The Missile Site Labor Commission." *ILR Research* 8 (1962): 11.

Parker, P. J. "Apollo 9 Tests Lunar Module." *Spaceflight* 11 (July 1969): 230–33.

"Pegasus Returning Meteoroid Flux Data." *Aviation Week and Space Technology*, 22 Feb. 1965, p. 28.

Rogers, John G. "What Life at Cape Kennedy Does to Marriage." *Parade*, 9 July 1969.

Rosen, Milton W. "Big Rockets." *International Science and Technology*, Dec. 1962, pp. 66–71.

Rutledge, Philip C. "Vertical Assembly Building—Design of Foundations." *Civil Engineering* 35 (Jan. 1965): 50–52.

"SA-9 Launch." *Aviation Week and Space Technology*, 22 Feb. 1965, p. 27.

"Saturnalia at Canaveral." *Newsweek*, 6 Nov. 1961.

"Saturn Flight Specs. Manned Shot Plan." *Aviation Week and Space Technology*, 30 Apr. 1962, p. 32.

"Saturn's Success." *Time*, 3 Nov. 1961.

Sherrod, Robert. "The Selling of the Astronauts." *Columbia Journalism Review*, May/June 1973, pp. 17–25.

Smith, Richard A. "Canaveral, Industry's Trial by Fire." *Fortune* 65 (June 1962): 135.

———. "Now It's an Agonizing Reappraisal of the Moon Race." *Fortune* 68 (Nov. 1963): 124–29, 270–80.

Sloan, James E., and Jack F. Underwood. "Systems Checkout for Apollo." *Astronautics and Aerospace Engineering* 1 (Mar. 1963): 37–40.

Taylor, Hal. "Big Moon Booster Decisions Looming." *Missiles and Rockets*, 28 Aug. 1961, pp. 14–15.

Tedesko, Anton. "Base for USA Manned Space Rockets (Structures for Assembly and Launching)." *International Association for Bridge and Structural Engineering Publications* 26 (1967): 535.

———. "Space Truss Braces Huge Building for Moon Rocket." *Engineering News-Record*, 6 Feb. 1964, pp. 24–27.

———. "Design of the Vertical Assembly Building." *Civil Engineering* 35 (Jan. 1965): 45–49.

"To Design for the Moon Age, Four Firms Work as One Team." *Engineering News-Record*, 6 Feb. 1964, pp. 46–48.

"To Strive, to Seek, to Find, and Not to Yield. . . ." *Time*, 3 Feb. 1967, pp. 13–16.

"To the Moon." *Time*, 18 July 1969, pp. 20–31.

"Tracking and Data Acquisition." *Spaceflight* 11 (June 1969): 190.

Trainer, James. "Titan III Plans Await DOD Approval." *Missiles and Rockets*, 14 May 1962, pp. 35–36.

Vonbun, Friedrich O. "Ground Tracking of Apollo." *Astronautics and Aeronautics* 4 (May 1966): 104–15.

von Braun, Wernher. "Exploring the Space Sea." *Ordnance* 49 (July–Aug. 1964): 50.

von Tiesenhausen, Georg. "Saturn Ground Support and Operations." *Astronautics* 5 (Dec. 1960): 30.

"Washington Roundup." *Aviation Week and Space Technology*, 18 June 1962, p. 25.

Congressional Documents

House Committee on Government Operations, Subcommittee. *Investigation of the Boeing–TIE Contract*. Hearing, 90th Cong., 2d sess., 15 July 1968.

House Committee on Post Office and Civil Service, Subcommittee on Manpower. *Decision of Comptroller General of the United States Regarding Contractor Technical Personnel*. H. Rept. 188, 89th Cong., 1st sess., 18 Mar. 1965.

House Committee on Science and Astronautics. *Apollo 13 Accident*. Hearing, 91st Cong., 2d sess., 16 June 1970.

———. *Cape Canaveral: The Hope of the Free World*. Print, 87th Cong., 2d sess., 24 May 1962.

———. *Equatorial Launch Sites—Mobile Sea Launch Capability*. H. Rept. 710, 87th Cong., 1st sess., 12 July 1961.

———. *Management and Operation of the Atlantic Missile Range*. Print, 86th Cong., 2d sess., 5 July 1960.

———. *Master Planning of NASA Installations*. H. Rept. 167, 89th Cong., 1st sess., 15 Mar. 1965.

———. *1962 NASA Authorization*. Hearings, pt. 1, 87th Cong., 1st sess., 13 Mar.–17 Apr. 1961.

———. *Report on Cape Canaveral Inspection*. Print, 86th Cong., 2d sess., 27 June 1960.

———. *Space, Missiles, and the Nation*. Rept. 2092 pursuant to H. Res. 133, 86th Cong., 2d sess., 5 July 1960.

———. *Transfer of the Development Operations Division of the Army Ballistic Missile Agency to the National Aeronautics and Space Administration*. Hearing on H.J. Res. 567, 86th Cong., 2d sess., 3 Feb. 1960.

House Committee on Science and Astronautics, Subcommittee on Manned Space Flight. *1968 NASA Authorization*. Hearings on H. R. 4450, H. R. 6470, pt. 2, 90th Cong., 1st sess., 14–21 Mar. 1967.

———. *1964 NASA Authorization*. Hearings on H. R. 5466, pts. 2a, 2b, 88th Cong., 1st sess., 6 Mar.–6 June 1963.

———. *1963 NASA Authorization*. Hearings on H. R. 10100, pt. 2, 87th Cong., 2d sess., 6 Mar.–10 Apr. 1962.

House Committee on Science and Astronautics, Subcommittee on NASA Oversight. *Apollo Program Pace and Progress*. Print, 90th Cong., 1st sess., 17 Mar. 1967.

———. *Engineering Management of Design and Construction of Facilities of NASA*. Print, 91st Cong., 1st sess., 11 Aug. 1969.

———. *Investigation into Apollo 204 Accident*. Hearings, 2 vols., 90th Cong., 1st sess., 10 Apr.–10 May 1967.

———. *NASA-DOD Relationship*. Print, 88th Cong., 1st sess., 26 Mar. 1964.

———. *Pacing Systems of the Apollo Program*. Print, 89th Cong., 1st sess., 15 Oct. 1965.

Senate Committee on Aeronautical and Space Sciences. *Amending the National Aeronautics and Space Administration Authorization Act for the Fiscal Year 1962*. S. Rept. 863 to accompany S. 2481, 87th Cong., 1st sess., 1 Sept. 1961.

———. *Apollo Accident*. Hearings, pts. 1–6, 90th Cong., 1st sess., 7 Feb.–9 May 1967.

———. *Apollo 204 Accident*. S. Rept. 956, 90th Cong., 2d sess., 30 Jan. 1968.

———. *NASA Authorization for Fiscal Year 1963*. Hearings on H. R. 11737, 87th Cong., 2d sess., 13–15 June 1962.

———. *NASA Authorization for Fiscal Year 1965*. Hearings on S. 2446, pt. 2, 88th Cong., 2d sess., 4–18 Mar. 1964.

———. *NASA Authorization for Fiscal Year 1966*. Hearings on S. 927, pt. 2, 89th Cong., 1st sess., 22–30 Mar. 1965.

Senate Committee on Aeronautical and Space Sciences, NASA Authorization Subcommittee. *NASA Authorization for Fiscal Year 1961*. Hearings on H. R. 10809, pt. 1, 86th Cong., 2d sess., 28–30 Mar. 1960.

Senate Committee on Appropriations, Subcommittee. *Independent Offices Appropriations, 1964*. Hearing on H. R. 8747, pt. 2, 88th Cong., 1st sess., 18 Oct. 1963.

Senate Committee on Government Operations, Permanent Subcommittee on Investigations. *Work Stoppage at Missile Bases*. Hearings pursuant to S. R. 69, pts. 1, 2, 87th Cong., 1st sess., 25 Apr.–9 June 1961.

———. *Work Stoppage at Missile Bases,* S. Rept. 1312, 87th Cong., 2d sess., 29 Mar. 1962.

Conference Papers

Aden, R. M. "Electrical Support Equipment for the Saturn V Launch Vehicle System." *Proceedings of 2d Space Congress*, Cocoa Beach, FL, 5–7 Apr. 1965.

Clements, J. S. "S-1C Stage Instrumentation Checkout Concepts at KSC." *Proceedings of 3d Space Congress*, Cocoa Beach, FL, 7–10 March 1966.

Debus, Kurt. "Some Design Problems Encountered in Construction of Launch Complex 39." Hermann Oberth Gesellschaft, Darmstadt, Germany, 25 June 1964.

———. "Trends and Problems in Instrumentation and Operations in NASA's Future Space Efforts." AFMTC-AFESD-AFCEA Symposium, Patrick A.F.B., FL, 6 Mar. 1962.

———. "Launch Operations for Saturn V/Apollo." 10th Annual Meeting of the American Astronautical Society, New York, 6 May 1964.

Deese, James H. "The Problem of Low Level Wind Distribution." Structural Engineers Councils of Florida, Tampa, FL, 9 Nov. 1964.

Duran, B. E. "Saturn I/IB Launch Vehicle Operational Status and Experience." Paper 680739, Society of Automotive Engineers, Los Angeles, 7–11 Oct. 1968.

Eudy, Glenn. "Saturn V Mechanical Ground Support Equipment." *Proceedings of 2d Space Congress*, Cocoa Beach, FL, 5–7 Apr. 1965.

George, W. V., and C. A. Stinson. "An Automated Telemetry Checkout Station for the Saturn V Systems." 1966 National Telemetering Conference, Boston, MA, 12 May 1966.

Goff, H. C., and J. M. Schabacker. "Apollo Spacecraft Integrated Checkout Planning." *Proceedings of 3d Space Congress*, Cocoa Beach, FL, 7–10 Mar. 1966.

Hope, J. R., and C. J. Neumann. "Probability of Tropical Cyclone Induced Winds at Cape Kennedy." *Proceedings of 5th Space Congress*, Cocoa Beach, FL, 11–14 Mar. 1968.

Jafferis, William. "Prelaunch and Launch Checkout Operations—Uprated Saturn and Saturn V Vehicles." Project SETE, New York, 24–28 July 1967.

Johansen, R. B. "Developments in On-Board and Ground Checkout Systems." AIAA paper 70-245, American Institute of Aeronautics and Astronautics Launch Operations Meeting, Cocoa Beach, FL, 2–4 Feb. 1970.

Knothe, Adolf H. "Range Safety—Do We Need It?" AIAA paper 70-249, American Institute of Aeronautics and Astronautics Launch Operations Meeting, Cocoa Beach, FL, 2–4 Feb. 1970.

Marshall, John M. "The Mobile Concept and Automated Checkout of the Apollo/Saturn V Space Vehicle." 22d International Astronautical Congress, Brussels, Sept. 1971.

Montgomery, R. M. "Range Safety of the Eastern Test Range." AIAA paper 70-246, AIAA Launch Operations Meeting, Cocoa Beach, FL, 2–4 Feb. 1970.

Moore, F. Brooks, and William Jafferis. "Apollo/Saturn Prelaunch Checkout Display Systems." IEEE Conference on Displays, Univ. of Loughborough, England, 7–10 Sept. 1971.

Petrone, Rocco. "Apollo/Saturn V Launch Operations." AIAA Third Annual Meeting, Boston, 29 Nov.–2 Dec. 1966.

———. "Ground Support Equipment and Launch Installations at John F. Kennedy Space Center, NASA, for the Manned Lunar Landing Program." 15th International Astronautical Congress, Warsaw, 1964.

———. "Saturn V/Apollo Launch Operations Plan." AIAA Space Flight Testing Conference, Cocoa Beach, FL, 18–20 Mar. 1963.

Richard, Ludie G. "Saturn V System Philosophies." *Proceedings of 2d Space Congress*, Cocoa Beach, FL, 5–7 Apr. 1965.

Robin, E. A. "Development and Utilization of Computer Test Programs for Checkout of Space Vehicle." *Proceedings of 4th Space Congress*, Cocoa Beach, FL, 3–6 Apr. 1967.

Rorex, James E., and Robert P. Eichelberger. "Digital Data Acquisition System in Saturn V." *Proceedings of 2d Space Congress*, Cocoa Beach, FL, 5–7 Apr. 1965.

Rudolph, Arthur. "Operational Experience with the Saturn V." AIAA paper 68-1003, AIAA 5th Annual Meeting, Philadelphia, 21–24 Oct. 1968.

Salvador, G., and R. W. Eddy. "Saturn IB Stage Launch Operations." *Proceedings of 5th Space Congress*, Cocoa Beach, FL, 11–14 Mar. 1968.

Speer, F. A. "Saturn I Flight Test Evaluation." AIAA paper 64-322, American Institute of Aeronautics and Astronautics, Washington, July 1964.

Taylor, G. H. "Operational Television System for Launch Complex 39 at the John F. Kennedy Space Center." *Proceedings of 5th Space Congress*, Cocoa Beach, FL, 11–14 Mar. 1968.

Taylor, T., Jr. "System Considerations for Establishing Prelaunch Checkout Effectiveness." *Proceedings of 2d Space Congress*, Cocoa Beach, FL, 5–7 Apr. 1965.

Tedesko, Anton. "Assembly and Launch Facilities for the Apollo Program, Merritt Island, Florida: Design of the Structure of the Vertical Assembly Building." ASCE Structural Engineering Conference, 21 Oct. 1964.

Thilges, J. N. "Range Safety, A Thorn in the Flesh." *Proceedings of 3d Space Congress*, Cocoa Beach, FL, 7–10 Mar. 1966.

von Tiesenhausen, Georg. "Ground Equipment to Support the Saturn Vehicle." American Rocket Society, Washington, Dec. 1960.

Technical Reports

Army Ballistic Missile Agency:

Dodd, R. P., and J. H. Deese. *Juno V (Saturn) Heavy Missile Launch Facility, 1st Phase Request, 2nd Phase Estimate.* Atlantic Missile Range, 14 Feb. 1959.

Hall, Charles J. *A Committee Study of Blast Potentials at the Saturn Launch Site*, Rep. No. DHM-TM-9-60. Redstone Arsenal, AL, Feb. 1960.

Hamilton, J. S., J. L. Fuller, and P. F. Keyes. *Juno V Transportation Feasibility Study*, Rep. No. DLMT-TM-58-58. Redstone Arsenal, AL, 5 Jan. 1959.

Koelle, H. H., F. L. Williams, W. G. Huber, and R. C. Callaway, Jr. *Juno V Space Vehicle Development Program (Phase I), Booster Feasibility Demonstration,* Rep. DSP-TM-10-58. Redstone Arsenal, AL, 13 Oct. 1958.

———. *Juno V Space Vehicle Development Program (Status Report)*, Rep. No. DSP-TM-11-58. Redstone Arsenal, AL, 15 Nov. 1958.

Organizational Manual, Army Ballistic Missile Agency, Redstone Arsenal, AL, 5 Mar. 1959.

Army Corps of Engineers:

Apollo Launch Complex 39 Facilities Handbook. South Atlantic Division, Canaveral District, undated.

Army Ordnance Missile Command:

Koelle, H. H., F. L. Williams, and W. G. Huber. *Saturn Systems Study*, Rep. No. DSP-TM-1-59. Redstone Arsenal, AL, 13 Mar. 1959.

Project Horizon: A. U.S. Army Study for the Establishment of a Lunar Military Outpost. Redstone Arsenal, AL, 8 June 1959.

Saturn Systems Study II, Rep. No. DSP-TM-13-59. Redstone Arsenal, AL, 13 Nov. 1959.

Air Force:

Air Force Eastern Test Range. *Range Safety Manual.* AFETR Manual 127-1. Patrick A.F.B., FL, 1 Sept. 1972.

Joint Air Force/NASA Hazards Analysis Board. *Safety and Design Considerations for Static Test and Launch of Large Space Vehicles.* Patrick Air Force Base, FL; Air Force Missile Test Center, 1 June 1961.

Headquarters, NASA:

Ad Hoc Task Group. *A Feasible Approach for an Early Manned Lunar Landing* (Fleming Committee Report). Washington, D.C., 16 June 1961.

Allen, William H., ed. *Dictionary of Technical Terms for Aerospace Use.* Washington, D.C.: GPO, 1965.

America's Next Decades in Space, A Report for the Space Task Group. Washington, D.C., Sept. 1969.

Apollo Inter-Center ICD Management Procedure, CM-001, 001-1B. Washington, D.C., Jan. 1969.

Logsdon, John M. *NASA's Implementation of the Lunar Landing Decision,* HHN-81 (comment edition). Washington, D.C., Aug. 1969.

NASA-DOD. *Joint Report on Facilities and Resources Required at Launch Site to Support NASA Manned Lunar Landing Program* (Debus-Davis Report). Cape Canaveral, FL, 31 July 1961.

NASA Special Committee on Space Technology. *Recommendations Regarding a National Civil Space Program* (Stever Committee Report). 28 Oct. 1958.

Office of Manned Space Flight. *Apollo Configuration Management Manual,* NPC 500-1. Washington, D.C., 18 May 1964.

———. *Apollo Program Flight Summary Report,* Apollo Missions AS-201 through Apollo 8. Washington, D.C., Jan. 1969.

———. *Apollo Program Development Plan,* Rep. No. M-D MA500. Washington, D.C., 1 Jan. 1966.

PERT: Program Evaluation and Review Technique, Handbook, NPC-101. Washington, D.C., 1 Sept. 1961.

"A Plan for Manned Lunar Landing" (Low Committee Report). Washington, D.C., 7 Feb. 1961.

Report of Apollo 204 Review Board to the Administrator, NASA, 5 Apr. 1967, 1 vol. plus 14 vols. of appendixes and a set of looseleaf color photographs.

Report of the Apollo 13 Review Board. Washington, D.C., 15 June 1970.

Report to the Administrator, NASA, on Saturn Development Plan, by Saturn Vehicle Team (Silverstein Committee Report). Washington, D.C., 15 Dec. 1959.

Kennedy Space Center:

Apollo/Saturn V Launch Operations Test and Checkout Requirements, AS-504 and All Subsequent Missions, K-V-051-01. KSC, FL, 1968.

Apollo/Saturn V MILA Facilities Descriptions, K-V-011, Coordination Draft. KSC, FL, 30 June 1965.

Brewster, H. D., and W. G. Hughes. *Lightning Protection for Saturn Launch Complex 39,* KSC, FL, 18 Oct. 1963.

Catalog of Launch Vehicle Tests, Saturn V, Apollo/Saturn V, Revision 1, GP-244. KSC, FL, 15 June 1966.

Hanna, George V. *Chronology of Work Stoppage and Related Events, KSC/AFETR through July 1965.* KSC, FL, Oct. 1965.

Index of KSC Apollo Tree Documents and other KSC Generated Documents in the KSC Library, GP-856. KSC, FL, 1970.

Jarrett, Francis E., Jr., and Robert A. Lindemann. *Historical Origins of NASA's Launch Operations Center to July 1, 1962,* KHM-1. Cocoa Beach, FL, Apr. 1964.

KSC Apollo/Saturn Program Development/Operations Plan, 100-39-0001, 2 vols. KSC, FL, 10 Oct. 1965.

Launch Support Equipment Engineering Division. *Development of Design Criteria for Saturn V Flame Deflector*, TR-174-D. KSC, FL, 1 June 1965.

———. *Saturn V Launch Support Equipment General Criteria and Description*, SP-4-37-D. KSC, FL, 15 Sept. 1964.

———. *Saturn V Electrical Ground Support Equipment for Launch Complex 39*, SP-96-D. KSC, FL, 21 Dec. 1964.

———. *Saturn V Launch Support Equipment General Criteria and Description*, SP-4-37-D, Revision. KSC, FL, 15 Sept. 1964.

Launch Vehicle Checkout Automation and Programming Office. *Apollo (Saturn) Automated Checkout*. KSC, FL, 23 Aug. 1974.

McMurran, W. R., ed. "The Evolution of Electronic Tracking, Optical, Telemetry and Command Systems at the Kennedy Space Center." KSC, FL, 17 Apr. 1973.

Public Affairs Office. *Kennedy Space Center Story*. KSC, FL, Dec. 1971.

Saturn Launch Vehicle Checkout Automation Development Plan, KSC-100-39-0007. Cocoa Beach, FL, 8 Aug. 1966.

Scheller, Donald R. "Management by Exception, Activation of Apollo/Saturn V Launch Complex 39." KSC, FL, 15 May 1967.

A Selective List of Acronyms and Abbreviations, GP-589 Revised. KSC, FL, July 1972.

Walter, George W. *Modifications to Bearing for Traction Support Rollers on Crawler-Transporters*. KSC, FL, 15 Dec. 1965.

"Weather Effects on Apollo/Saturn V Operations, Apollo 4 through Apollo 13," 630-44-0001. KSC, FL, 27 July 1970.

Manned Spacecraft Center:

Analysis of Apollo 12 Lightning Incident, MSC-01540. Houston, TX, Feb. 1970.

Atlantic Missile Range Operations: Facilities, 1959–1964. Houston, TX, 15 Apr. 1963.

John F. Kennedy Space Center, NASA Fiscal Year 1963 Estimates, Apollo Mission Support Facilities. Cape Canaveral, FL; Florida Operations, 1963.

Launch Operations Center:

Concepts for Support Service at the Merritt Island Launch Area. Cape Canaveral, FL, 6 May 1963.

Criteria for Design Pad "A" Launch Complex 39. Cape Canaveral, FL, 19 Dec. 1962.

Criteria for Launch Complex 39, Crawler Transfer System and Utilities. Cape Canaveral, FL, 5 Sept. 1962.

Deese, James. "Concept Development of Saturn Service Structure No. II." Cape Canaveral, FL, Apr. 1963.

Saturn C-5 Facilities Evaluation for Complex 39, LTR-1-2. Cape Canaveral, FL, 10 Sept. 1962.

Raffaelli, A. R. *Introduction to Lightning*, LT1R-2-DE-62-6. Cape Canaveral, FL, 14 Dec. 1962.

Marshall Space Flight Center:

Buchanan, Donald D., and George W. Walter. *Appraisal of Transfer Modes for Saturn C-5 Mobile Systems*, Rep. No. MIN-LOD-DH-9-62. Huntsville, AL, 11 June 1962.

Catalog of Systems Tests for Saturn S-1 Stage. Huntsville, AL, 1 Sept. 1961.

Chrysler Corporation Space Division. *Saturn IB Vehicle Handbook*. Huntsville, AL, 25 July 1966.

Duren, O. K. *Interim Report on Future Saturn Launch Site*, Rep. No. MIN-LOD-DL-1-61. Huntsville, AL, 10 May 1961.

Gwinn, Ralph T., and Kenneth J. Dean. "*Consolidated Instrumentation Plan for Saturn Vehicle SA-1*," Rep. No. MTP-LOD-61-36.2a. Huntsville, AL, 25 Oct. 1961.

Launch Countdown, Saturn Vehicle SA-1, Rep. No. MTP-LOD-61-35.2. Redstone Arsenal, AL, 30 Oct. 1961.

Launch Facilities and Support Equipment Office, Launch Operations Directorate. *A Preliminary Study of Launch Facility Requirements for the C-4 Space Vehicle*. Huntsville, AL, Oct. 1961.

———. *Project Saturn, C-1, C-2 Comparison*. Rep. No. M-MS-G-113-60. Huntsville, AL, 21 Jan. 1961.

MSFC Automation Plan. Huntsville, AL, 8 May 1962 (revised, 1 June 1964).

Results of the Saturn I Launch Vehicle Test Flights, Rep. No. MPR-SAT-FE-66-9. Huntsville, AL, 1 Aug. 1961.

Richard, Ludie G., and Charles O. Brooks. *The Saturn Systems Automation Plan*. Huntsville, AL, 15 Sept. 1961.

SA-1 Vehicle Data Book, Rep. No. MTP-MS and M-E-61-3. Huntsville, AL, 26 June 1961.

Software for IU 201 at MSFC, SA-201 at KSC, SA-501 at KSC, Rep. No. R-DIR-64-1. Huntsville, AL, 1 Dec. 1964.

Sparks, Owen L., comp. *Preliminary Concepts of Launch Facilities for Manned Lunar Landing Program*, Rep. No. MIN-LOD-DL-3-61. Huntsville, AL, 1 Aug. 1961.

———. *Preliminary Feasibility Study on Offshore Launch Facilities for Space Vehicles*, Rep. No. IN-LOD-DL-1-60 Interim. Huntsville, AL, 29 July 1960.

Technical Information Summary, AS-501, Apollo Saturn V Flight Vehicle, R-ASTR-S-67-65. Huntsville, AL, 15 Sept. 1967.

Walter, George W. *Saturn Mobile (Canal) Concept Flame Deflector and Launcher/Transporter Emplacement Evaluation*, Rep. No. MIN-LOD-2-62. Huntsville, AL, Feb. 1962.

———. *Saturn Mobile (Rail) Concept: An Examination of Rail Transfer Systems for a Launcher/Transporter*, Rep. No. MIN-LOD-DH-3-62. Huntsville, AL, 3 Apr. 1962.

Contractor:

Beech Aircraft Corporation. *Saturn C-5 Propellant Transportation Optimization Study, ER-13539*. Boulder, CO, 25 June 1962.

Boeing Atlantic Test Center. *Launch Complex 39 GSE Systems Descriptions*. KSC, FL, 3 Aug. 1965.

Brown Engineering Co. *An Evaluation of an Enclosed Concept for a C-5 Vertical Assembly Building*. Huntsville, AL, 2 Apr. 1962.

———. *An Evaluation of an Open Concept for a C-5 Vertical Assembly Building*. Huntsville, AL, 2 Apr. 1962.

———. *Fixed Pad Concept of Launch Complex 39 for the Saturn C-5 Vehicle*. Huntsville, AL, 28 Sept. 1962.

Chrysler Corp. Space Division. *Saturn I/IB Automation Orientation, HSE-R 115*. Huntsville, AL, undated.

Culver, A. J., and K. G. Baird. "ERS Recovery Plan." KSC, FL: Boeing Atlantic Test Center Management Systems Staff, Apr. 1966.

General Electric Co. Apollo Support Department. *Systems Description of Saturn V Launch Vehicle Ground Electrical Support Equipment at Vehicle Launch Facility 39-1*. Daytona Beach, FL, 27 Sept. 1965.

Green, Arthur W., Lewis E. Williamson, Robert P. Dell, and Reed B. Jenkins. *Reliability Study of Saturn SA-3 Pre-Launch Operations*. Washington, D.C.: ARINC Research Corporation, 3 Jan. 1963.

Hartsfield, Annie May, Mary Alice Griffin, and Charles M. Grigg, eds. *Summary Report NASA Impact on Brevard County*. Tallahassee, FL: Institute of Social Research, Florida State University, 1966.

The Martin Co. *Rescue and Escape Systems from Tall Structures (RESTS)*. Denver, CO, Oct. 1963.
———. *Saturn C-2 Operational Modes Study, Summary Report*. Baltimore, MD, June 1961.
———. *Saturn C-3 Launch Facilities Study*, Rep. No. ER 12125, 3 vols. Baltimore, MD, Dec. 1961.
Maurice H. Connell and Associates. *Alternate "I" Heavy Missile Service Structure*. Miami, FL, 24 Apr. 1959.
———. *Heavy Missile Launch Facility Criteria*. Miami, FL, 15 Mar. 1959.
———. *Heavy Missile Service Structure Criteria*. Miami, FL, undated.
———. *Siting Study and Recommendation, Saturn Staging Building and Service Structure, Complex 34 AFMTC*. Miami, FL, Jan. 1960.
———. *Saturn C-5 Launch Facilities, Study of Rail Systems for Vertical Transporter/Launcher Concept*. Huntsville, AL, May 1962.
Sutherland, L. C. *Sonic and Vibration Environments for Ground Facilities—A Design Manual*, report WR 68-2. Wyle Laboratories, 1968.
TRW Space Technology Laboratories. *A Study of the KSC Safety Program*. Cape Canaveral, FL, May 1965.

Interviews

Aden, Robert, MSFC Astrionics, Electrical, by Benson, 30 Oct. 1974.
Alexander, William, Washington, D.C., by James Frangie, 8 Aug. 1969.
Atkins, John R., KSC Safety Off., by Faherty, 5 Sept., 5 Nov. 1973, 29 May 1974.

Barfus, Armond, KSC Support Ops., Development Testing Lab., by Benson, 5 Dec. 1973.
Bertram, Emil, KSC Launch Ops., Requirements and Resources Off., by Benson, 28 Sept.,* 15 Nov. 1973.*
Bidgood, Clarence, Washington, D.C., by Frangie, 14 Nov. 1968, 13 Aug. 1969.
Black, Dugald, KSC Dep. Dir., Support Ops., by Benson, 28 Feb. 1974.
Bobik, Joseph, KSC Spacecraft Ops., Quality Surveillance Div., by Benson and Faherty, 26 June 1974.
Boylston, Clifford, Brown Engineering Co., Huntsville, AL, by Benson, 21 July 1972.
Brewster, Heyward, KSC Design Engineering, Design Documentation Br., by Benson, 30 Nov. 1973.*
Brown, Joseph Andrew, KSC Design Engineering, Architectural Sec., by Faherty, 25 Sept. 1973.
Bruns, Rudolf, KSC Information Systems, Computer Systems Div., by Frangie, 22 Aug. 1969; by Benson, 3 Jan. 1974.
Bryan, Frank, KSC Launch Vehicle Ops. Engineering Staff, by Benson, 19 Dec. 1973.
Buchanan, Donald, KSC Design Engineering, by Frangie, 5 Sept. 1969; by Benson, 22 Sept., 7 Nov., 28 Nov. 1972; 4 Oct. 1974.
Burke, J. F., Chief, KSC Saturn/Apollo Facilities Br., by Frangie, 23 Apr. 1969.

Carlson, Norman, KSC Launch Vehicle Ops., Test Ops. Br., by John Marshall, 16 Dec. 1970; by Benson, 5 Sept. 1974.*
Carothers, Ralph Dale, KSC Spacecraft Ops., Preflight Ops. Br., by Benson, 14 June 1974.*
Chambers, Milton, KSC Launch Vehicle Ops., Gyro and Stabilizer Systems Br., by Benson, 19 Aug. 1974.*

*Indicates telephone interview.

Chandler, William, KSC Launch Vehicle Ops., Electrical Systems Br. (LCC), by Benson, 20 Nov. 1973.*

Chauvin, Clarence, KSC Spacecraft Ops. CSM Test Staff, by Benson, 2 Apr.,* 23 May, 6–7 June 1974; by Faherty, 24 June 1974.

Childers, Frank, KSC Information Systems, Quality Surveillance Off., by Faherty, 7 Nov. 1972, 18 Mar. 1973.

Clark, Raymond, KSC Design Engineering, by Benson, 30 June 1972.

Claybourne, John P., KSC Sciences and Applications Project Off., Earth Resources, by Benson, 5 Jan. 1973.

Clearman, William T., Jr., Cocoa Beach, FL, by Benson, 5 Jan., 13 Sept.,* 26 Oct. 1973.*

Cochran, Harold, KSC Spacecraft Ops., Communications and R. F. Sec., by Benson, 7 June 1974.*

Collner, Joseph D., KSC Spacecraft Ops., Communications and R. F. Sec., by Benson, 7 June 1974.

Corbett, Belzoni A., Jr., KSC Information Systems, by Benson, 11 Jan. 1974.

Corn, Graydon F., KSC Launch Vehicle Ops., Propellants Br., by Benson, 23 July 1973,* 22 Apr. 1974.*

Crawford, Harvey, KSC Spacecraft Ops., Environmental Control and Cryogenics Sec., by Benson, 28 June 1974.*

Crunk, Henry, KSC Launch Vehicle Ops., Mechanical and Propulsion Systems, by Benson, 12 Apr. 1973.*

Davidson, James, KSC Launch Vehicle Ops., Electrical Systems, by Benson, 31 July 1973.*

Davis, Edwin, KSC Design Engineering, Launch Accessories, by Benson, 18 Jan. 1973.*

Debus, Kurt H., KSC Center Dir., by James Covington, 22 Aug. 1969; by Benson and Faherty, 16 May 1972.

Deese, James, KSC Design Engineering, Systems Analysis, by Benson, 10 May 1972, 4 Oct. 1973; by Faherty, 16 Mar. 1973.

Dodd, Richard P., KSC Design Engineering, Project Integration Off., by Benson, 13 Aug. 1974.

Donnelly, Paul, KSC Assoc. Dir. for Ops., by John Marshall, 19 Nov. 1970; by Benson and Faherty, 19 June 1974.*

Duggan, Orton L., KSC Apollo Off., by James Grimwood, 13 Nov. 1969.

Duren, Olin K., MSFC Astronautical Lab., Materials, by Benson, 29 Mar., 16 May 1972,* 1 Nov. 1974.*

Edwards, Marion, KSC Launch Vehicle Ops., Launch Instrumentation, by Benson, 7 May 1973.

Ely, George, KSC Launch Vehicle Ops., Flight Control, by Benson, 8 July 1974.

Enlow, Roger, KSC Technical Support, by Benson, 5 Dec. 1973.*

Ernst, Lloyd, KSC Design Engineering, LC-39 Area Management Br., by Faherty, 8 Nov. 1973.

Fannin, Edward, KSC Launch Vehicle Ops., Mechanical and Propulsion, by Benson, 23 July 1973.*

Finn, James E., Design Engineering, Cables and Special Power Sec., by Faherty, 16 Jan. 1973.

Fiorenza, Vincent, GE Space Div., Apollo and Ground Systems, KSC, by Benson, 25 Apr. 1974.

Foster, Leroy, General Electric, Daytona Beach, FL, by Benson, 30 Apr. 1974.

Fowler, Calvin, GE Space Div., Apollo and Ground Systems, KSC, by Benson, 25 Apr. 1974.

Fulton, James, KSC Design Engineering, by Benson, 26 Oct. 1973.*

Gassman, Marvin, Apollo-Skylab Program Off., Configuration Management Br., by Benson, 8 Nov. 1973.
Gibbs, Asa, Satellite Beach, FL, by James Covington, 7 Aug. 1969.
Glaser, William, KSC Launch Vehicle Ops., Telemetry, by Benson, 7 May 1973.
Goldsmith, James, MSFC Procurement Off., Saturn Div., by Benson, 6 Nov. 1974.*
Gooch, Harold, KSC Administration, Labor Relations, by Faherty, 1 July 1974.
Gramer, Russell, KSC Installation Support, Quality Surveillance Div., by Faherty, 19, 21 Sept. 1972, 19 July 1973.
Greenfield, Terry, KSC Design Engineering, Digital Electronics, by Benson, 3 Jan. 1974.
Gruene, Hans, KSC Dir. of Launch Vehicle Ops., by John Marshall, 19 Nov. 1970; by Benson and Faherty, 10 May 1972.

Haggard, Ken M., Lockheed, Personnel Industrial Relations, by Faherty, 16 July 1973.
Hahn, Richard, KSC Design Engineering, Analysis, by Benson, 3 Dec. 1973.
Hand, Larry, KSC Design Engineering, Communications, by Benson, 2 Oct. 1973.
Hangartner, James, KSC Spacecraft Ops., Mechanical Systems, by Benson, 3 July 1974.*
Hansel, John, KSC Quality Control, by Faherty, 11 July 1973.
Harris, Gordon, KSC Chief, Public Affairs, by Benson, 12 Apr. 1974.
Harris, Steven, KSC Design Engineering, Field Engineering Off., by Benson, 24 Oct. 1973.*
Henschel, Charles F., KSC/NASA Test Ops. Off., by John Marshall, 17 Nov. 1970.
Horn, Frank W., Jr., KSC Apollo-Skylab Programs, by Benson, 28 Sept. 1973.
Horner, William J., Jr., KSC Security Off., by Faherty, 1 July 1974.
Huffman, Bobby R., KSC Launch Vehicle Ops., Launch Instrumentation Systems Div., by Benson, 7 May 1973.
Hughes, R. Bradley, KSC Information Systems Engineering Application, by Benson, 11 Jan. 1974.
Humphrey, John T., KSC Launch Vehicle Ops., Propulsion and Vehicle Mechanical, by Benson, 2 Aug. 1974.*

Jafferis, William, KSC Launch Vehicle Ops., Systems Engineering, by Benson, 19 Dec. 1973, 22 Jan. 1974.*
Jelen, Wilfred G., KSC Information Systems Data, by Benson, 15 Jan. 1974.
Jenke, Richard, Huntsville, AL, by Benson, 29 Oct. 1974.
Johnson, Charles, Florida Dept. of Commerce, Employment Service, Cocoa, FL, by Benson, 8 Apr. 1974.*
Johnson, Edwin C., KSC Spacecraft Ops., CSM and Payloads Project Engineering, by Benson, 10 June 1974.*
Johnson, Harold G., KSC Support Ops., Planning and Contract, by Benson, 17 Dec. 1973.
Jones, James, Information and Measurement Systems, Test Analysis Sec., by Faherty, 8 Apr. 1973.
Joralan, Albert, KSC Design Engineering Data Systems, by Benson, 3 Jan. 1974.

Kapryan, Walter, KSC Dir. of Launch Ops., by Benson, 25 Apr. 1974.
Kaufman, James R., KSC Administration, Manpower Utilization, by Benson, 27 Mar. 1974.*

Lamberth, Horace, KSC Spacecraft Ops., Fluid Systems, by Benson, 24 Apr. 1974.*
Lang, J. Robert, KSC Spacecraft Ops., Environmental Control and Cryogenic, by Benson, 7 June 1974.*
Lealman, Roy, KSC Launch Vehicle Ops., Electrical G and C Systems, by Benson and Faherty, 27 June 1974.
Leet, Joel, KSC Shuttle Project Planning, by Benson, 8 Nov. 1973, 13 Feb. 1974.
Lloyd, Russell, KSC Support Ops., by Benson, 13 Feb. 1974.
Lowell, Albert, General Electric, Daytona Beach, FL, by Benson, 16 Apr. 1974.

McCafferty, Riley, Johnson Space Center, Crew Training and Simulation, by Ivan Ertel, 28 Jan. 1971; by Benson, 20 Feb. 1974.
McKnight, James N., KSC Spacecraft Ops., Preflight, by Benson, 26 June 1974.*

Malley, George, Chief Counsel, Langley Research Center, by Faherty, 6 Nov. 1973.*
Marsh, Thomas, KSC Launch Vehicle Ops., Propulsion and Vehicle, by Benson, 7 May 1973.
Mathews, Edward, JSC Manager, Space Shuttle Systems Integration, by Benson, 19 Feb. 1974.
Matthews, George D., KSC Information Systems, Telemetry, by Benson, 11 Jan. 1974.
Medlock, Joe, KSC Launch Vehicle Ops., Checkout, Automation and Programming Off., by Benson, 7 Oct. 1974.
Melton, Lewis, KSC Administration, Resources and Financial Management, by Covington, 18 Feb. 1969.
Montgomery, Ann, KSC Spacecraft Ops., Preflight Ops., by Benson, 26 June 1974.
Moore, Robert T., KSC Information Systems, Planning and Technical Support, by Benson, 27 Sept. 1973.
Morris, Owen, JSC Manager, Apollo Spacecraft Program Off., by Benson, Grimwood, and Courtney Brooks, 20 Dec. 1972.
Moser, Robert, KSC Test Planning Off., by Benson, 30 Mar., 18 July 1973,* 17 Apr. 1974.*
Moxley, Paul, KSC Spacecraft Ops., Propulsion Systems, by Benson, 19 July 1974.
Mrazek, William, MSFC Assoc. Dir. for Engineering, by Benson, 2 Aug. 1972.*
Murphy, John, KSC Apollo-Skylab Program Off., Launch Vehicle and Workshop Br., by Benson, 9 Nov. 1973.

Nazaro, Ron, IBM, KSC, by Faherty, 8 July 1974.
Newall, Robert, KSC Launch Vehicle Ops., S-IC Systems, by Benson, 31 July 1973.*
Norwalk, William, Hqs. Auditing Div., by Benson, 23 Jan. 1974.*

O'Conner, H. J., Mgr., Wildlife Refuge, by Faherty, 31 May 1972.
O'Hara, Alfred D., KSC Launch Vehicle Ops. Management, by Benson, 14 Jan. 1974.*
Owens, Lester, KSC Design Engineering Systems, by Benson, 12 Apr., 21 Nov. 1972.

Page, George F., KSC Spacecraft Ops., by Benson, 28 Jan. 1974.*
Pantoliano, Thomas, KSC Launch Vehicle Ops., Mechanical and Propulsion, by Benson, 18 Apr. 1973.*
Parker, Clarence C., KSC Installation Support, by James L. Frangie, 14 Feb. 1969; by Benson and Faherty, 2 Feb. 1972.
Parsons, Walter, KSC Design Engineering Systems, by Benson, 21 Jan. 1974.
Paul, Henry C., KSC, Chief, Checkout Automation and Programming Off., by John Marshall, 9 Dec. 1970.
Petrone, Rocco, NASA Hq., Apollo Program Dir., by Eugene Emme and Tom Ray, 17 Sept. 1970; by Benson, Emme, and Faherty, 25 May 1972.
Pickett, Andrew, KSC Shuttle Projects Off., by Benson, 26 July 1973.*
Policicchio, Mrs. Caroline, KSC, by James Covington, 4 Aug. 1969.
Poppel, Theodor, KSC Design Engineering, Field Engineering Off., by Benson, 12 Jan. 1973,* 24 Jan. 1973.
Porcher, Arthur G., KSC Design Engineering, by Benson, 28 Apr. 1972.*
Potate, John, NASA Office of Manned Space Flight, by Benson, Faherty, and Ray, 25 May 1972.
Potter, John, KSC Design Engineering, Field Engineering Off., by Benson, 24 Oct. 1973.
Preston, G. Merritt, KSC Manager, Shuttle Project, by Benson and Faherty, 12 Dec. 1973, by Benson, 22 Jan. 1974.*
Proffitt, Richard C., KSC Spacecraft Ops., Launch Complex 39, by John Marshall, 1 Dec. 1970; by Faherty, 20 June 1974.

Ragusa, James, KSC, Off. of the Dep. Director, by Benson, 11 Sept. 1974.*

Redfield, Marvin, NASA Hqs., Advanced Development, by Benson, Faherty, and Ray, 25 May 1972.

Reyes, Raul Ernest, KSC Spacecraft Ops. Preflight, by Faherty, 19 Jan., 3 June, 30 Oct. 1973, 24 June 1974.

Renaud, Fred, KSC, Bendix Launch Support, by Faherty, 4 Apr., 16 May 1973.

Richard, Ludie, MSFC Dep. Dir., Science and Engineering, by Benson, 12 Dec. 1973,* 30 Oct. 1974.

Rigell, Isom, KSC Launch Vehicle Ops., by Benson, 3 Dec. 1973.*

Roberts, John T., KSC Design Engineering, Utilities Sec., by Benson, 1 Nov. 1973.*

Rosen, Milton W., NASA Hqs., by Barton Hacker and Eugene Emme, 14 Nov. 1969; by Benson and Faherty, 25 May 1972.

Rowland, R. D., Asst. to the President, Hays International Corp., Birmingham, AL, by Benson, 25 July 1972.*

Russell, Labrada, KSC Installation Support, Librarian, by Benson, 15 Mar. 1973.

Sasseen, George T., KSC Spacecraft Ops., Engineering, by Benson, 26 July 1973,* 4 Feb. 1974*; by Faherty, 8 July 1974.

Scholz, Arthur, KSC, Boeing Aerospace Co., Field Ops. and Support, by Benson, 18 June 1974.

Seully, Edward J., McDonnell-Douglas Astronautics Co., by Faherty, 20 Apr. 1973.

Shea, Joseph, at Washington, D.C., by Eugene Emme, 6 May 1970.

Sherrer, Leroy, KSC Launch Vehicle Ops. Contractor Technical Management, by Benson, 25 July 1973.

Siebeneichen, Paul, KSC Community Relations Off., by Faherty, 29 Jan. 1973.*

Sieck, Robert, KSC Spacecraft Ops., Shuttle Project, by Benson, 4 Apr. 1974.*

Smith, Jackie E., KSC Spacecraft Ops., Experiments, by Benson, 4 June 1974.

Smith, Richard G., MSFC, Manager Saturn Program Off., by Benson, 29 Oct. 1974.

Sparkman, Orval, KSC Design Engineering, Mechanical Design, by Benson, 15 Dec. 1972, 13 June 1974.*

Sparks, Owen L., MSFC, Performance and Flight Mechanics, by Benson, 31 Mar. 1972.

Stringer, M. S., KSC Internal Review Staff, by James Frangie, 19 Dec. 1968.

Stein, Martin, URSAM Project Architect for LCC, by Covington, 8 Aug. 1969.

Thompson, John, KSC Launch Vehicle Ops., Checkout, Automation and Programming Off., by Benson, 7 Oct. 1974.

Twigg, John M., KSC Launch Vehicle Ops., Skylab and Space Shuttle, by John Marshall, 23 Nov. 1970.

von Tiesenhausen, Georg, by Benson, 29 Mar. 1972, 20 July 1973.

Wagner, Walter, KSC Apollo-Skylab Programs, Configuration Management, by Faherty, 7 Aug. 1973; by Benson, 21 Sept. 1973.

Walter, George, KSC Design Engineering, Structures, by Benson, 7 Nov. 1972, 26 Jan. 1973.

Walton, Thomas, KSC Design Engineering, LPS Systems, by John Marshall, 17 Dec. 1970; by Benson, 23 Jan. 1974.*

Wasileski, Chester, KSC Design Engineering, Facilities and Systems, by Benson, 14 Sept., 14 Dec. 1972.

Wedding, Michael A., Chief, Checkout Equipment Br., Automation—Spacecraft, by John Marshall, 11 Dec. 1970.

Wendt, F. Gunter, North American Rockwell Test Management, by Faherty, 18 June 1973.

White, James, KSC Design Engineering, Electrical and Electronic Design, by Benson, 9 May 1973.

Whiteside, Carl, KSC Launch Vehicle Ops., Electrical G and C, by Benson, 4 Jan.,* 29 Aug. 1974.

Widick, Herman K., KSC Spacecraft Ops., LM and Skylab Test, by John Marshall, 15 Dec. 1970; by Benson, 23 May 1974.

Wiley, Alfred N., KSC Spacecraft Ops., by Benson, 31 Oct. 1973, 13 Feb. 1974.

Williams, Francis L., NASA Hq. Off. of Analysis and Evaluation, by Benson, 6 Apr. 1972.*

Williams, Grady, KSC Dep. Dir. for Design Engineering, by Benson, 29 Mar. 1973.

Wills, Tom, KSC Design Engineering, Mechanical Design, by Benson, 28 Nov. 1973.

Wojtalik, Fred, MSFC Astrionics Lab., Guidance and Control, by Benson, 30 Oct. 1974.

Yates, Maj. Gen. Donald N. (USAF, Ret.), by Faherty, 17 Sept. 1973.

Youmans, Randell E., KSC Launch Vehicle Ops., Test Ops., by John Marshall, 5 Feb. 1971.

Zeiler, Albert, KSC Design Engineering, Mechanical Design, by F. E. Jarrett and W. Lockyer, 11 Aug. 1970; by Benson, 24 Mar., 11 July, 24 Aug. 1972, 23 July 1973*; by Faherty, 24 Aug. 1972.

INDEX